ADVANCES IN ELECTROCHEMISTRY AND ELECTROCHEMICAL ENGINEERING

VOLUME ELEVEN

ADVANCES IN ELECTROCHEMISTRY AND ELECTROCHEMICAL ENGINEERING

•

EDITORS

HEINZ GERISCHER, *Fritz Haber Institute,*
Berlin-Dahlem, Germany
CHARLES W. TOBIAS, *University of California,*
Berkeley, California

ADVISORY BOARD

E. J. CAIRNS, *General Motors Research Laboratory,*
Warren, Michigan
P. DELAHAY, *New York University,*
New York, New York
M. FLEISCHMANN, *University of Southampton,*
Southampton, Great Britain
N. IBL, *Federal Institute of Technology,*
Zurich, Switzerland
J. A. A. KETELAAR, *University of Amsterdam,*
The Netherlands
J. KORYTA, *Academie of Sciences,*
Praha, Czechoslovakia
R. PARSONS, *University of Bristol,*
Bristol, England
C. WAGNER, *Max-Planck-Institut,*
Göttingen, Germany
E. B. YEAGER, *Case-Western Reserve University,*
Cleveland, Ohio

ADVANCES IN ELECTROCHEMISTRY AND ELECTROCHEMICAL ENGINEERING

•

VOLUME ELEVEN

Edited by
HEINZ GERISCHER

*Fritz-Haber Institute
Berlin-Dahlem, Germany*

and

CHARLES W. TOBIAS

University of California, Berkeley

A Wiley-Interscience Publication

JOHN WILEY & SONS, New York • Chichester • Brisbane • Toronto

Copyright © 1978, by John Wiley & Sons, Inc.

All rights reserved. Published simultaneously in Canada.

Reproduction or translation of any part of this work beyond that permitted by Sections 107 or 108 of the 1976 United States Copyright Act without the permission of the copyright owner is unlawful. Requests for permission or further information should be addressed to the Permissions Department, John Wiley & Sons, Inc.

Library of Congress Catalog Card Number: 61-15021

ISBN 0-471-87528-7

Printed in the United States of America

10 9 8 7 6 5 4 3 2 1

Introduction to the Series

This series was planned to make available authoritative reviews in the area of electrochemical phenomena, and to bridge the gap between electrochemistry as a part of physical chemistry and electrochemical engineering. Increased interest in electrochemistry and its applications renders this publication timely.

Chapters will vary considerably in their method of approach and length. Some chapters will be definitive while others, dealing with unsettled problems in rapidly evolving areas, will have to be somewhat tentative. Coverage over a number of volumes should be quite comprehensive. Emphasis will be placed in successive volumes either on the physicochemical or the engineering approach, although some volumes may cover both aspects. Specific problems in electroanalytical chemistry will not be included, except for certain theoretical questions not usually considered by analytical chemists.

1961   PAUL DELAHAY
                                                     CHARLES W. TOBIAS

We regret very much that our colleague Paul Delahay, because of interests and obligations in other areas, has decided to resign after fifteen years as co-editor of the Advances. The favorable acceptance of the first nine volumes and an increasing interest in electrochemical science and technology, in our opinion, well justify continuation of the series. We expect to maintain the high standards set in the past and plan no departure from the purpose stated in the Introduction to the Series.

1976   HEINZ GERISCHER
                                              CHARLES W. TOBIAS

Preface

Electron transfer from an electrode to an electrolyte solution where the solvent is the sole interacting partner is the simplest electrode reaction. This process, however, can be studied in aqueous solution only with highly excited electrons such as those generated by light absorption. Although photoelectron emission from an interface into a vacuum has been known for a long time, this phenomenon has attracted the attention of electrochemists only relatively recently. A number of investigations devoted to this subject have revealed important new aspects of the thermodynamics and kinetics of electrochemical reactions and have also established a link between electrochemistry and radiation chemistry. Pleskov and Rotenberg, both recognized authorities in this field, provide in their chapter a valuable perspective of the experimental and theoretical results obtained in recent years, mostly in the Soviet Union.

The formation of metal monolayers by electrodeposition on other metallic substrates has created much interest among electrochemists in recent years. This phenomenon, which occurs much more frequently than has generally been assumed, should have a significant impact on our conceptual understanding of the mechanism of nucleation of new phases. In Kolb's chapter the relevant experimental techniques are reviewed, with special emphasis on modern spectroscopic methods that greatly extend the capabilities of traditional electrochemical tools. Paralleling chemisorption experiments in the gas phase, the thermodynamic interpretation of the electrochemical results provides information on the dependence of chemisorption energies on the degree of surface coverage and on the nature of the substrate. Results of these electrochemical studies on single crystals form a bridge to LEED studies of superstructures on surfaces in a vacuum.

The chapter on the zinc electrode by McBreen and Cairns focuses on the chemistry and electrochemistry of zinc relevant to its use in electrochemical energy conversion. Detailed treatment is given of anodic and cathodic processes at this electrode, with emphasis on behavior in alkaline electrolytes. A description of the complex behavior of zinc on repeated alternate anodic and cathodic current passage is followed by an evaluation of the status of various practical cells employing this electrode. Although significant progress has been made in recent years in the understanding of the intricate behavior of zinc electrodes, it is clearly evident that further research efforts must be devoted to the clarification of certain puzzling phenomena that impede their performance. Considering the need for efficient, low-cost cells in energy conversion and storage, the potential benefits of advancing our knowledge of the behavior of zinc electrodes may indeed be large.

Flow-through porous electrodes have received considerable attention in recent years because of their potential usefulness in technological applications. Newman and Tiedemann's chapter provides a definitive treatment of the performance of these electrodes, emphasizing the analogy between mass transfer in packed-bed chemical reactors and porous electrodes. To evaluate the applicable flow rates, the effects of ohmic drop, axial diffusion and dispersion, and electrode depth must be considered. High conversion ideally requires operation at the limiting current, and the ohmic drop within the electrode limits the maximum allowable flow rate. Quantitative comparison is made of the performance in various configurations of electrodes relative to flow direction. Flow-through electrodes are well suited to a variety of practical applications, including the removal of metal ions from dilute streams, and to the execution of electro-organic synthesis processes in which mass transfer may severely limit reactor capacity.

HEINZ GERISCHER
CHARLES W. TOBIAS

Berlin, Germany
Berkeley, California
December 1977

Contributors to Volume 11

Elton J. Cairns
    General Motors Research Laboratories, Warren, Michigan

Dieter M. Kolb
    Fritz Haber Institute, Berlin-Dahlem, Germany

James McBreen
    Brookhaven National Laboratory, Upton, Long Island, New York

John S. Newman
    Department of Chemical Engineering, University of California, Berkeley, California

Yuri V. Pleskov
    Institute of Electrochemistry, Academy of Sciences of the U.S.S.R., Moscow

Z. A. Rotenberg
    Institute of Electrochemistry, Academy of Sciences of the U.S.S.R., Moscow

William Tiedemann
    Corporate Applied Research Group, Globe Union, Inc., Milwaukee, Wisconsin

Contents

Photoemission of Electrons from Metals into Electrolyte Solutions as a Method for Investigating Double-Layer and Electrode Kinetics, *by* Yu. V. Pleskov and Z. A. Rotenberg 1

Physical and Electrochemical Properties of Metal Monolayers on Metallic Substrates, *by* D. M. Kolb 125

The Zinc Electrode, *by* J. McBreen and E. J. Cairns 273

Flow-through Porous Electrodes, *by* J. S. Newman and W. Tiedemann 353

Index 439

ADVANCES IN ELECTROCHEMISTRY AND ELECTROCHEMICAL ENGINEERING

VOLUME ELEVEN

Photoemission of Electrons from Metals into Electrolyte Solutions as a Method for Investigating Double-layer and Electrode Kinetics

YURI V. PLESKOV and Z. A. ROTENBERG
Institute of Electrochemistry
Academy of Sciences of the U.S.S.R.
Moscow

1.	Introduction	3
	1.1. Brief Historical Background	3
	1.2 Main Features of Photoemission into Electrolytes	8
2.	Theory	11
	2.1. Photoemission Theory	11
	2.2. General Theory of Photodiffusion Currents	25
3.	Techniques of Photoemission Experiments	28
4.	Basic Laws of Photoemission	33
	4.1. Five-halves Law	34
	4.2. Independence of Photoemission Threshold from the Nature of Metal	37
	4.3. Dependence of Photocurrent on Scavenger Concentration	41
	4.4. Work Function and Energy State of Electrons in Electrolyte Solutions	44
5.	Structure of Electrical Double Layer and Photoemission	48
	5.1. General	48
	5.2. Measurements of ψ'-Potential	51

5.3.	Determination of Zero Charge Potential	57
5.4.	Thickness of Dense Part of Electrical Double Layer	63
5.5.	Specific Adsorption Studies	64
5.6.	Adsorption Layers of Large Thickness	71

6. Kinetics of Electrochemical and Chemical Reactions Connected with Photoemission of Electrons ... 72

   6.1. Kinetics of Electrochemical Conversion of Atomic Hydrogen ... 73

      6.1.1. Formation and Removal of Hydrogen from Solution ... 73

      6.1.2. Basic Kinetic Laws of Electrochemical Removal of Atomic Hydrogen ... 79

      6.1.3. Mechanism of Electrode Reactions Involving Atomic Hydrogen ... 87

   6.2. Electrochemical Reactions Following Chemical Transformations in the Bulk of the Solution ... 97

      6.2.1. Kinetic Equations ... 97

      6.2.2. Electrooxidation and Homogeneous Decomposition of the Anion Radical NO_3^{2-} ... 99

      6.2.3. Chemical and Electrochemical Reactions Involving CO_2^- and $CH_3^{\cdot}$ Radicals ... 104

      6.2.4. Homogeneous Chemical Reactions Involving $H^{\cdot}$ and $OH^{\cdot}$ Radicals ... 105

7. Conclusions ... 111
8. Acknowledgments ... 113
9. List of Symbols ... 113
10. References ... 115

1. INTRODUCTION

1.1. Brief Historical Background

The history of the studies on the so-called photovoltaic phenomena at the metal-electrolyte interface dates back over 130 years. In 1839 Becquerel (33) detected electric currents when one of two identical electrodes immersed in dilute acid solutions was illuminated. This phenomenon, subsequently termed the "Becquerel effect", had been the object for detailed investigations. Numerous studies that followed Becquerel's pioneer paper were carried out with diverse systems containing either the metal-electrolyte or the metal-vacuum interface. Among the works of the second group, Hertz's investigations (86) are the most fundamental ones, which resulted in 1887 in the discovery of the outer photoeffect (i.e., electron photoemission) at the metal-vacuum interface. This phenomenon, thoroughly studied by Stoletov (161) in the following year, consists in the transition of electrons, after having absorbed a light quantum, from the metal into the surrounding medium where they exist for a short time in the form of free (quasifree) electrons. Subsequently, the electron photoemission drew great attention and has continued to be the object of intensive investigations both in theoretical and experimental aspects (8,63, 81,159). The results obtained in these studies had a decisive influence on the development of quantum theory of light, solid-state physics, and other problems of modern physical sciences. Hertz's discovery gave a powerful impetus to the production of vacuum photocells and other instruments.

The successs that accompanied the creation of vacuum photocells and their subsequent improvements for a long time distracted the attention of those engaged in the study of electrochemical photocells. Nevertheless, diverse photosensitive electrochemical systems have been developed since then, and theories have been advanced to explain the generation of the photoresponse in such systems. A classification of the electrochemical photocells with respect to their operation mechanisms is given in a paper (58) that reviews the early works on the photovoltaic systems as well as in a later survey (42). All photocells can be roughly devided into three main groups by the nature of the processes taking place in them.

In the first group we can include those systems where the photocurrent is generated as a result of the absorption of light by the solution leading to homogeneous photochemical reactions. In this case, the potential (current) variation at the electrode is caused by the formation of excited molecules, free radicals, or other photolysis products that might undergo reduction or oxidation at the electrode. As a matter of fact, the electrode itself does not participate in the photoprocess but acts as a mere detector for photochemical reactions. Such systems are reviewed elsewhere in the literature (95), and we do not examine them here. The second group of the photoprocesses that take place on semiconductor electrodes (78,120), metals coated with oxide films or dyes (94), and insulating electrodes (116) also does not fall within the scope of this review. Here the photocurrent is produced mainly by the inner photoeffect, that is, generation of minority current carriers in the semiconductor (electrons in the conduction band or holes in the valence band) that participate in the electrode reaction. Finally, in the third group we can include the photovoltaic systems consisting of metals with a clean surface in contact with a nonabsorbing electrolyte. Precisely, it is on such systems that Becquerel conducted his experiments.

A large number of papers are devoted to the study of the Becquerel effect, of which the paper of Sihvonen (158) deserves special attention, with its suggestion of several possible mechanisms for explaining the generation of photocurrent: (a) emission of electrons from the metal into solutions under the action of light (especially in the form of solvated electrons), (b) photodischarge of dissolved species, and (c) decomposition of adsorbed molecules under the action of light with the formation of cations and anions on the surface that ultimately pass into the solution.

Audubert (9), who studied the Becquerel effect on platinum, copper, and mercury, believed that the photoresponse generation mechanism is very similar to the outer photoeffect in vacuum discovered by Hertz. Later he revised his viewpoint and suggested that the photocurrent is generated as a result of photolysis of water molecules adsorbed on the electrode surface (10,11).

The effect of light was also observed during the decomposition of alkali metal amalgam in aqueous solution (43). The rate of this process depended on the polarization of light. When the electric field vector of the electromagnetic wave was directed normal to the metal surface, then the light accelerated the reaction much faster than when it was oriented parallel to the surface. These facts are consistent with the photoemission measurements in vacuum, and hence can serve as an experimental evidence in support of Sihvonen's hypothesis (158) that photoelectrons pass from the metal into the solution.

Deserving special mention are the works by Bowden (46) and Hillson and Rideal (93), who investigated the effect of light on the rate of electrochemical evolution of hydrogen and oxygen. For explaining the observed effects, Hillson and Rideal (93) proposed that the illumination of the electrode leads to activation of some hydrogen atoms adsorbed on the surface, thus accelerating the recombination or electrochemical desorption of hydrogen. Such an explanation is probably tenable for metals with medium or high surface coverage by atomic hydrogen. According to these authors, in the anodic processes light is responsible for the disintegration of the surface oxide molecules, thus giving rise to active oxygen, which prior to deactivation is capable of reacting with the radicals on the surface.

The numerous explanations proposed show that the Becquerel effect is a complex phenomenon consisting of several effects diverse in nature. Only today it has been possible to divide it into its "elementary components".

In the early 1960s the interest toward photoelectrochemistry has been revived mainly in the results obtained by three investigators, each advancing his own theory for interpreting the nature of photocurrent at metal electrodes. Berg and colleagues studied the action of light on the mercury dropping electrode using the conventional polarographic techniques (35-42). He attributed the photocurrent mainly to the fact that the absorption of light by a metal increases the energy of electrons inside the metal, and thus there is every possibility for the electrode reaction to procede on such a "hot" electrode. Heyrovsky and colleagues (87-92) explained the photoeffect by the decomposition of a surface charge-transfer complex formed between the electrode metal and the solvent (water, ethanol, etc.) or a species in solution under the effect of light. The chemical bond in such a complex may be polarized differently depending on the electron-donor or electron-acceptor nature of the adsorbate. On absorbing light by the metal, this bond is ruptured, and the bond electrons migrate to the electrode (anodic photocurrent) or to the adsorbed molecule that leaves the electrode surface (cathodic photocurrent). Such concepts that the photodesorption and photodecomposition of surface compounds are elementary acts of photovoltaic processes were proposed by Veselovsky quite early (165).

In 1963 Barker and Gardner (25) suggested that a major role in this phenomenon belongs to the photoemission of electrons from metals into solutions and their subsequent interaction with the solvent and dissolved substances.[+] The idea of photoemission of electrons from metals into electrolyte as we have seen, is not a new concept; nevertheless, only Barker and his collaborators succeeded first in demonstrating (by indirect but sufficiently reliable methods) the photoemission of electrons into solutions and pointing out the range of applicability of these concepts. Their results instigated further progress in the experimental and theoretical study of photoemission. Thus a quantum mechanical theory was developed to explain the photoemission into electrolytes, and investigations were undertaken and are still being carried out to study this phenomenon in different countries: (a) the United States (60,62,155), the Soviet Union (52,55,103,127), Argentina (44,135), Japan (96), Australia (155), and Italy (56). These three theories, which we review to explain the generation of photocurrents in electrochemical systems, as is evident now from numerous studies, are not mutually exclusive. One or the other mechanism may predominate, depending on the experimental conditions.

[+] The earlier investigations of the Harwell group are briefly reviewed elsewhere (26).

1.2. Main Features of Photoemission into Electrolytes

In contrast to the photoemission of electrons from metals into vacuum, the photoemission of electrons into electrolytes is simply the first stage of a complex multistep process. Its scheme was first formulated in 1965 (25,28) (Fig. 1). It includes the following successive stages (a) photoemission proper (i.e., the transition of electrons through the interface after absorption of a light quantum), (b) thermalization and solvation (hydration in aqueous solution) of emitted electrons, leading to a reduction in their initial energy to the level of thermal kinetic energy and subsequent formation of solvated electrons[+] [this stage usually completes in a time interval of the order 10^{-11} s (2,126)], and (c) entrance of solvated electrons, formed in solution, into chemical reaction with the solutes capable of capturing electrons (electron scavengers). For instance, H_3O^+, NO_3^- ions or N_2O, O_2, CO_2 molecules may serve as scavengers. Especially, there are many scavengers among organic substances (5,124). If there are no scavengers in the solution, then the solvated electrons return to the electrode at a comparatively rapid rate, and thus the resultant stationary photocurrent in the system will be zero.[++]

[+] The formation of hydrated electrons in water was first suggested by Platzman (125) in 1953. In particular, solvated electrons are produced as an intermediate in many radiation-chemical and photochemical reactions. Different methods for the preparation of solutions of solvated electrons, their properties and reactivity are reviewed elsewhere (4,5,85,124).

[++] Solvated electrons can also react with the solvent molecules. The rate of this reaction, however, is usually so small that it cannot give rise to noticeable photocurrents.

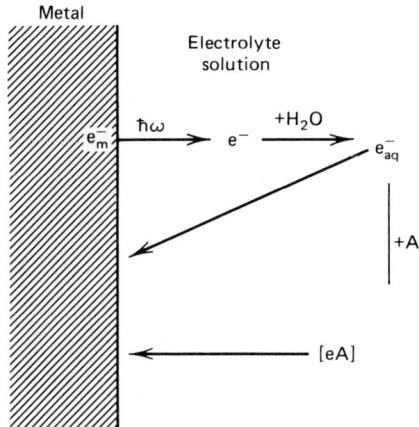

Fig. 1. Scheme of total photoprocess.

The situation is far simpler in the case of emission of electrons from a metal into vacuum. Under the influence of the electrical field, an electron there, after leaving the metal, freely moves from the photocathode (emitter) to the anode (collector). Therefore, we discuss those special features of photoemission from a metal into an electrolyte that distinguish it from the photoemission into vacuum.

An electrical double layer exists at the metal-electrolyte interface, and all of the external potential applied to the electrode is concentrated in this layer, thus giving rise to an additional photoemission parameter, namely, the electrode potential. In sufficiently strong electrolyte solutions the potential remains practically constant outside the dense part of the double layer (if we discard the ohmic potential drop in the bulk of the solution). However, in the metal-vacuum system the potential drop between the cathode and the anode is stretched over the entire space between the electrodes. If the electrical field near the cathode is not sufficiently strong as to create autoelectronic emission[+], the electron work function does not depend on the external potential difference applied. On the other hand, if a potential difference ϕ is applied to the metal-electrolyte system, the energy level of an electron in solution at a distance exceeding the thickness of the double layer changes by $e\phi$ (from now onward, $e>0$ is the absolute value of the electron charge). In other words, the electron work function must change as follows:

$$w_{ms}(\phi) = w_{ms}(0) + e\phi \qquad (1)$$

where $w_{ms}(0)$ is the work function describing the electron emission from metal into solution at the electrode potential taken to be zero, and $w_{ms}(\phi)$ is the work function at a potential ϕ.

Unlike the emission into vacuum, in photoemission into electrolyte, the electrons fall into a condensed medium and thus gain some energy as a result of the electron-medium interaction. Therefore, if we compare the state of the metal surface contacting the electrolyte with that in vacuum under comparable conditions (e.g., when the metal surface is not charged), we find that the electron work function differs in these two cases.

[+] Autoelectronic emission is usually observable at field strength of the order of 10^7 V/cm (63). Such fields can be created only with the help of special emitters. The Schottky effect (variation in the height of the potential barrier at the metal-vacuum interface under the effect of an external field) also begins to exert influence only in strong fields. Yet, the intensity of the electrical field near the cathode in standard vacuum photoemission experiments usually does not exceed a few hundred volts per centimeter.

Due to the electrical double layer and the presence of a condensed medium the nature of the motion of electrons near the electrode differs from the motion in vacuum where the electrons are subjected to two forces, namely, the force of the driving field and the image forces. The electrons emitted into electrolytes enter into collective interaction with the solvent and solute molecules. We can also stipulate that the image forces in the electrolyte are partially or totally screened by the ions in the solution (see below). All of these facts are reflected in the quantum-mechanical description of electron motion. In the subsequent pages we show that these specific features are responsible for the difference in the dependence of photocurrents on the radiation frequency between photoemission into electrolytes and into vacuum.

Photoemission into vacuum is not accompanied by any chemical reaction; it is a purely physical effect. Photoemission into electrolytes however, as already mentioned, can be observed only when the electrolyte contains electron scavengers. A condition essential for the stationary photocurrent is that the solvated electrons must undergo a chemical reaction with the scavengers. This feature makes the photoemission look like the usual electrochemical reaction, and this similarity is exhibited in a number of common basic laws.

2. THEORY

2.1. Photoemission Theory

Einstein laid the foundations of the modern theory of light-with-matter interaction in his paper (65) published in 1905.

He gave a theoretical explanation of the empirical laws of the then known photoeffect. The phenomenological theory proposed by Fowler (68) in the 1930s acted as the main lever for further developments in this field. This theory is based on the Sommerfeld model of free electrons in metals (160) and expresses the photocurrent I as a parabolic function of the frequency of light ω:

$$I = \frac{aA_o}{2} \frac{\hbar}{k^2} (\omega-\omega_o)^2 \qquad (2)$$

when $\hbar(\omega-\omega_o) \gg kT$, where A_o is the Sommerfeld constant, $\hbar$ is the Planck constant, k is the Boltzmann constant, ω_o is the threshold frequency of the photoeffect (the so-called red boundary), and T is the absolute temperature. The coefficient a accounts for the fraction of metal electrons excited by light absorption.

Fowler's parabolic law is in satisfactory agreement with the experimental data on the photoemission of the electrons from metals into vacuum (63,81). The specific properties of the interface and the medium into which the photoelectrons are transferred were not considered in the formulation of this law, which even today is successfully applied in emission electronics for determining the emissive properties of metals. The assumptions stipulated by Fowler in his theory in the form of postulates were eventually proved by the modern theory of photoeffect.

A quantum-mechanical theory was developed by Tamm and Shubin (162), Mitchell (117,118), and others. In the late 1960s Brodsky and Gurevich developed a general approach for describing the photoemission phenomenon, including the photoemission at the metal-electrolyte interface. Their theoretical description is based on the so-called threshold approximation (32). Such a general description is possible because at radiation frequencies rather close to the threshold, the: (a) kinetic energy of electrons emitted from the metal is less than the energy parameters characterizing the internal structure of the metal and (b) emitted electron wavelength exceeds the thickness of the transition region at the interface. Thus under certain conditions one may neglect the structure of the interface and the electronic structure of the metal if the interest is focused only on the frequency dependence of the photocurrent near the threshold. For a reader interested in the complete and rigorous treatment of the problem, we suggest the original papers (48-51), as our aim in this work is merely to outline the physical content of the topic.

The total photoemission current flowing in a direction normal to a microscopically uniform metal-electrolyte solution interface is as usual calculated by the following formula:

$$I = \int j_x(E_i, P_{\shortparallel}, \omega) \frac{1}{\exp\{(E_i - \tilde{\mu})/kT\} + 1} \rho(E_i, P_{\shortparallel}) dE_i dp_{\shortparallel} \qquad (3)$$

where E_i and $p_{\shortparallel} = \{p_y, p_z\}$ are the energy and initial electron-momentum components parallel to the surface (x = 0), $j_x(E_i, P_{\shortparallel}, \omega)$ is the partial photocurrent density outside the metal corresponding to the initial values of E_i and $p_{\shortparallel}$. The second expression under the integral sign describes the Fermi distribution of electrons inside the metal, and $\tilde{\mu}$ denotes the electrochemical potential of the electron. It is convenient to take the energy of a delocalized electron in the electrolyte solution as the reference point for the energy.[+] Then the electrochemical potential of an electron in the metal $\tilde{\mu}$ with respect to the solution is equal to the metal-solution work function but with the opposite sign: $-w_{ms}$. The last factor in Eq. 3 denotes the density-of-states

---

[+] The physical nature of such a level is discussed in Section 4.4.

function in the metal.

The range of integration of Eq. 3, that is, the range of admissible values of E_i and $p_\shortparallel$, is determined from the law of conservation of energy:

$$p = \sqrt{2m(E_i + \hbar\omega) - p_\shortparallel^2} \qquad (4)$$

where p is the x-component of the momentum of an electron emitted into vacuum and m is the mass of the electron. By virtue of the translational symmetry in the interface plane, conservation of the values of the tangential components of the momentum $p_\shortparallel = \{p_y, p_z\}$ is accounted for in Eq. 4; in other words, they remain equal to their initial value inside the metal. For a given frequency ω only those initial values of E_i and $p_\shortparallel$ are admissible for which p is real. In the contrary case (p is imaginary), the electron does not leave the metal and the current j_x for these values of E_i and $p_\shortparallel$ is zero. Thus the variables of integration in Eq. 3 must satisfy the following inequalities:

$$2m(E_i + \hbar\omega) > p_\shortparallel^2$$

$$\text{or} \quad 0 \leqslant p_\shortparallel^2 \leqslant 2m(E_i + \hbar\omega)$$

$$\text{and} \quad -\hbar\omega < E_i < \infty \qquad (5)$$

The steady-state current $j_x(E_i, p_\shortparallel, \omega)$ outside the metal averaged in the yz-plane is calculated by the well-known quantum-mechanical expression:

$$j_x = \frac{i\hbar}{2m}\left\langle \Psi_f \frac{\partial \Psi_f^*}{\partial x} - \Psi_f^* \frac{\partial \Psi_f}{\partial x} \right\rangle \qquad (6)$$

where Ψ_f is the wave function of the emitted electron in the final state outside the metal far away from the interface and Ψ_f^* is its complex conjugate function. According to Eq. 6, for calculating the partial photocurrent j_x it is necessary to determine the wave function of an electron outside the metal that has to be matched to the corresponding wave functions inside the metal and in the transition region at the interface with the medium into which electrons are emitted.

A schematic representation of the electron energy levels in the metal, both in the solution and at the interface, is shown in Fig. 2. In this figure δ denotes the thickness of the interface region where the surface forces decay; it corresponds to the thickness of the electrical double layer, which in concentrated electrolyte solutions is close to 1 Å (for a more detailed discussion, see Section 5.4.). The motion of an electron for $x > \delta$ is approximately described by the one-dimensional effective potential V (x), which accounts only for the field in the solution. On the other hand, the field of the doubel layer and other forces that influences the motion of an electron in the vicinity of the surface is concentrated within the transition region $0 < x < \delta$.

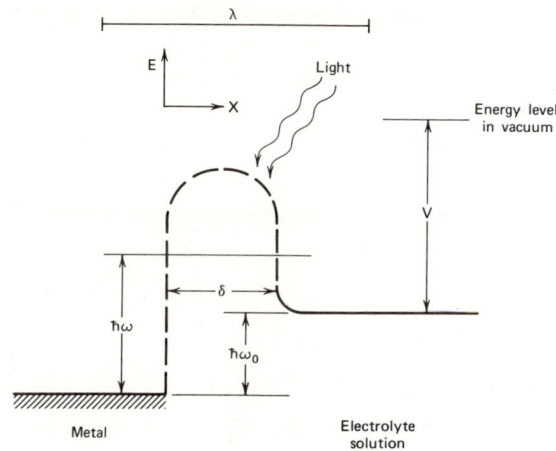

Fig. 2. Energy diagram for illuminated metal surface.

The Schrödinger equation, which must be satisfied by the unknown function Ψ_f for $x > \delta$, is

$$\left\{ \frac{\hbar^2}{2m} \Delta + E_f - V(x) \right\} \Psi_f (x,y,z) = 0 \qquad (7)$$

where $E_f = E_i + \hbar\omega$. With the reference point taken for the energy, the potential $V(x)$ vanishes as $x \to \infty$. As transverse momentum $p_\parallel = \{p_y, p_z\}$ is conserved when the electron penetrates through the interface, it is convenient to express the function Ψ_f in the form

$$\Psi_f = \exp\left[\frac{i}{\hbar}(yp_y + zp_z)\right] \Psi(x)$$

Then from Eq. 4 we find that $\Psi(x)$ is given by the following equation:

$$\left\{ \frac{\partial^2}{\partial x^2} + \frac{p^2}{\hbar^2} - \frac{2m}{\hbar^2} V(x) \right\} \Psi(x,p) = 0 \qquad (8)$$

When $x \to \infty$ the solution $\Psi(x,p)$ must describe an outgoing wave propagating from the metal surface. Let $f(x,p)$ be the solution that describes such a wave and fulfills the boundary condition

$$\lim_{x \to \infty} j_x \left[f(x,p)\right] = \frac{p}{m}$$

at infinity. If the potential $V(x)$ tends to zero sufficiently rapidly when $x \to \infty$, say by the exponential law, then

$$\lim_{x \to \infty} f(x,p) = \exp(ipx/\hbar)$$

apart from a phase factor. Hence by means of the boundary condition, that has to be obeyed by $\Psi_f(x,p)$ at infinity, it follows that Ψ_f must be proportional to $f(x,p)$:

$$\Psi(x,p) = L(p)\, f(x,p)$$

Substituting this into Eq. 6, we obtain for the partial current

$$\dot{j}_x [\Psi_f] = \frac{p}{m} |L(p)|^2 \qquad (9)$$

and substituting this into Eq. 3 we obtain for the photocurrent

$$I = \frac{1}{m} \int p\, |L(p)|^2 \left(\exp\{(E_i - \tilde{\mu})/kT\} + 1 \right)^{-1} \rho(E_i, p_\parallel)\, dE_i\, dp_\parallel \qquad (10)$$

This formula is as general as the boundary condition at infinity on which it solely relies, and the main problem of the theory is to calculate the square modulus of $L(p)$.

Since we started from Eq. 7, which is valid only for $x \geq \delta$, $L(p)$ is to be found from the boundary condition at $x = \delta$, that is, by matching $L(p)f(x,p)$ to the wave function in the region $x \leq \delta$ (see Fig. 2 and discussion following Eq. 6). Consequently, $L(p)$ in general depends on the internal electronic structure of the metal, the structure of the interface, and the potential $V(x)$ within the solution (since $f(x,p)$ also depends on it). The examination of the question as to which of these influences may be ruled out within the threshold approximation is considerably facilitated by the fact that it suffices to match just the wave function (and not also its derivative), since one is interested in the momentum dependence of $|L(p)|^2$ only. Thus we expect that the influence of the transition region becomes negligible if its extent is sufficiently small, since in that case for matching the solution $\Psi(x)$ of the Eq. 8 at $x \approx \delta$ to the wave function for $x < \delta$ we may use the value of $\Psi(x)$ for $x = 0$ instead of $x = \delta$. For a quantitative estimate we take $|\Psi(0) - \Psi(\delta)| / |\Psi(\delta)| \ll 1$ as a criterion for the validity of such an approximation and calculate $\Psi(0)$ by means of Taylor expansion around $x = \delta$:

$$\Psi(o) = \Psi(\delta) - \delta \left.\frac{d\Psi}{dx}\right|_{x=\delta} + \frac{1}{2}\delta^2 \left.\frac{d^2\Psi}{dx^2}\right|_{x=\xi} ; \quad 0 \leq \xi \leq \delta$$

If $\Psi(x)$ obeys a Schrödinger equation such as Eq. 8 in $0 \leq x \leq \delta$ [denoting the potential in this region by $U(x)$], it follows from the Taylor expansion that we may neglect the transition region, if

$$\delta \left|\frac{d \ln \Psi}{dx}\right|_{x=\delta} <<1 ; \quad \frac{m\delta^2}{\hbar^2}\left|U\right|_{max} <<1 \qquad (11)$$

If the potential $V(x)$ outside the transition region is negligible then $\Psi_f(x,p) \sim \exp(ipx/\hbar)$ for $x > \delta$ and

$$\delta \left|\frac{d \ln \Psi}{dx}\right|_{x \tilde{=} \delta} \quad \frac{p\delta}{\hbar} <<1$$

This inequality can be given a direct physical interpretation: the thickness of the transition region δ must be considerably less than the de Broglie wavelength $\lambda = (\hbar/p)$ of the emitted electron.[+]

Besides this condition (Eq. 11), the threshold approximation implies another inequality that follows from the behavior of electrons inside the metal ($x < 0$). The final energy of electrons

---

[+] To give some estimate, the de Broglie wavelength of an emitted electron is about 2 to 6 Å if its energy is in the range 0.1-1 eV. It should be pointed out that the model solution of the problem for the surface barrier of triangular shape (corresponding to the potential distribution in the double layer in a concentrated electrolyte) showed (55) that the condition << ("much less than") in the inequalities Eq. 11 can be understood as < ("less than") in most cases of interest, since it is not these parameters that are actually compared, but the exponential functions whose powers they represent.

inside the metal is contained in the respective Schrödinger equation for $x < 0$ in the form of the sum $E_f + V_m(x)$ with a large interaction potential inside the metal V_m^+. The characteristic scale of electron interaction inside the metal is of the same order as that of the kinetic energy of electrons on the Fermi surface (E_F). Consequently, if the quantity E_f always remains far less than E_F, the wave function $\Psi(x)$ in the inner region must not noticeably vary with E_f provided there are no isolated levels (narrow bands) or resonances (like plasmons) in the threshold region. Evidently, the same holds true at the boundary $x = 0$. Hence it follows that $\Psi(x)$ is approximately independent of E_f and also on $p \leqslant (2 m E_f)^{1/2}$, if

$$\frac{E_f}{E_F} \ll 1 \qquad (12)$$

Since the final energies of the electrons giving the major contribution into the photocurrent do not exceed $\hbar(\omega-\omega_o)$, Eq. 12 can be rewritten as

$$\frac{\hbar(\omega-\omega_o)}{E_F} \ll 1 \qquad (13)$$

which along with the existence of the threshold frequency ω_o clearly reveals the meaning of the "threshold" approximation.

Assuming that Eqs. 11 and 12 are satisfied, we get within the threshold approximation the boundary condition:

$$\Psi_f(\delta,p) \simeq \Psi_f(o,p) = \text{const}$$

+ The potential V_m by the order of magnitude is equal to the depth of the "square potential well" in the metal. It is usually about 7-10 eV, but the energy E_f does not as a rule exceed 1 eV.

Then, as shown by Brodsky and Gurevich (51), the square modulus of L(p) is given by

$$|L(p)|^2 = \frac{|\Lambda|^2}{|f(o,p)|^2} \quad (14)$$

where the constant Λ depends not on the momentum p but only on the properties of the metal.

This threshold approach gives the solution apart from some constant $|\Lambda|^2$, whose determination would have been equivalent to the complete solution of the problem. However, the constancy of $|\Lambda|^2$ that is, its independence of p, is sufficient for finding L(p), and consequently the frequency dependence of the total current I, which is now a function of the interaction forces in the solution only.

The total photoemission current I can be found by substituting the explicit form of $|L(p)|$ into Eq. 10 and then integrating it with respect to the initial electron states in the metal given by Eq. 5. The upper limit of integration with respect to energy is ∞ and the lower limit is $-\hbar\omega$. Considerable simplifications can be introduced here with the help of the second threshold condition (Eq. 13). Since $\hbar(\omega-\omega_0)$ is far less than the kinetic energy of electrons in the metal, the major contribution to the emission current is due to electrons with initial energies close to the Fermi surface. Therefore, provided there is no prominent structure in the density of states near the Fermi surface, we can assume that the electron density-of-states in Eq. 10 is a constant:

$$\rho(E_i, p_{\|}) = \rho(E_i, p_{\|})\Big|_{E_i=\tilde{\mu},\ p_{\|}=0} = \rho_o$$

and thus take it outside the integral. As L(p) depends only on p, using Eq. 4 we can integrate Eq. 10 with respect to p by introducing the cylindrical coordinates $|p_{\|}| = (p_y^2 + p_z^2)^{1/2}$ and

Θ arctan (p_y/p_z). Thus we obtain (62)

$$I = 2\pi|\Lambda|^2 ekT\rho_o \int_o^\infty \frac{p}{|f(o,p)|^2} \ln\left(1+\exp\left\{\frac{\tilde{\mu}+h-E_f}{kT}\right\}\right) dE_f \quad (15)$$

where $E_f = (p^2/2m)$.

At ordinary temperatures, metals exhibit strong degeneracy, and thus we can put $T = 0$ for radiation frequencies for which $kT \ll \hbar(\omega-\omega_o) \ll E_F$. Then with the help of the relationship

$$\lim_{T \to 0} \left[kT\ln(1 + e^{y/kT})\right] = y \cdot \Theta(y)$$

where $\Theta(y)$ is a "step" function ($\Theta = 1$ for $y > 0$, and zero for $y < 0$), we obtain

$$I = 2\pi|\Lambda|^2 e\rho_o \int_o^{\hbar(\omega-\omega_o)} \frac{p}{|f(o,p)|^2} (\hbar\omega - \hbar\omega_o - E_f) dE_f \quad (16)$$

Thus after substituting the explicit function $|f(o,p)|^2$ into Eq. 16 we can easily find the photoemission current.

Photoemission of electrons into highly concentrated electrolytes is the simplest and also most important case. All of the potential drop here is concentrated within the dense part of the double layer. If the double layer is thin enough to be totally included in the transition region δ, we can assume that $V(x) = 0$ outside this region. In such an approximation the solution $f(x,p)$ coincides with the asymptotic expression

$$\lim_{x \to \infty} f(x,p) = \exp(ipx/\hbar)$$

in the entire region $x > \delta$. Then we have

$$|f(o,p)|^2 = 1 \quad (17)$$

Substituting Eq. 17 into the general formula (Eq. 16) we can calculate the total current

$$I = 2\pi |\Lambda|^2 (2m)^{1/2} e\rho_o \int_0^{\hbar(\omega-\omega_o)} x^{1/2} (\hbar\omega - \hbar\omega_o - x) \, dx$$

or finally

$$I = \begin{cases} A(\hbar\omega - \omega_o)^{5/2} & \text{for } \omega \geqslant \omega_o \\ 0 & \text{for } \omega < \omega_o \end{cases} \quad (18)$$

This law of photoemission into electrolytes given by Eq. 18, in contrast to the Fowler's parabolic law, is called the "five-halves law" (48).[+] If the long-range image forces are taken into consideration in the threshold approximation, we once again obtain the parabolic law (48). Therefore, the main cause for the discrepancy between these two laws (Fowler's parabolic law and the five-halves law) is that there are no image forces in the emission of electrons into electrolytes, although they do exist in the case of emission into vacuum. According to Brodsky and Gurevich (48,51), the image forces in the emission into electrolytes are screened off by the ions in the electrolyte solution.

The quantum-mechanical description of the screening-off effect requires a very complex many-body problem to be solved and cannot be consistently given in terms of an one-particle description of electron motion in a medium. Therefore, we confine ourselves here to qualitative remarks. The theory under consideration is based on the time-independent Schrödinger equation (Eq. 7), and the electron is described by a wave function in the

---

[+] It may be noted that formulas formally resembling the five-halves law were obtained earlier by Mitchell (117) and Adawi (1a) by using a particular model for the barrier (photoemission through a rectangular surface barrier).

form of a monochromatic wave (E_f is fixed). Formally, this corresponds to a strictly time-independent electron-density distribution outside the electrode emitter. Taking into account the quantum nature of the photoemission process, we can assume that in the approximation considered there exists a steady-state probability flow in the direction $x \to \infty$ rather than a classical "escape" of single electrons. Therefore, when the "transition" period corresponding to the beginning of the experiment is over, in the given approximation the photoemission process should not involve spatial changes of the charge-distribution density. Thus since there are ions present in solution, they should undergo rearrangement, which will screen off this charge.

Of course, the electron wave function is actually not strictly monochromatic and corresponds to a wave packet. Therefore, the screening-off time is not infinite (as in the case of a monochromatic wave), but still lesss than the time τ of the passing of this wave packet through the interface. However, as before, the time τ is much greater than the time of the passing of the classical "point" charge through the dense part of the double layer. In this case, just as in dynamic charge screening in metals, collective ion motions should arise. It is these collective motions (and not, e.g., the classical ions diffusion), that lead to effective elimination of long-distance forces.[+]

[+] This reasoning agrees with the estimation $\tau \simeq 1/v$, where $v = 10^7 \div 10^8$ cm·s^{-1} is the velocity of the emitted electron and $1 \simeq 10^{-6}$ cm is the characteristic size of the wave packet, whence $\tau = 10^{-13} \div 10^{-14}$ s. The time characterizing the collective ion motions is $\tau_i = (4\pi e^2 N/M)^{-1/2}$, where N is the number of charged particles in unit volume and M their mass. At $N \simeq 10^{19} \div 10^{20}$ cm^{-3} and $M \simeq 10^3 - 10^4$ m_o, τ_i proves to be of the order of τ.

For zero absolute temperature frequency ω_o in Eq. 18 is the "red boundary" of photoeffect. Under real experimental conditions (T > 0), according to the general expression (Eq. 15), a photocurrent can exist also if $\omega < \omega_o$, because the metal contains electrons with an energy greater than that of the Fermi surface. However, the number of such electrons exponentially decreases with the energy, and the photocurrent must decrease likewise with decreasing frequency ω in the subthreshold region. As in the case of Fowler's law (Eq. 2), for $\hbar(\omega-\omega_o) \gg kT$ the photocurrent can be described with sufficient accuracy by the expression (Eq. 18) for $T \neq 0$. This gives a comparatively simple method for the determination of the threshold frequency ω_o with the help of Eq. 18 by extrapolating to $I \to 0$.

It is easy to find the dependence of the photoemission current into the electrolyte on the electrode potential from the relationship between the work function and the electrode potential (Eq. 1): $w_{ms}(\phi) = w_{ms}(o) + e\phi$. Substituting the work function w_{ms} into Eq. 18 we obtain the photocurrent-voltage relationship:

$$I = A (\hbar\omega - \hbar\omega_o - e\phi)^{5/2} \qquad (19)$$

The work function $\hbar\omega_o$ belongs to the electrode potential used as the zero of reference (e.g., the reference electrode potential). From Eq. 19 it follows that the photoemission law predicts an identical functional dependence of the emission current on the electrode potential and the quantum energy of the light. Eq. 19 is extensively used for plotting the experimental photocurrent versus potential curves for electrochemical systems, but ceases to be valid if the condition of the threshold approximation is violated. This occurs, for example, in the case of adsorption on the electrode of sufficiently bulky organic molecules or of diluted electrolyte solutions (for more details, see Sections 5.2., 5.4., and 5.6.).

The formula for (one-photon) photoemission current (Eq. 19) applies accordingly to Brodsky and Gurevich (51), to surface photoexcitation as well as to bulk photoexcitation[+] provided the light frequency is sufficiently close to the threshold. In what follows we do not go into the details of the nature of photoexcitation. Similarly, many-photon emission observed when the electrode is illuminated by a strong laser beam (14,27,102) and some other types of excitation [e.g., participation of surface plasmons (152), "photoemission of holes" (80)] do not fall within the scope of our review.

2.2. General Theory of Photodiffusion Currents

A complete description of the photoemission phenomena includes also the processes that occur with the photoelectrons after they have entered the electrolyte. These are: (a) thermalization and solvation, (b) reaction of solvated electrons with scavengers, (c) further conversion of the products of these reactions in the bulk of the solution and at the electrode, and (d) return and capture of unreacted solvated electrons by the electrode surface. A quantitative analysis of these phenomena is generally carried out by solving the respective diffusion equations (84) or by studying the equivalent electrical circuits containing the elements describing diffusion (20,28). Therefore, the photocurrents associated with the chemical and electrochemical transformations of solvated electrons were termed the "photodiffusion currents" (84).

[+] The works of Barker, Bottura, et al. (23), Babenko, Benderskii, et al. (12-14), and Gerischer (79) have shown that the bulk photoexcitation makes a large contribution into the measured photocurrent. The bulk photoeffect is dominant in semiconductor electrodes (83).

Application of diffusion equations for describing the photodiffusion currents is generally believed to be justifiable as the characteristic thickness of the region in which these processes take place (a few tens of Ångstrøms) exceeds the "diffusion step" of the particle (which is close to the intermolecular distance) in order of magnitude.

The resultant photocurrent in the system j, according to the scheme shown in Fig. 1, is

$$j = I - I_e \pm I' \qquad (20)$$

where I is the emission current, I_e is the current produced by the return of solvated electrons back to the electrode, and I' is the current of reduction ("plus" sign) or oxidation ("minus" sign) of the products [eA] of scavenger-electron interaction. Under steady-state conditions the current I' is proportional to the number of electrons trapped by the scavengers; thus Eq. 20 can formally be expressed as

$$j = \nu (I - I_e) \qquad (21)$$

where ν is the stoichiometric coefficient showing the total number of electrons transferred through the interface as a result of capture of one electron by a scavenger. For example, if all of the products [eA] are reduced at the electrode, then $\nu = 2$, for one-electron electrooxidation we have $\nu = 0$. If the products [eA] do not participate in any of the electrochemical reactions, then $\nu = 1$.[+]

[+] In some cases the value of ν may exceed 2, for instance, $\nu = 8$ for the photoemission of electrons into sulfur hexafluoride solutions at negative potentials (56).

The current I_e produced by the return of solvated electrons back to the electrode is found by solving the diffusion equation for solvated electrons. Under steady-state conditions this equation is (84)[+]

$$D_e \frac{d^2 c_e}{dx^2} - k_A c_A c_e + \Phi(x) = 0 \qquad (22)$$

where c_e is the local concentration of solvated electrons, k_A is the rate constant of their interaction with the scavengers, c_A is the scavenger concentration in the solution, and D_e is the diffusion coefficient of solvated electrons. The "source function" $\Phi(x)$ describes the rate of deposition of solvated electrons in the solution as a result of thermalization and solvation of photoelectrons. From the law of conservation of charge, we have

$$e \int_0^\infty \Phi(x) dx = I \qquad (23)$$

The solution of Eq. 22 under the boundary conditions $c_e(\infty) = 0$ and $k_s c_e(0) = D_e (dc_e/dx)_{x=0}$ with reference to the fact that $I_e = eD_e (dc_e/dx)_{x=0}$ gives an expression for the resultant photocurrent j (84):

$$j = \nu \left[I - \frac{ek_s}{QD_e + k_s} \int_0^\infty \Phi(x) \exp(-Qx) dx \right] \qquad (24)$$

where k_s is the rate constant of capture of an electron by the metal surface, $Q^{-1} = (D_e/k_A c_A)^{1/2}$ is the thickness of the layer

[+] Here the simplest and most real case is considered when all of the emitted electrons are solvated well before they react with the scavengers or return to the electrode. A more general case including the reactions of "dry" electrons is discussed elsewhere (127).

through which the electron must wander before reacting with the scavenger.

Equation 24 fully describes the photocurrent in the system as a function of the scavenger concentration and the kinetic constants of the solvated electron reactions. For high scavenger concentration, when $Qx \gg 1$ in the region where $\Phi(x)$ differs from zero, the integral in Eq. 24 tends to zero; thus

$$j \simeq \nu I \qquad (25)$$

For sufficiently low scavenger concentrations, when $Qx \ll 1$ and the electrons are captured by the surface at a very fast rate ($k_s \gg D_e Q$), we have

$$j = \nu x_o Q I \qquad (26)$$

where x_o denotes the mean distance from the emitter electrode surface at which the solvated electrons are formed. It is determined as the first moment of the function $\Phi(x)$:

$$x_o = \frac{e}{I} \int_o^\infty x\Phi(x)\,dx$$

From Eqs. 25 and 26 it follows that the experimentally measured photocurrent is equal to the emission current only for sufficiently high scavenger concentrations. In the low concentration range the photocurrent, forming only a part of the emission current, is a linear function of the square root of the scavenger concentration.

3. TECHNIQUES OF PHOTOEMISSION EXPERIMENTS

Usually the photoemission phenomena in the metal-electrolyte system are investigated by illuminating the electrode with light of wavelength of 250 to 450 nm (quantum energy of 2.5 to 5.0 eV). Mercury or xenon high-pressure lamps with a continuous or linear spectrum are used as a light source. The desired wavelength band is obtained by means of sufficiently high-transmission monochromator or with the help of interference filters. The width of the spectral band obtained by these filters is 5 to 10 nm. The cell and the focusing optics are made of quartz.

Regarding the cell illumination method and accordingly the measurement techniques, the equipment used in the photoemission studies can be devided into direct current (DC), alternating current (AC), and pulse installations. In the equipment of the first type the light intensity is maintained constant and DC ammeters are used. Under usual illumination intensities the order of measured photocurrent densities lies in the range of 10^{-6} to 10^{-7} A/cm^2.

Modulation of light intensity, in principle, permits: (a) investigation of the photocurrent relaxation processes taking place in an electrochemical cell (e.g., diffusion of reagents or of the reaction intermediates) and (b) application of AC amplifiers, which in some respects are more convenient to handle. As the estimates show, in case (a) frequencies above 10 kHz have to be used in the experiments, whereas in case (b) it is sufficient to change the light intensity at a frequency of the order of 10 to 100 Hz, for example, with the help of a rotating disc chopper placed on the path of the light beam or by modulating the current fed to the lamp. Selective amplifiers tuned to a preset modulation frequency are normally used for measuring the photocurrent (or photopotential). An installation of this type is described by De Levie and colleagues (61,62). In these experiments a dropping mercury electrode was used as the photocathode, and the photocurrent was measured at a predetermined instant of lifetime of the drop. The components of the electrode impedance were measured along with the photocurrent for checking the state of the electrode surface.

The block scheme of this installation is shown in Fig. 3. Two disc choppers are placed on the path of the light beam from the mercury lamp. One of them modulates the light with a frequency of about 16 Hz. With the help of an auxiliary light source and a photodiode it also generates the reference pulse of the same frequency for the synchronous detector. The second chopper (frequency of 0.5 Hz) is intended for timing the recorder and monitoring the drop lifetime. The stylus contacts the tape at a specific time instant after the preceeding drop has been released.

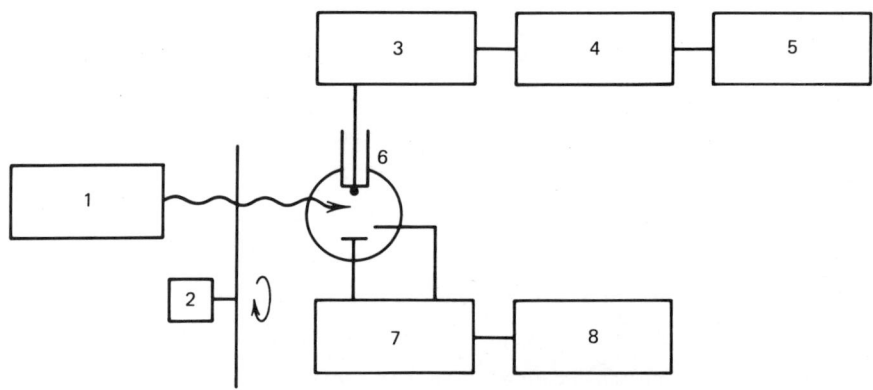

Fig. 3. Installation with modulated illumination (62): (1) light source, (2) light beam chopper, (3) current amplifier, (4) synchronous detector, (5) recorder, (6) cell, (7) potentiostat, (8) potential monitor.

Barker, Gardner, et al., (28) studied the dropping mercury electrode with square wave modulated light of a mercury lamp at a frequency of 225 Hz. A square-wave polarograph was used for polarizing the cell and measuring the photocurrent. This instrument is vastly superior to a simple AC amplifier for the detection of small modulated photocurrent signals because its noise level is low as it contains a lock-in amplifier. The flash-photolysis techniques (130) are applied in the pulse-measurement method. Flash lamps giving a flash of 10^{-4} to 10^{-3} s duration with an energy up to 100 J are used as the light source.

The quantity that fully characterizes the photoemission process is the quantum yield. In most experiments, however, the measurements were only of comparative nature, that is, the photocurrents were measured in relative units. For reliable photocurrent measurements it is necessary to maintain the electrode potential at a constant level irrespective of the illumination intensity. It is easy to realize this condition under steady-state illumination or low-frequency light modulation by using a normal potentiostat. But in high-frequency and flash measurements, the electrode impedance (determined by the high-frequency capacity of the double layer) is so low that it is difficult to create a potentiostat with sufficiently low input impedance. Therefore, the measurements are in such experiments usually taken in galvanostatic rather than in potentiostatic regime; in other words, the photopotential is measured with a constant current flowing through the electrochemical cell.[+] It is not

---

[+] Coulostatic mode (zero cell current and thus constant electrode charge) is a particular case of this regime. Delahay and colleagues (60,155) employed a coulostatic flash unit for studying the photoemission from mercury pool and dropping mercury electrodes.

difficult to calculate the photocurrent magnitude from the photopotential values provided one knows the electrode differential capacity.

On comparing these three methods, we find that the stationary illumination technique is the simplest for exact measurements of the photocurrent, but it is not suitable when there are appreciable dark currents (of course, it does not give any information on the relaxation processes). The flash method is not very precise and, in addition, there usually exists a certain amount of indeterminacy in this technique arising from the inaccuracies in the measured double layer capacities. The most universal method, probably, is that of the modulated illumination.

The majority of photoemission experiments was carried out with dropping or stationary mercury electrodes in aqueous solutions. Presently, electrodes of solid metals or semiconductors are finding increasing interest.

The species used as the scavengers for solvated electrons must: (a) react at a sufficiently high rate with the solvated electrons [rate constants are tabulated elsewhere (5,85,124)], (b) readily dissolve in the solvent used, and (c) not enter into electrochemical reactions (in darkness) in the investigated potential range. Moreover, the products of electron capture by the scavenger must not undergo oxidation at the electrode; in case of reduction, the reaction must proceed at a sufficiently fast rate (i.e., it must proceed in diffusion regime) in the investigated potential range. Otherwise, this reaction will distort the potential dependence of the photoemission current and thus create difficulties in the interpretation of the obtained curves. Futhermore, it is often desirable that the scavenger is not strongly (specifically) adsorbed at the electrode-solution interface in the investigated potential range, otherwise complications

and uncertainties may arise caused either by the capture of
hydrated electrons by adsorbed scavenger molecules (or ions) or
by light-induced electron transfer from the electrode to the
adsorbed entities [due to mechanisms of the types suggested by
Heyrovsky and colleagues (87-92) and Berg and colleagues
(35-42)].

Thus the selection of a suitable scavenger is determined by
the physical and chemical properties of the system under investigation (electrode material, solvent, and composition of the solution), and also by the objective of the investigation. The most
common scavengers used are H_3O^+ and N_2O.

4. BASIC LAW OF PHOTOEMISSION

After the three hypotheses described in the preceeding sections
had been formulated and a quantitative photoemission theory had
been developed, the subsequent experimental investigations were
directed at distinguishing the photoeffects of different kinds.
We can outline the basic criteria for discriminating the photoemission proper from other photovoltaic effects: (a) the dependence of the photocurrent on the presence and the concentration
of scavengers for solvated electrons, (b) characteristic dependence of the photocurrent on the electrode potential; relationship between the threshold potential and the quantum energy predicted by the formula (Eq. 19), and (c) independence of the
threshold potential on the nature of the electrode material.
For the time being, we can say that the experimental investigations have shown the photoemission of electrons to be the
dominant element of the electrode photoeffect in numerous studied
systems.

The classical experiment that convincingly demonstrated the emission nature of photocurrent is the investigation carried out by Barker, Gardner, et al. (28), who studied the photocurrent as a function of the scavenger concentration. By slightly disrupting the chronological order, we first consider the role of the electrode potential [items (b) and (c)].

4.1. Five-halves Law

Attempts for a quantitative interpretation of the experimental photocurrent curves in terms of Fowler's law did not lead to any successful results; no straight lines were obtained in $j^{0.5} - \phi$ coordinates as required by Eq. 2 after substituting the potential dependence of the work function (Eq. 1) into the formula (Eq. 2). However, if these curves are expressed in terms of the five-halves law (Eq. 19), that is, plotted in $j^{0.4} - \phi$ coordinates, it is possible to obtain straight lines (Fig. 4) whose intersection with the abscissa gives the photoemission threshold potential ϕ_o (for a given quantum energy $\hbar\omega$). A qualitative verification by the graphical method is corroborated by the statistical treatment of the experimental results. Such a computer-aided verification was first carried out in 1968 (147). Later, Benderskii, Babenko, et al. (34) applied the least-square methods to verify a large number of experimental $j - \phi$ curves obtained under diverse conditions, and have shown that the most probable value of the exponent of $(\hbar\omega - \hbar\omega_o - e\phi)$ for representing the experimental results in the form of a power law is close to 5/2. Setting aside certain controversial points arising as a result of lack of high scavenger concentration in many cases, placing the measured photocurrent j appreciably below the level of emission current,[+] we may assert that in those cases where the conditions of the threshold approximation (Eqs. 11 and 12) hold

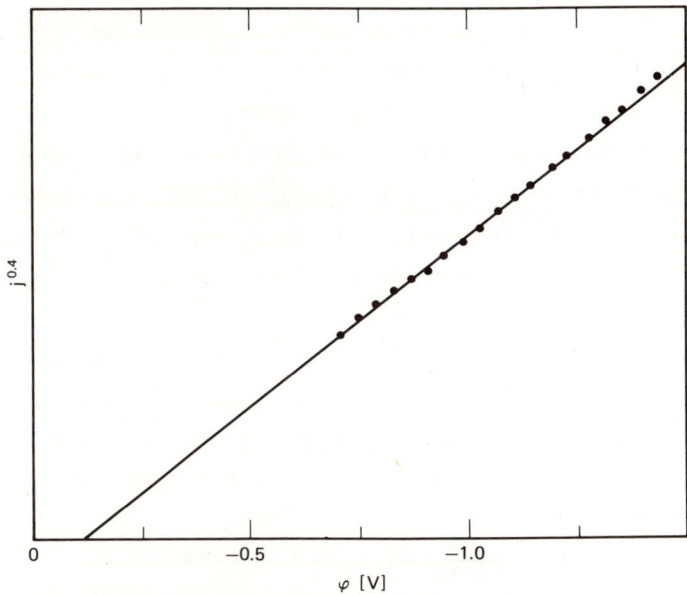

Fig. 4. Five-halves law. Lead, 1M HCl. Here and in the following figures the photocurrent is expressed in relative units and potentials are given against SCE. Measurements pertain to a wavelength of 365 nm if not mentioned otherwise.

[+] In this case the current j was found to be a function of the mean solvation length x_o, which in turn depends on the potential ϕ. Thus the additional $j - \phi$-dependence of current that arises as a consequence is responsible for the approximate cubic law observed in experiments (23) in place of the five-halves law. For details, see Section 4.3.

valid, the potential dependence or the quantum energy dependence of the photoemission rate obeys the five-halves law.

The experimental $j - \phi$ curves may deviate from the five-halves law in two cases: (a) if Eq. 11 is not fulfilled, for example, in dilute electrolyte solutions in which the thickness of the diffuse electric layer exceeds the wavelength λ of the emitted electron, or on electrodes coated with adsorbed layers thicker than λ and (b) when the electrode reaction of the reduced scavenger is slow and its rate constant depends on the potential.

These discrepancies were utilized for investigating the double layer structure and the electrode kinetics (see Sections 5 and 6, respectively).

From Eq. 19 it follows that as the influences of the quantum energy ($\hbar\omega$) and potential (ϕ) on the photocurrent are additive, any change in $\hbar\omega$ must entail the same change in the magnitude if the threshold potential ϕ_o (experimentally determined by extra-

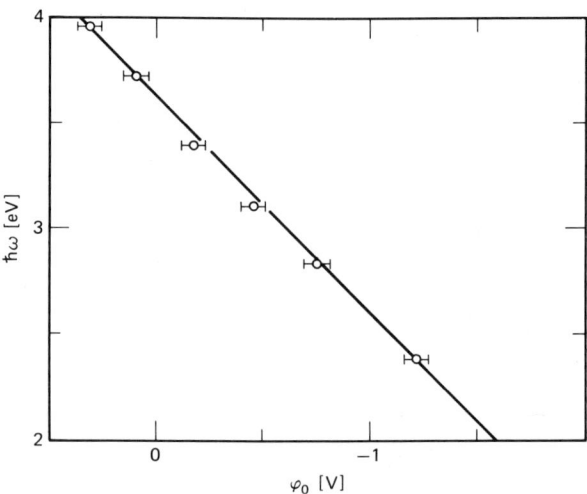

Fig. 5. Threshold potential versus quantum energy (151). Lead electrode, 1M HCl.

polating the curve $j^{0.4} - \phi$ to the point of intersection with the abscissa axis). This is exemplified by the plot in Fig. 5, and indeed, $d(e\phi_o) / d(\hbar\omega)$ is practically equal to unity.[+]

4.2. Independence of Photoemission Threshold from the Nature of Metal

The "red boundary" of the photoeffect in the photoemission into vacuum depends on the work function. The photoemission method for the measurment of work function is precisely based on this fact. Analogously, one may be led to believe that the photoemission threshold potential at the metal-electrolyte interface also depends on the nature of the metal and the points of extrapolation of $j^{0.4} - \phi$ lines to $j = 0$ for different metals would differ by the difference in their "vacuum" work function (or more exactly, the zero charge potentials).

However, the metal-electrolyte interface differs considerably from that of metal-vacuum; an additional variable, the electrode potential, acts in the metal-electrolyte system. The surface of a metal in vacuum, if it is not in contact with another metal, is not charged. On the contrary, a metal immersed in some electrolyte always acquires some potential and hence a charge that, in principle, may vary. According to Eq. 1, the electron work function varies with the potential applied. Therefore, the work function here must be referred to some definite potential. Electrons in metal electrodes, kept under the same potential,

---

[+] Here too (see footnote on p. 35) the low scavenger concentration (i.e., $j < 1$) complicates the situation: $d(e\phi_o) / d(\hbar\omega)$ becomes less than unity, a fact first observed by De Levie and Kreuser (62).

have identical Fermi energies, that is, their electrochemical potentials are equal. Thus the work function for the transport of an electron from a metal into an electrolyte solution does not depend on the nature of the metal but is uniquely determined by its electrode potential.

The difference in the metal nature is exhibited only in the height and the shape of the potential barrier at the interface, that is, in the transition region. But, as we have seen, the barrier properties are not very significant for the energetics of photoemission in the threshold approximation. It depends only on the difference between the initial and the final levels of the electron energy and not on the potential distribution at distances being small compared with the de Broglie wavelength of the electron. Therefore, the threshold of photoemission from metal into electrolyte does not depend on the nature of the metal (144).[+]

Figure 6 shows the $j^{0.4} - \phi$ curves obtained at mercury, lead, indium, and bismuth electrodes (151). Despite the large difference [almost 0.5 V (14,113)] in the zero charge potentials of these metals (and also in the "vacuum" work functions) their threshold potentials differ by not more than 50 mV, which is within the standard deviation range of experimental error. The independence of the threshold potential on the nature of the metal has been experimentally verified on a number of liquid

---

[+] This, in any case, is true for metals with a simple dispersion law, for which the "photoemission" work function $w_p = \hbar\omega_o$ is equal to the "thermodynamical" value w_t (which is found, e.g., from the contact potential difference measurements). In a more general case, w_p may exceed w_t for some metals (147). The photoemission from such electrodes has not yet been investigated.

(144) and solid (103,168) metals. We must emphasize here that this principle is equally applicable to certain electrodes whose zero charge potential varies as a result of adsorption. To the extent that Eq. 11 is valid (i.e., the thickness of the adsorption layer is less than the de Broglie wavelength of electron), the photoemission threshold potential is not sensitive to adsorption. The $j^{0.4} - \phi$ curves plotted for bismuth in solution containing halogen ions (which shift the zero charge potential of bismuth) are shown in Fig. 7 as an example. The similarity of the plots in Figs. 6 and 7 is self-evident and needs no

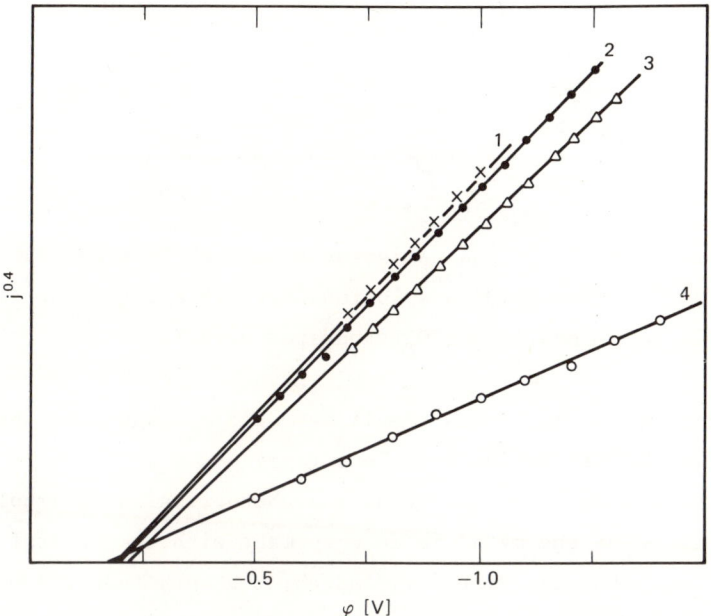

Fig. 6. Independence of photoemission threshold on the metal nature (151): (1) indium, 0.1N KF, (2) bismuth, 0.5N KCl, (3) lead, 0.5N Na_2SO_4, and (4) mercury, 0.5N KF. Scavenger N_2O.

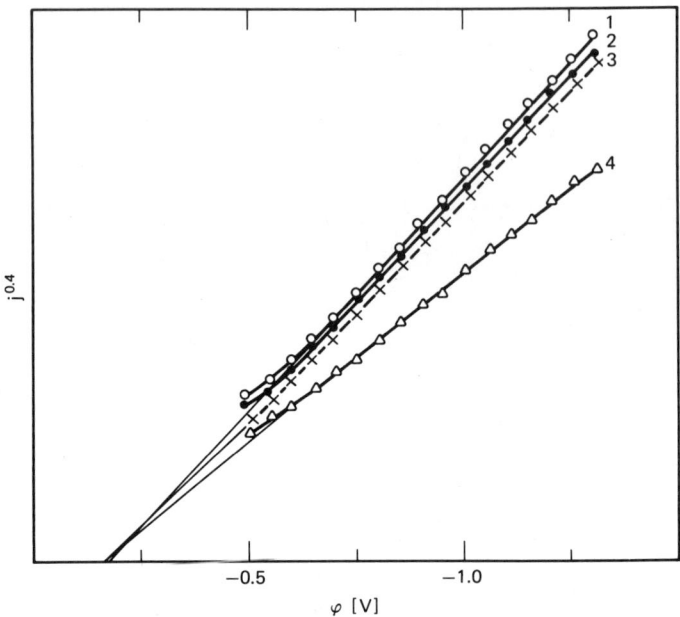

Fig. 7. Independence of photoemission threshold from adsorption (148). Bismuth, 0.5N solutions of: (1) KI, (2) KBr, (3) KCl, and (4) Na_2SO_4 saturated with N_2O.

further comment. An analogous result was also obtained for the mercury electrode in halide solutions (145).

The principle of independence of the photoemission threshold from the nature of the metal is in agreement with the general laws of electrochemical kinetics. Indeed, the work function of a metal is not explicitly contained in the kinetic equation of slow discharge if the electrode potential is expressed in reference to some arbitrarily chosen but fixed electrode (70,72,123, 163). Considerable difficulties are encountered in demonstrating this principle by direct experiments because, above all, in such reactions (e.g., the hydrogen ion discharge), the chemical bonds are ruptured and the reaction products are adsorbed on the electrode surface. The rate of these processes evidently depends on the nature of the metal. But for the simplest electrode reaction, reduction of anions, it has been established (of course, only after introducing the Ψ'-correction) that the electron transfer not accompanied by adsorption is indeed identical for different metals (77).

The photoemission of electrons at the metal-electrolyte interface can be regarded as the simplest charge-transfer reaction, which, unlike the usual electrode processes, is not accompanied by the formation or rupture of chemical bonds. For this very reason it proves to be a very convenient means for investigating certain fundamental questions in electrochemical kinetics.

4.3. Dependence of Photocurrent on Scavenger Concentration

One important criterion that distinguishes the photoemission from other photoeffects at the metal-electrolyte interface is the dependence of the photocurrent on the scavenger concentration c_A. According to Eq. 26 the relationship $j \sim \sqrt{c_A}$ holds valid in the low scavenger concentration range. A deviation from this square-root law must be observed with further increase in c_A and ultimately the photocurrent on attaining the value νI must cease to depend on c_A. Such a dependence of the photocurrent on the scavenger concentration was first detected by Barker, Gardner, et al. (28) on a dropping mercury electrode. Subsequently, it was confirmed on other metals, such as lead, bismuth, and indium (138,146,148). Figure 8 shows the experimental curves plotted for a lead electrode in solutions containing hydrogen ion as the scavenger of hydrated electrons (138).

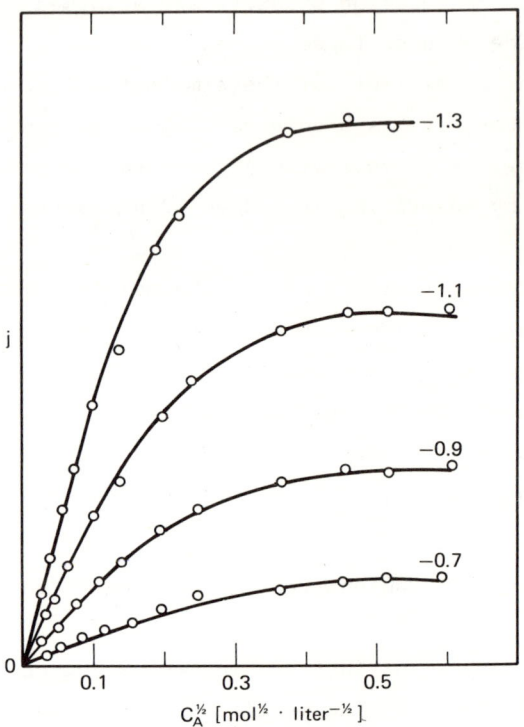

Fig. 8. Photocurrent versus square root of scavenger concentration (138). Lead, KCl + HCl solution (total concentration 1 mol/liter). Potentials are shown on the curves.

In the absence of scavengers, however, the so-called residual photocurrents may be observed (28,40); however, their nature remains an obscure, enigmatic puzzle. Barker and McKeown (29) believe that these residual photocurrents are generated by the photoelectron emission with subsequent capture of hydrated electrons mainly by the traces of scavenging impurities inevitably present in the electrolyte. As experimental evidence in support of this viewpoint, they mentioned the suppression of the residual photocurrents by dissolved CO_2 (competing scavenger), which after capturing the electrons undergo oxidation at the electrode (at potentials more positive than - 1.3 V). The concentration of these uncontrollable impurities is appreciably diminished if the solution is irradiated by UV light in the presence of SO_3^{2-}, which photoreduces the impurity to an inert state with respect to the solvated electron.

The study of photocurrent versus scavenger concentration is not limited to a detection of the emission nature of the photocurrent. Such experiments carried out on various metals at different potentials and quantum energies, under external fields and scavenger-concentration gradients, yield valuable information on the thermalization and solvation of emitted electrons and on further transformations of solvated electrons in electrolyte solutions. We briefly examine some problems that have been resolved by photoemission studies on the behavior of excess electrons in polar solutions.

According to Eq. 24, the photocurrent versus the scavenger-concentration relationship gives a method for constructing the shape of the "source function" $\Phi(x)$. The accuracy available today for the photocurrent measurements, however, is not adequate for completely solving this problem. Nonetheless, the first and second initial moments of $\Phi(x)$ have already been determined. A comparison of their values for different models narrows down the class of functions suitable for describing the deposition of solvated electrons originating from photoelectron emission into the solution (148).

The mean hydration length x_o has been reported to be in the range of 20 to 40 Å in (28,105,148). It is determined on the

basis of Eqs. 25 and 26 by comparing the photocurrents for low
and high scavenger concentrations. The magnitude of x_o increases
with the growth in the initial photoelectron energy (28,138).
The second moment, which is also derived from the $j - c_A$ dependence (148), suggests that the $\Phi(x)$ is close in shape to the
"theoretical" energy distribution of electrons in the metal.

The kinetic parameters k_A and D_e necessary for calculating
x_o by Eq. 26 are generally determined from radiation chemical
measurements, but they can also be obtained from photoemission
measurements when the scavenger undergoes electrochemical reduction in darkness (136). The measurements of photodiffusion
currents in solutions containing different scavengers find
extensive application in the determination of relative rate
constants for electron capture by scavengers (24,28,143).

Flash techniques, including laser illumination (27,102), will
undoubtedly find increasing application in the future developments of this field. With the help of very short-duration flashes
it would probably be possible to find the absolute rate constants
of solvated electron capture by scavengers (31) and by metal surface, as well as to directly determine the distance at which the
photoelectrons loose their initial energy in polar media.

4.4. Work Function and Energy State of Electrons in Electrolyte Solutions

The electron work function for the metal-electrolyte solution
system can be found from the relationship $w_{ms} = \hbar\omega_o = \hbar\omega - e\phi_o$
as the energy of light quantum $\hbar\omega$ and the corresponding threshold
potential ϕ_o are known. This work function belongs to some arbitrary but fixed potential with respect to which the threshold

potential ϕ_o is reckoned. For a given quantum energy we have $\hbar\omega - e\phi_o = $ const, and thus for any potential we can write

$$w_{ms}(o) = e\phi + \text{const} \qquad (27)$$

The constant in this relationship can be easily found from the $j^{0.4} - \phi$ line. For a quantum energy of 3.4 eV in aqueous solutions we have $\phi_o = -0.05 \pm 0.05$ V.[+] Consequently, $\hbar\omega - e\phi_o = 3.4 \pm 0.1$ eV. Thus the dependence of the work function on the electrode potential (against a saturated calomal reference electrode) has the following explicit form:

$$w_{ms} = 3.4 + e\phi \qquad (28)$$

This equation is universal (for aqueous solutions) and does not depend on the nature of the metal. It can be used for determining the relationship between the work function (for an electron emitted into electrolyte solution) and the zero charge potential of the metal. It is free of those constraints that are imposed on Frumkin's well-known linear relationship connecting the work function (for an electron emitted from metal into vacuum) to the zero charge potential of the metal (73). This latter relationship may cease to be a linear function if the orientation of solvent

---

[+] This very value of the threshold potential is obtained in the photocurrent measurements in high-concentration scavenger solutions (when $j = \nu I$) or from the measurements in low-concentration scavenger solutions (see Figs. 6 and 7) with proper correction introduced (141), with saturated calomel electrode used as the reference electrode.

dipoles at the metal-solution interface and the deformation of
the electronic "cloud" on the metal surface contacting the electrolyte depend on the nature of the metal.

The photoemission studies give a means for estimating the
energy of interaction of excess electron with a polar medium.
For this purpose it is necessary to compare the electron work
function for emission from metal into vacuum with that for emission into solution at zero charge potential. For mercury, the
work function for electron emission into solution at zero charge
potential (- 0.43 V) is 3.0 ± 0.1 eV (129), whereas for emission
into vacuum it is - 4.5 eV (82). The difference between these
values is to be attributed mainly to the energy gain as a result
of the electron-solvent interaction. The energy of such an interaction, calculated from the Volta potential in the metal-solution
system (52,64), is about 1.25 eV. This value is appreciably
(i.e., ~ 0.25 eV) less than the electron hydration energy, which
is about - 1.5 eV (135). Therefore, we can conclude that the
photoemission act proper produces in the initial stage not fully
hydrated (in a general case, solvated) electrons.

Here it is necessary to briefly examine the energy state of
excess electrons in polar solution. An electron emitted into the
solution, provided it does not lose its initial energy during the
thermalization process, moves with such a high velocity that only
the electronic polarization, but not the orientation polarization
of the medium, can follow. Such an electron (generally termed the
"dry" or "quasifree", but more appropriately referred to as a
"delocalized" electron) reminds us of an electron in the conductance band of a semiconductor crystal. The energy value of
1.25 eV determines the position of the lower edge of the "conductance band" in the solution with respect to the energy of

a free electron in vacuum.[+] As the kinetic energy of the electron decreases, it begins to interact with the orientation polarization of the medium. The end product of such an interaction is the solvated (hydrated) electron whose energy level lies well below the lower edge of the conductance band. This difference, as already mentioned, amounts to 0.2 to 0.3 eV for water. With the help of the photoemission measurements it was possible to determine the interaction energy of electrons with various solvents (64).

In the concluding paragraph of this section we deal with the role of photoemission investigations in the development of some general concepts on the electrode reaction mechanisms. The advances in radiation chemistry in the 1960s resulted in the discovery of solvated electrons in a number of solvents (including water) and in the investigation of their properties. These studies prompted some electrochemists (7,101,133,167) to stipulate an assumption that the thermoemission of electrons from the cathode into solution might be the primary act for many cathodic processes. The hydrated electrons formed in the vicinity of the electrode surface then enter into homogeneous chemical reaction with the depolarizers in the solution. Walker (166) was the first to attempt an experimental detection of hydrated electrons near the cathode surface by spectroscopic methods. Later it was found (57,131) that the effects observed by Walker were caused by the potential dependence of the reflectivity of the metal surface rather than by hydrated electrons in the solution.

[+] Since a solution, in contrast to a crystal, is a disordered medium, the lower boundary of its conductance band is not very distinct (119).

The electron work function value found from the photoemission measurements in aqueous solutions can be used to quantitatively estimate the probability of thermoemission of electrons into these solutions (47). This probability proved to be low in magnitude, and thus the contribution of the "emission" mechanism to the cathodic currents observed in aqueous solutions, even under high negative potentials, definitely is almost negligible. Noticeable thermoemission currents can be expected only at potentials more negative than - 3 V. But at these potentials the probability of the "usual" (i.e., the electrochemical) cathodic process mechanism (e.g., reduction of water molecules or alkali metal ions) is higher by several orders.

Indeed, in some nonaqueous solutions, such as hexamethylphosphoric triamide, for which the attained cathodic overvoltage is as high as 3 V, the electrochemical generation of solvated electrons at different metal cathodes has been observed (3,110, 111). Nonetheless, hydrogen is evolved in hexamethylphosphoric triamide from acid solutions at considerably lower overvoltage (~ 2 V) than the generation of electrons. Hence hydrogen evolution takes place via the usual proton discharge mechanism rather than by the "emission" mechanism.

5. STRUCTURE OF ELECTRICAL DOUBLE LAYER AND PHOTOEMISSION

5.1. General

The area where photoemission and its accompanying processes take place contains the electrical double layer. Therefore, it is natural to suppose that the structure of the double layer must influence both the photoemission act proper and the processes associated with the formation and subsequent behavior of solvated electrons.

The general key role that the electrical double layer plays in electrode processes was first noticed by Frumkin as early as 1933 in his attempts to explain the effect of double-layer structure on the hydrogen overvoltage (69). Quantitatively, this effect is expressed by introducing the Ψ'-potential into the kinetic equation of the reaction to describe the variation in concentration of reacting ions at the electrode and the influence of the potential drop in the Helmholtz double layer on the activation energy (59,75). Photoemission can be considered as the simplest case of electron transfer across the interface. Therefore, the role of the electrical double layer in photoemission phenomena and in the electrochemical kinetics is similar in many respects.

First, the double layer may influence the emission act proper. It is true that if the thickness of the double layer is less than the de Broglie wavelength of the emitted electrons, the effect of the structure of the interface on photoemission, as shown in the previous pages, can be discarded. But in dilute solutions of electrolytes, the thickness of the diffuse layer is comparable with the de Broglie wavelength, and in some cases far exceeds it. In such situations, the double-layer structure has a noticeable effect on the photoemission proper. It can be quantitatively described by assuming some specific model for the double layer and then calculating the photoemisssion current for a potential barrier of a given shape at the emitter surface. In the subsequent pages we examine three cases: (a) the diffuse layer, (b) adsorption layer of large thickness, and (c) the Helmholtz layer comparable in thickness with the electron wavelength.

Second, the double layer may influence the processes in solutions initiated by the photoemission act, and this effect may consist mainly in the following:

1. The structure of the double layer has a strong effect on the drift of electrons in the electrical field when they return to the electrode after hydration and on their interaction with charged scavengers. Besides, the field of the diffuse layer may as well influence the drift of charged products of electron capture by scavengers (e.g., NO_3^{2-}).
2. The concentration of charged scavengers undergoes variations in the diffuse double-layer electric field. The near-the-electrode concentration of charged scavengers under high absolute values of the Ψ'-potential may considerably differ from the concentration in the bulk of the solution. An attempt was made elsewhere (44) to account for this effect quantitatively.
3. For electrode reactions, slower than the diffusion process, in which the products of electron scavenging take part, the rate constants of the heterogeneous processes enter into the expression for the photocurrent. As they depend essentially on the Ψ'-potential, the structure of the electrical double-layer will have an influence on the photocurrent.

Third, when the thickness of the diffuse layer is comparable with the distance at which solvated electrons are deposited in the solution the diffusion back to the electrode is affected. At solvation length of 20 to 30 Å, the double layer may be expected to exert considerable influence in dilute electrolyte solutions.

5.2. Measurements of Ψ'-Potential

A quantitative theory of photoemission that accounts for the finite size of the electrical double layer (of both its dense and diffuse parts) has been suggested by Brodsky, Sheberstov, et al. (53,54,157). It is based on solving the Schrödinger equation for the motion of an electron in a potential field characteristic for the electrical double layer. Figure 9 shows the schematic representation of the potential distribution V(x) in the usual one-dimensional model of the double layer divided into its dense and diffuse parts (59,75) (for the case of $\Psi' < 0$). The potential distribution in the diffuse layer is calculated in the Gouy-Chapman theory (59).

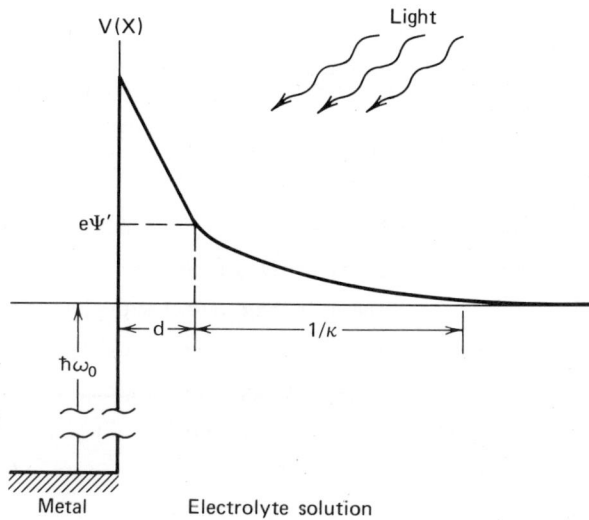

Fig. 9. Energy diagram for the photoemission in dilute electrolyte solutions.

The quantitative effect of the double layer on emission is characterized by the numerical values of the dimensionless parameter $(2m|e\Psi'|)^{1/2} / \kappa\hbar$, which is the ratio of the effective thickness of the diffuse layer to the de Broglie wavelength of the emitted electron possessing the energy $|e\Psi'|$. When the inequality

$$\frac{(2m|e\Psi'|)^{1/2}}{\kappa\hbar} \gg 1 \qquad (29)$$

is satisfied, the diffuse layer is nonpermeable to the emitted electrons; that is, the electrons with an energy less than $|e\Psi'|$ cannot tunnel through it. Numerical estimates show that this inequality (Eq. 29) is satisfied only for sufficiently dilute solutions (< 0.1M) when the Debye screening length[+] exceeds 20 to 30 Å. In more concentrated solutions the electron tunnels freely through the potential barrier of the diffuse layer, and the Ψ'-potential has no significant effect on the photoemission act.

At negative Ψ'-potentials, when Eq. 29 holds true, the expression

$$I = A\left[\hbar\omega - \hbar\omega_o - e(\phi-\Psi')\right]^{5/2} \qquad (30)$$

was derived for the photoemission current. It differs from Eq. 19 in that it contains the potential difference $\phi - \Psi'$ in place of the potential ϕ alone. This result has a direct physical meaning. Equation 29 shows that the potential $V(x)$ varies very slowly at distances of the order of one electron wavelength, and thus the diffuse part of the double layer plays the same role of a "well

---

[+] The electron wavelength λ corresponding to an energy of 0.15 eV is approximately 5 Å and in 0.01M solution of 1:1 electrolyte, the Ψ'-potential for negative surface charges reaches approximately − 150 mV.

with a flat bottom" like the uncharged bulk of the solution in concentrated solutions (where the Ψ'-potential can be neglected). In other words, when the electrode potential varies by ϕ, the bottom of the potential well, and consequently, the work function for the electron emission from metal into solution changes by $e(\phi-\Psi')$ but not by $e\phi$ as in concentrated solutions.

Equation 30 is strictly applicable only for $\Psi' < 0$. If $\Psi' > 0$, the potential well has to be substituted for the potential barrier. Although this case is not considered quantitatively in the theory, we can yet assume that, in passing from the metal into the potential well, the emitted electron gains an additional amount of energy of the order of $|e\Psi'|$.

A comparison of Eqs. 30 and 19 clearly suggests that a consideration of the effect of the diffuse layer on the photoemission proper is equivalent to the introduction of the so-called Frumkin Ψ'-correction into the kinetic equation of the electrode reaction.

With the help of Eq. 30 we can directly determine the Ψ'-potential in dilute solutions from the shift in the photocurrent curve with respect to a similar curve for concentrated solution (where $\Psi' = 0$). For this purpose, the coefficient A in Eq. 30 should not depend on the Ψ'-potential. This condition is fulfilled for sufficiently high concentrations of scavengers[+] (usually of the order of 1 mol/liter) when all of the emitted electrons are captured in the solution. However, this is very difficult to realize in practice, and thus we have to reconcile experiments with lower scavenger concentrations where A may be a function of the electrolyte concentration owing to the effect

---

[+] Of course, the scavengers must be noncharged so as not to influence the ionic concentration of the solution.

of Ψ'-potential on the secondary process (above all, on the drift of hydrated electrons). Nevertheless, it is possible even in this case to choose such a pair of solutions of different electrolyte concentrations that the "side" effect of Ψ'-potential in both solutions may be the same, and thus we may apply Eq. 30 for the determination of the difference in the Ψ'-potentials in such solutions, $\Delta\Psi'$. The dependence of $\Delta\Psi'$ on the potential of a mercury electrode in potassium fluoride solutions of concentrations 10^{-2} and 10^{-3} mol/liter determined by this method is

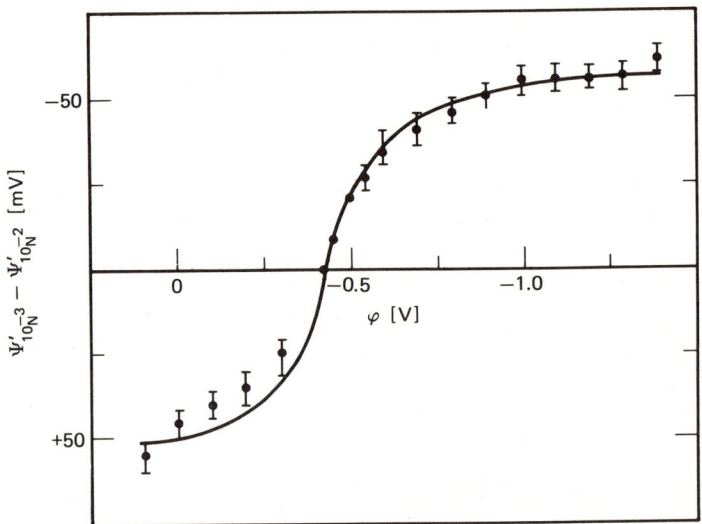

Fig. 10. Difference of Ψ'-potentials in 0.001N and 0.01N KF solutions as a function of the potential of mercury electrode (112). Solid line, calculated according to Gouy-Chapman theory.

exemplified in Fig. 10.[+] It is in satisfactory agreement with
the predictions of the Gouy-Chapman theory (solid line in the
figure) (112).

In dilute electrolyte solution under sufficiently negative
potentials the electric field of the diffuse layer effectively
retards the return of hydrated electrons back to the electrode,
and hence we may expect that the limiting photocurrent could be
attained at far less scavenger concentrations than in strong
solution for which $\Psi' = 0$. This, especially, is true of posi-
tively charged scavengers (e.g., hydrogen ions) whose near-the-
electrode concentration in the field of negative Ψ'-potential
may be considerably higher than in the bulk of the solution.
Under these conditions it is now possible to determine the
absolute value of Ψ'-potential in dilute solutions by comparing the $j^{0.4} - \phi$ curve for this solution with the curve for the
concentrated electrolyte solution, because in both cases the
photocurrent is equal to the emission current. This method was
applied for measuring the Ψ'-potential in dilute acid solutions
(0.1 to 0.001N) as a function of the electrolyte concentration
(Fig. 11). Again these results are in good agreement with the
Gouy-Chapman theory (150).

[+] Nitrous oxide was used as the scavenger. Its concentration in
saturated solution (0.025 mol/liter) under a pressure of 1 at
is not sufficient for capturing all the electrons emitted into
the solution. Nonetheless, we can show that in such dilute electrolyte solutions the variation in the electrolyte concentration
has no effect on the secondary process but affects only the
photoemission proper.

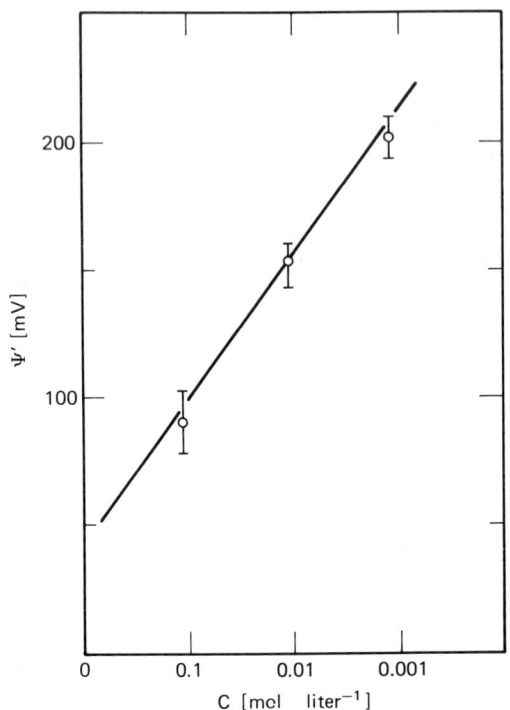

Fig. 11. Ψ'-potential versus electrolyte (HCl) concentration (150). Mercury electrode, − 1.1 V. Solid line, calculated according to Gouy-Chapman theory.

It must be stressed here that the photoemission method is applied for determining not the local Ψ'-potential (e.g., as in the kinetic investigations), but precisely the mean potential in the outer Helmholtz plane as the emitted electron is strongly delocalized in this plane. Such averaged values are calculated in the Gouy-Chapman theory.

As the photoemission method is based on the quantum-mechanical treatment of electron transitions through the interface, it essentially varies from the differential capacity method widely used for measuring the Ψ'-potential. Thus the photoemission measurements give one more evidence to demonstrate the validity of Gouy-Chapman theory, at least, in the concentration range of 10^{-2} to 10^{-3}N for the 1:1 electrolytes in the absence of any specific adsorption.

5.3. Determination of Zero Charge Potential

In addition to the photoemission proper, the structure of the double layer has significant influence, as already pointed out, on the other stages of the processes accompanying the photoemission act. The theory of motion of solvated electrons in the diffuse layer field was analyzed in detail by Rotenberg and Gurevich (139,140). It was found that the retardation of the motion of hydrated electrons on their return path to the electrode meets striking similarities with the processes of electroreduction of anions in dilute electrolytes (76) whose discharge is strongly inhibited at negative Ψ'-potentials. All of these facts again demonstrate that a close interrelationship exists between photoelectron emission at the metal-electrolyte interface and the "usual" electrochemical reactions.

Without going into the details of the explicit functional dependence of j on Ψ' or on the concentration of an indifferent electrolyte, we consider the fact that $\Psi' = 0$ at the zero charge potential in the absence of specific adsorption. Therefore, all phenomena associated with the effect of electrolyte concentrations on the photocurrent as a result of the variations in Ψ'-potential vanish at the zero charge potential. As a consequence, the photocurrent curves at different electrolyte concentrations (in the low concentration range) must meet at this potential.

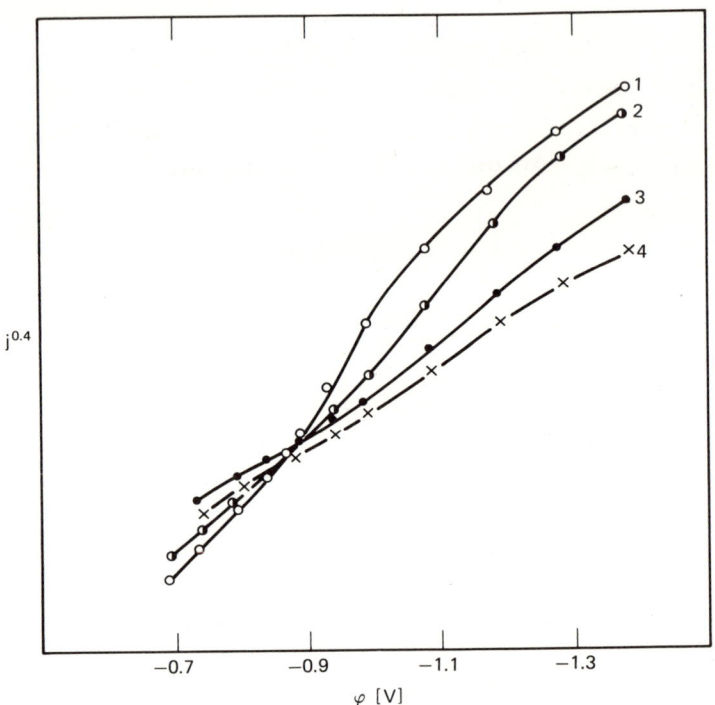

Fig. 12A. Determination of zero charge potentials (151).
(A) lead: (1) 0.001N HCl, (2) 0.001N HCl + 0.01N NaCl, (3) 0.001N HCl + 0.1N NaCl1 (4) 0.001N HCl + 0.23N NaCl.

Figure 12 shows the $j^{0.4} - \phi$ curves plotted for acid solutions at various concentrations of an indifferent electrolyte on lead and cadmium electrodes (132,137,146,151). From these plots it is clear that for all metals the curves at different concentrations indeed intersect near the zero charge potential. In the negative surface charge region, the photocurrent monotonically decreases

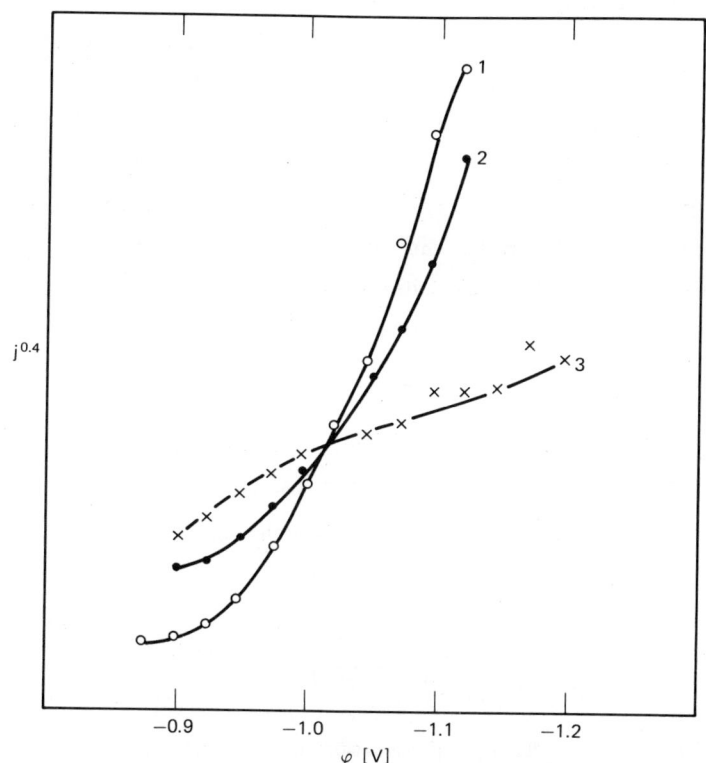

Fig. 12B. Determination of zero charge potentials (151). (B) cadmium: (1) 0.001N HCl, (2) 0.001N HCl + 0.01N KCl, (3) 0.001N HCl + 0.1N KCl.

with growth in the total concentration of the electrolyte. Such a behavior pattern shows the Ψ'-potential to have a significant influence mainly on the photodiffusion processes (variation in the migration of electrons), and the effect of Ψ'-potential on the emission proper plays a secondary role.

Therefore it is evident that the zero charge potential is in some respects a "singled-out" point on the photocurrent curve, and this is vividly exhibited in dilute solutions. By the intersection of the photocurrent curves for different electrolyte concentrations at a constant scavenger concentration, we can quantitatively determine the zero charge potential. In Table I, the zero charge potentials determined by the photoemission method are compared with the results of capacitance measurements.

Our attention is drawn to the fact that in the absence of any specific adsorption, the zero charge potentials found by the photoemission method are in quantitative agreement with the differential capacity measurements.[+]

Another method for measuring the zero charge potential (128) is based on the application of Eq. 30. The function $\Psi'(\phi)$ has an inflexion at the zero charge potential (see Fig. 10). If Eq. 30 is differentiated with respect to $-\phi$, assuming that A does not depend on ϕ, we obtain

$$\frac{dI^{0.4}}{d(-\phi)} = eA^{0.4} \left(1 - \frac{d\Psi'}{d\phi}\right) \tag{31}$$

As the derivative $d\Psi'/d\phi$ has a maximum at the zero charge potential, the derivative $d(I^{0.4})/d(-\phi)$ will accordingly have a minimum. Thus we can estimate the zero charge potential from the minimum of the $d(I^{0.4})/d(-\phi)$ curve. This method is applicable

---

[+] The zero charge potentials determined by the capacitance method were measured in KF and Na_2SO_4 solutions not containing any surface active ions. In the photoemission measurements the specific adsorption of Cl^- ions somewhat shifts the zero charge potentials of lead and cadmium toward the negative side.

TABLE I

Zero Charge Potential of Metals

Metal	Solutions	Zero charge potential (vs. SCE)	
		Photoemission (149,151)	Capacitance (74,113)
Hg	0.001N KF + N_2O 0.01N KF + N_2O	−0.43	−0.43
Pb	0.001N HCl 0.001N HCl + 0.01N KCl	−0.85±0.02	−0.81±0.02
Bi	0.001N HCl 0.001N HCl + 0.01N KCl	−0.64	−0.64±0.02
Bi	0.001N KNO_3 0.001N KNO_3 + 0.01N KF	−0.63	−
Cd	0.001N HCl 0.001N HCl + 0.01 KCl	−1.03	−1.00±0.02
In	0.001N HCl 0.001N HCl + 0.01N KCl	−0.90±0.03	−0.90±0.02
Sb	0.001N KNO_3 0.001N KNO_3 + 0.01N KCl	−0.4±0.02	−0.4

only to systems for which Eq. 30 is valid to a sufficient degree of accuracy, that is, when the coefficient A is independent of ϕ. This condition, as already mentioned, is fulfilled for dilute

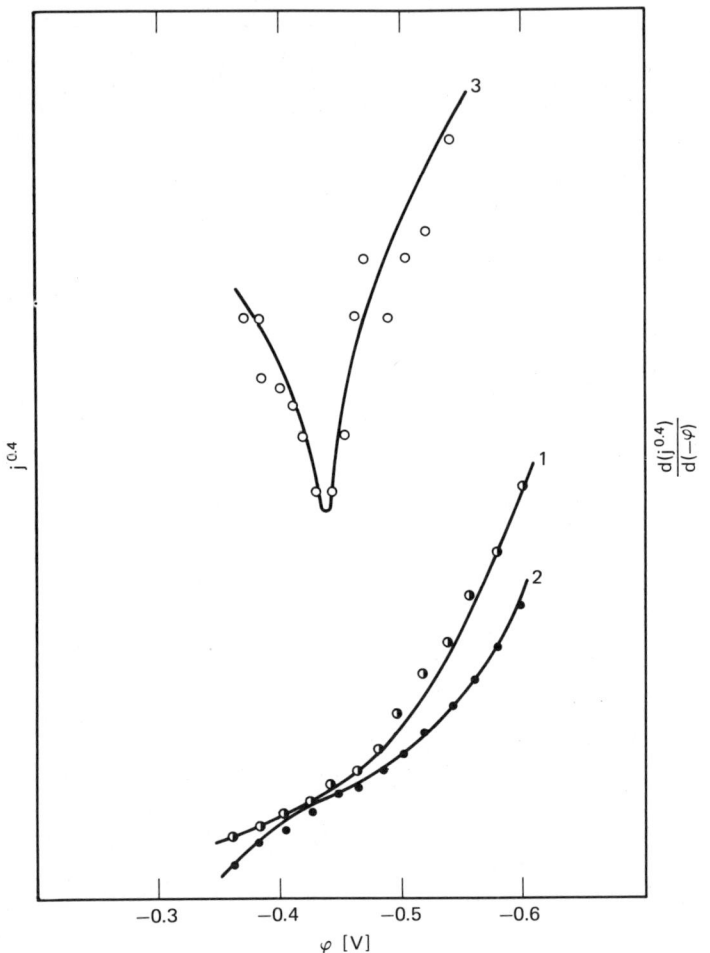

Fig. 13. Determination of zero charge potential of mercury (128): (1) 0.001N HCl, (2) 0.001N HCl + 0.009N KCl, (3) 0.01N KCl solution saturated with N_2O.

electrolyte solutions saturated with nitrous oxide. The $d(I^{0.4})/d(-\phi)$ curve is shown in Fig. 13 for a 0.01N KCl solution saturated with N_2O. The minimum potential on the curve practically coincides with the zero charge potential of mercury. This figure also shows the $j^{0.4} - \phi$ curves for two dilute electrolyte solutions. They intersect at the zero charge potential. It is obvious that both methods give very close results.

5.4. Thickness of Dense Part of Electrical Double Layer

In the preceding pages it was assumed that the dense part of the electrical double layer does not play a significant role in the photoemission process. This assumption is based on the fact that the wavelength of the emitted electron exceeds the thickness of the region at the interface where the potential drop mainly occurs. In some experiments, however, a deviation from the five-halves law was observed that can be attributed to the violation of Eq. 11, where the dense part of the double layer is of not negligible thickness.

Basing on the theoretical calculations of the photoelectron tunneling through the potential barrier of the dense part (53, 156), Lakomov, Rotenberg, et al. (112) have estimated that the thickness of the Helmholtz layer on the mercury electrode in 0.1M NaF under potentials more negative than the zero charge potential is about 2 Å. This value is appreciably less than the total thickness of the Helmholtz layer which is the sum of the cation radius and the diameter of a water molecule as a "water interlayer" exists near the metal surface. Yet in the true sense of the estimate, the obtained thickness is not the distance between the metal surface and the outer Helmholtz plane but characterizes the region where the potential drops down rapidly.

According to the model calculation (109) a major part of the potential drop (up to 90%) takes place in the water molecules layer (as this layer has a lower dielectric permeability). Further decay in the potential right up to the outer Helmholtz plane takes place at far less rate. Therefore, the estimate given for the thickness of the dense layer is by no means in contradiction with the concept of a "water interlayer" attached to the electrode surface.

5.5. Specific Adsorption Studies

At least two effects must be taken into consideration in studying specific adsorption of ions and molecules on the electrode:

1. Specific adsorption of charged particles gives rise to a potential change in the outer Helmholtz plane; we accordingly denote this potential by the symbol Ψ'.
2. Specifically adsorbed ions or molecules block a certain part of the electrode surface, so that it may lead essentially to a reduction in the emission current from the areas covered with the adsorbate. Furthermore, we cannot altogether rule out the fact that the adsorbed particles may increase the quantum yield of the photoeffect, in which case the adsorption of particles on the electrode surface may be accompanied by an increase in the photocurrent.

Both of these cases were analyzed in detail from a theoretical viewpoint by Sheberstov, Brodsky, and Gurevich (53,54,157), who used a model in which the specifically adsorbed particles form a regular structure on the metal surface. The interaction of the adsorbed particles with an electron is characterized by a scattering length. Under these assumptions it is possible to interlink the phenomenologically introduced one-dimensional Ψ'-potential with individual characteristics of the adsorbed particles, surface coverage and geometry of the surface structure.

We here examine in brief outline form only the qualitative aspects of the problem. The first of these effects [item (1)], namely, change in the Ψ'-potential and its effect on the photocurrent, is most apparent during the specific adsorption of halogen anions on the electrode, its coverage by these anions usually being small.

Because of the wave properties of the emitted electrons and the fact that photodiffusion phenomena take place well remote from the surface, the effect of adsorption of halogen anions on the photocurrent in terms of the Ψ'-potential might be well exhibited only in dilute solutions. All of those concepts of the effect of Ψ'-potential on the photocurrent that apply when there is no specific adsorption are also applicable to this case as well. We may recall here that the work function for the electron emission from metal into electrolyte and the threshold potential ϕ_o are not influenced by the adsorption of anions on the electrode surface in concentrated electrolyte solutions (see Section 4.2.) insofar as the thickness of the diffuse layer is sufficiently small and the threshold approximation condition (Eq. 11) is satisfied.

The photocurrent curves, recorded on a mercury electrode in 0.01M KF, KCl, KBr, and KI solutions, are shown in Fig. 14 (112). For well-negative potentials, when the specific adsorption can be neglected, the photocurrent curves coincide for all these four solutions. A reduction in the photocurrent is observed, as compared with the current in KF solutions, in the region of specific adsorption of halogen ions. This is consistent with the adsorption activity of anions. The point where this reduction in photocurrent onsets in the series Cl^-, Br^-, I^- shifts toward the negative potential region. In the range of more positive potentials the photocurrents for Br^- and I^- show an increase as compared with that of F^-.

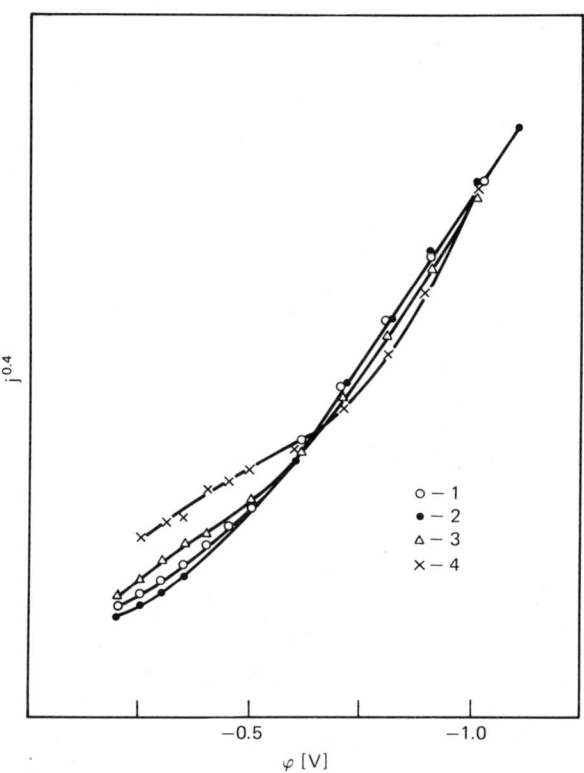

Fig. 14. Effect of halogen ion adsorption on photocurrent of mercury electrode (145): (1) KF, (2) KCl, (3) KBr, (4) KI (0.01N solutions).

The decrease in photocurrent observed in the adsorption of halides is associated with the effect of Ψ'-potential on the photoemission. Had the influence of halides consisted only in this Ψ'-effect, by the corresponding shift in the photocurrent curves it would have been possible to determine the "adsorptional" Ψ'-potential and to calculate the potential at the outer Helmholtz plane. Unfortunately, as Fig. 14 shows, a counter-influence is superimposed on this effect, and it dominates under positive potentials. The nature of the increasing photocurrent is in all probability correlated with the dual influence of the specific adsorption of anions. For negatively charged surfaces, when all of the emitted electrons practically are captured in the solution because of the action of the diffuse layer field, the photocurrent is very close in magnitude to the emission current, and the Ψ'-potential variation, as a result of adsorption of halides, exerts considerable influence mainly on the emission proper. An inverse situation is observed close to the zero charge potential under relatively low scavenger concentrations (e.g., in solutions saturated with nitrous oxide), namely, when most of the emitted electrons return to the electrode. Therefore, the effect of anion adsorption, accompanied by a shift in the Ψ'-potential toward more negative values, on the "back" current of the electrons to the electrode is more pronounced and prevails over the hampering effect of adsorption on the photoemission.

A more complex picture of the adsorbate effect on the photocurrent can be observed in the case of tetrabutylammonium (TBA) cation under large surface coverages (112,145). In concentrated solutions where the potential at the outer Helmholtz plane shows little variation, the adsorption of TBA at the electrode, if it at all affects the magnitude of the photocurrent, takes place mainly through the screening of the surface. In dilute solutions, besides this effect, the Ψ'-potential variations play an important role. Indeed, the change in the photocurrent during the adsorption of TBA on mercury and lead electrodes is appreciably less in concentrated than dilute solutions. Hence on these metals TBA influences the photocurrent largely through the variations in the Ψ'-potential. This conclusion is in qualitative agreement with the experimental results obtained for the effect of TBA on the hydrogen overvoltage (106). In the "ordinary" discharge potential range, where the hydrogen evolution overpotential depends on the magnitude of Ψ'-potential, it increases in the presence of TBA. In a barrierless process, where the Ψ'-potential has no effect on the overvoltage, there is practically no such TBA effect either. Thus both of these electron-transfer processes – hydrogen ion discharge and the photoemission act – are almost not inhibited by the particles adsorbed on the surface. The effect of these particles is largely due to the changes in Ψ'-potential.

Without any regard for the detailed mechanism of the effect of TBA on the photocurrent, the photoemission measurements can be used for quantitatively investigating the TBA adsorption. Figure 15 shows the curves obtained for a mercury electrode in a solution containing different amounts of TBA. Since the electrolyte is sufficiently concentrated, the five-halves law is preserved in the TBA adsorption region with the same threshold potential as in a "pure" solution. At potentials more negative than -1.1 V, TBA undergoes desorption. The $\Theta - \phi$ curves for different TBA concentrations (Θ = surface coverage) can be calculated in this potential range using the assumption of the additivity of photoemission currents from free areas on the electrode surface and those covered with the adsorbate.

Quantitatively they resemble similar curves obtained by the differential capacitance method (145).

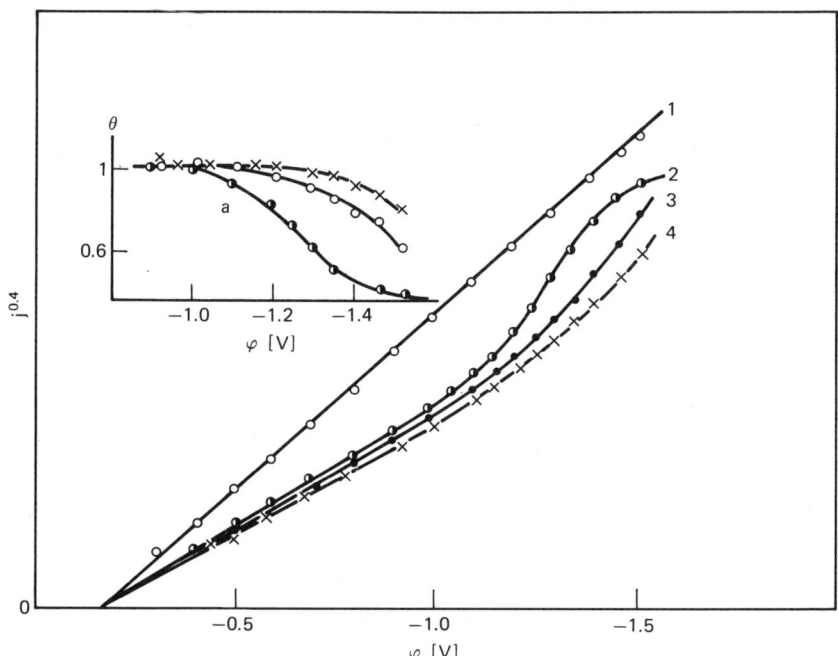

Fig. 15. Effect of TBA adsorption on photocurrent of mercury electrode (145). 0.1N KF saturated with N_2O; TBA concentration: (1) 0, (2) 10^{-3}, (3) 10^{-4}, (4) 10^{-5} mol/liter; (a) calculated potential dependence of surface coverage for − 1.3 V.

There are apparently other causes besides those mentioned on p. 64 to explain the effect of specific adsorption on the photoemission processes. The j − φ curve on a platinum electrode in alkaline solution saturated with N_2O is shown in Fig. 16 as an illustration. It draws our attention to the fact that the photocurrent shows a sharp drop in the potential range more negative than − 0.6 V, precisely the region where the adsorption of hydrogen takes place on platinum. It is really interesting to note that exactly in this potential region the electroreduction of N_2O is inhibited (6). Thus far the mechanism for the action of adsorbed hydrogen on photoemission from platinum electrode is obscure. We can only suppose that hydrogen under moderate surface coverages acts as an acceptor for electrons in metals, causing the subsurface layer of platinum to be depleted in electrons and the photoemission intensity to drop down.

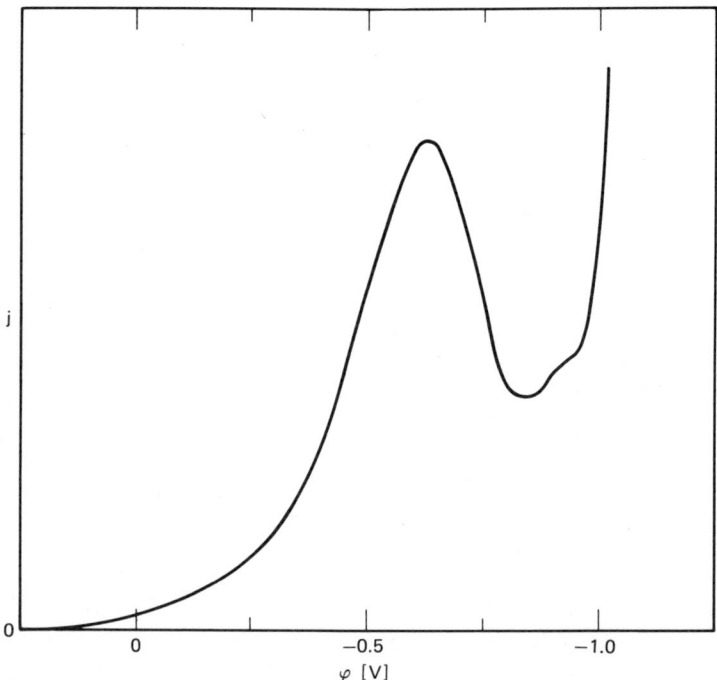

Fig. 16. Photocurrent on platinum electrode (6), 0.2N KOH saturated with N_2O.

5.6. Adsorption Layers of Large Thickness

The threshold approximation condition (Eq. 11) is violated for the adsorption of organic molecules with long hydrocarbon chains (of the order of 10 Å) so that the adsorption layer formed by them on the electrode surface (coverage close to unity) is sufficiently thick. For example, during the adsorption of cetyl alcohol and palmitic and myristic acids on mercury electrode, the photocurrent diminishes by 100 times (145). Here the five-halves law does not hold true, and the relationship between the photocurrent and potential was determined by model calculations for tunneling of electrons through a thick potential barrier of the adsorbed layer. The photoemission current was found to be a function of both thickness and dielectric permeability of the adsorbed layer and also of the difference between the work function for electron emission into the electrolyte and into this layer (156).

A combination of the photoemission and the capacitance measurements is a convenient method for determining the characteristics of the adsorbed layers. Since both the photocurrent and the differential capacitance depend on the thickness d_{ads} and the dielectric permeability ε_{ads} of the layer, so by a simultaneous photocurrent and capacitance measurement we can determine d_{ads} and ε_{ads} separately (66). Table II illustrates the results obtained in this way.

The values of d_{ads} proved to be very close to the length of the hydrocarbon chains in the investigated molecules calculated from the length of C-C bonds and the angles between them. This shows that the hydrocarbon chains are orientated normally to the electrode surface in such "condensed" layers. The reduction in the dielectric permeability with growing chain length probably shows that the content of water in the adsorbed layer decreases due to its increased hydrophobicity.

The results cited in this section form only a minor part of the material accumulated to date. We made an attempt to give manyfold examples for illustrating the possibilities of the photoemission method in double-layer studies.

TABLE II

Characteristics of Adsorbed Layer (Alcohol on Mercury) (66)

Alcohol	d_{ads}, Å	ε_{ads}
Hexyl alcohol	9.3	4.1
Octyl alcohol	11.7	3.2
Decyl alcohol	12.7	–
Cetyl alcohol	18.0	2.0

6. KINETICS OF ELECTROCHEMICAL AND CHEMICAL REACTIONS CONNECTED WITH PHOTOEMISSION OF ELECTRONS

Various processes arising in the solution after the capture of solvated electrons by the scavenger and leading to the formation of free radicals in the near-the-electrode electrolyte layer end

in electrode reactions. In the simplest case these are purely
electrochemical reactions not complicated by the homogeneous
transformations of the free radicals. An example is the electrochemical removal of atomic hydrogen from solutions. During the
emission of electrons into solutions containing NO_3^-, NO_2^-, CO_2,
and many organic molecules, the electrode reactions are usually
accompanied by chemical reactions in the bulk of the solution.
Here, the photoelectron emission acts as a source of free
radicals near the electrode surface, and at the same time it can
be used as a measuring instrument in the quantitative investigations of their reactions. Today many of these reactions have
been studied in sufficient detail, and a review of their special
features constitutes the subject matter for this section.

6.1. Kinetics of Electrochemical Transformation of Atomic Hydrogen

6.1.1. FORMATION OF ATOMIC HYDROGEN AND ITS REMOVAL FROM
SOLUTION. Atomic hydrogen, with its interesting electrochemical
properties, has since long been attracting the attention of
electrochemists. This is quite understandable as it is an intermediate in the thoroughly investigated reaction of cathodic
hydrogen evolution. During the photoemission process, hydrogen
atoms are generated in the electrolyte solution as a result of
scavenging of hydrated electrons by hydrogen ions:

$$e_{aq}^- + H_3O^+ \longrightarrow H + H_2O \tag{A}$$

Hydrogen atoms either recombine to molecules or enter into
electrochemical reactions on approaching the electrode surface.

Recombination is hardly a real way for the removal of hydrogen from the solution, because as a second-order reaction it proceeds at a very slow rate at the low atomic hydrogen concentrations usually achieved in the photoemission experiments. Therefore, practically the whole amount of atomic hydrogen formed in the solution diffuses to the electrode and is removed there via the electrochemical mechanism.

In a wide potential range hydrogen atoms undergo electrochemical desorption at an electrode that inadequately adsorbs atomic hydrogen:

$$H + HS^- + e^- \longrightarrow H_2 + S^- \qquad (B)$$

where HS stands for the proton source that might be hydrogen ions or water molecules in aqueous solutions.[+] At more positive potentials, in addition to such processes, ionization of atomic hydrogen is also possible:

$$H + H_2O \longrightarrow H_3O^+ + e^- \qquad (C)$$

As does any other cathodic reaction, this reduces the photocurrent. This reaction, first noticed by Barker, Gardner, et al. (28) on mercury electrode, was subsequently studied in detail on both mercury and some solid electrodes (142, 149, 150).

Here photoemission acts as a source of atomic hydrogen in the vicinity of the electrode surface. As the hydrogen formation rate is easily controlled in such experiments, the photoemission

---

[+] Krishtalik (107) was the first to point out the possibility of removing hydrogen via the electrochemial mechanism involving water molecules in acid solutions.

method is more convenient than the methods used earlier (15,71,114).[+]

With respect to the electrode reactions (Reactions B and C), the photocurrent j measured in the experiments is determined by the following equations:

$$j = j_o + \vec{j}_H - \overleftarrow{j}_H \qquad (32)$$

where $j_o = I - I_e$ is the emission current minus the reverse current of returning electrons not captured by scavengers, $\vec{j}_H$ is the cathodic current of Reaction B, and $\overleftarrow{j}_H$ is the anodic current of Reaction C. In principle, two mechanisms are possible for Reactions B and C, namely via the adsorption of hydrogen atoms on the electrode surface with the formation of H_{ads} or directly from the dissolved state bypassing the adsorbed state. If both these steps are taken into account, under steady-state conditions the following equations hold:

$$j_o = (k_o + \vec{k}_1 + \overleftarrow{k}_1) \, c_H(o) \qquad (33)$$

$$k_o c_H(o) = (\vec{k}_2 + \overleftarrow{k}_2)\Theta \qquad (34)$$

$$j = j_o + (\vec{k}_1 - \overleftarrow{k}_1) \, c_H(o) + (\vec{k}_2 - \overleftarrow{k}_2) \qquad (35)$$

where k_o is the adsorption rate constant expressed, as the

[+] The photoemission method of generating atomic hydrogen near the electrode surface is similar to the method developed by Levina and Kalish (114) in which the atomic hydrogen is produced during the glow discharge and then diffuses to the electrode through a thin layer of solution.

other kinetic constants, in electrical units, $\vec{k}_1$ and $\overleftarrow{k}_1$ are the rate constants of the cathodic and anodic removal of atomic hydrogen directly from the solution, $\vec{k}_2$ and $\overleftarrow{k}_2$ are similar rate constants for the adsorbed hydrogen, $c_H(o)$ is the near-the-electrode concentration of atomic hydrogen, and Θ is the surface coverage by hydrogen. We have neglected the recombination of adsorbed hydrogen atoms on the electrode surface in Eqs. 33 - 35, what is justified in the case of an electrode with high hydrogen overvoltage. If the Reaction B proceeds with the simultaneous participation of water molecules and hydrogen ions, then the rate constants $\vec{k}_1$ and $\vec{k}_2$ can be formally divided into two terms $\vec{k}_i = \vec{k}_i' + X\vec{k}_i''$, where $\vec{k}_i'$ and $\vec{k}_i''$ are the rate constants of hydrogen removal with the participation of water molecules and of hydrogen ions respectively, X is the molar fraction of H_3O^+.

On eliminating Θ and $c_H(o)$ from Eqs. 33 - 35, we obtain

$$j = 2 j_o \left[\frac{\vec{k}_1}{k_o + \vec{k}_1 + \overleftarrow{k}_1} + \frac{k_o \vec{k}_2}{(k_o + \vec{k}_1 + \overleftarrow{k}_1)(\vec{k}_2 + \overleftarrow{k}_2)} \right] \quad (36)$$

In the limiting case where it is possible to neglect the oxidation reactions ($\overleftarrow{k}_1 = \overleftarrow{k}_2 = 0$), $j = 2 j_o$, we obtain the usual dependence (the five-halves law in concentrated solution) for the photocurrent.[+] In a general case, as the rate constants in Eq. 36 are functions of the potential, the $j - \phi$ dependence is more complicated. Such an equation would be difficult to verify in experiments and thus it is more convenient to consider the limiting cases for comparing the theory with the experiment.

[+] Such a potential dependence of the photocurrent is also observed when the rate constants in Eq. 36 do not depend on the potential, which is true, for instance, for activationless processes.

Reactions B and C proceed bypassing the adsorption step $(k_o = 0)$; then

$$j = \frac{2 j_o \vec{k}_1}{\vec{k}_1 + \overleftarrow{k}_1}$$

The potential dependence of the rate constants $\vec{k}_i$ and $\overleftarrow{k}_i$ can be expressed by the following equations (59,75):

$$\vec{k}_i = \vec{k}_{io} \exp \frac{-\alpha_i F\phi}{RT}, \quad \overleftarrow{k}_i = \overleftarrow{k}_{io} \exp \frac{\beta_i F\phi}{RT}$$

In this case the $j - \phi$ dependence takes the form:[+]

$$\phi = \frac{2.3 RT}{(\alpha_1 + \beta_1)F} \left(\log \frac{2 j_o - j}{j} + \log \frac{\vec{k}'_{1o} + x\vec{k}''_{1o}}{\overleftarrow{k}_{1o}} \right) \quad (37)$$

Atomic hydrogen enters into both reactions only from the adsorbed state ($\vec{k}_1 = \overleftarrow{k}_1 = 0$). From Eq. 36 we obtain

$$\phi = \frac{2.3 RT}{(\alpha_2 + \beta_2)F} \left(\log \frac{2 j_o - j}{j} + \log \frac{\vec{k}'_{2o} + x\vec{k}''_{2o}}{\overleftarrow{k}_{2o}} \right) \quad (38)$$

Reaction B proceeds via the adsorption step, while Reaction C does not ($\vec{k}_1 = \overleftarrow{k}_2 = 0$), then

$$\phi = \frac{2.3 RT}{\beta_1 F} \left(\log \frac{2 j_o - j}{j} + \log \frac{k_o}{\overleftarrow{k}_{1o}} \right) \quad (39)$$

[+] In deriving Eq. 37, as well as Eqs. 38 and 40, it was assumed that the transfer coefficient α_i is independent on whether Reaction B involves water molecules or hydrogen ions. In a general case, the transfer coefficient α_i might prove to be different for the two mechanisms of the electrochemical desorption.

We assume here that the rate constant of adsorption k_o is independent of the potential.

Reaction B takes place bypassing the adsorbed state, while reaction C proceeds via H_{ads} ($\vec{k}_2 = \overleftarrow{k}_1 = 0$); hence we have

$$\phi = \frac{2.3\ RT}{\alpha_1 F} \left(\log \frac{2\ j_o - j}{j} + \log \frac{k'_{1o} + Xk''_{1o}}{k_o} \right) \qquad (40)$$

Equations 37 - 40 cover all of the real mechanisms for the removal of atomic hydrogen. In their form they are similar to the kinetic equations for the slow discharge except that in place of the discharge current, these equations contain the dimensionless parameter $(2\ j_o-j)/j$, which equals the ratio of current for hydrogen oxidation $\overleftarrow{j}_H$ to the current for its reduction $\vec{j}_H$. The structure of Eqs. 37 - 40 is such that a linear relation must exist between ϕ and $\log\{(2\ j_o-j)/j\}$; moreover, the slope of the line characterizes the transfer coefficients for the respective reactions.

These equations can be easily modified if we take into account the effect of the electrical double-layer structure on the rate of the electrode processes. For this purpose it is sufficient to introduce the Ψ'-potential into the expression for the rate constants $\vec{k}_i$ and $\overleftarrow{k}_i$, and then express the near-the-electrode concentration of hydrogen ions (more exactly, the molar fraction X) in terms of Ψ'-potential. This question is given detailed consideration in the subsequent pages, where we review the experimental results.

6.1.2. BASIC KINETIC LAWS OF ELECTROCHEMICAL REMOVAL OF ATOMIC HYDROGEN.

Potential Dependence

Figures 17 and 18 show the potential dependence of the photocurrent in $j^{0.4} - \phi$ coordinates for mercury and bismuth obtained in different solutions containing hydrogen ions as a scavenger.

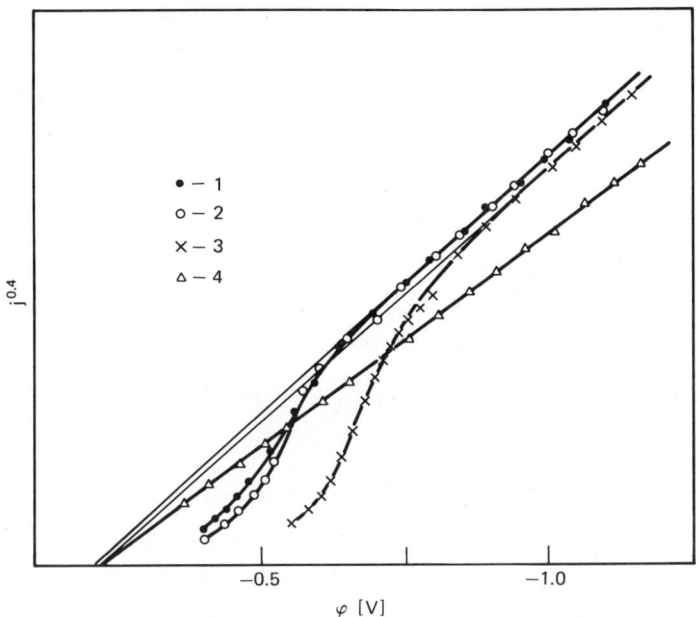

Fig. 17. "Hydrogen drop" of photocurrent on mercury electrode (142). Solutions: (1) 0.1N KCl, (2) 0.1N KBr, (3) 0.1N KI, (4) 0.1N KCl + 10^{-4}M TBA containing 0.01 mol/liter of HCl.

When there is no appreciable specific adsorption (e.g., in chloride solutions), the curve for the mercury electrode shows a deviation from the linear law at potentials more positive than -0.6 to -0.7 V. On the bismuth electrode the reduction in photocurrent is observed at potentials more positive than -0.8 V.[+]

For an experimental verification of the validity of Eqs. 37 – 40, in addition to the potential dependence of the photocurrent, j, we must ascertain the current $2 j_o$ in the absence of any oxidation of atomic hydrogen. This can be determined either by extrapolating the linear segment of the $j^{0.4} - \phi$ curve to the positive potential region (where the linear dependence is violated) or by directly measuring the photocurrent with such scavengers whose products of interaction with hydrated electrons do not undergo oxidation at the electrode. Nitrous oxide, for example, is such a scavenger that gives rise to an OH radical after the capture of an electron. This OH radical is easily reduced at the electrode in the entire potential range under consideration. By comparing the photocurrents for hydrogen ions and N_2O, we can calculate the

---

[+] A reduction in the photocurrent due to anodic oxidation of hydrogen atoms was observed also on antimony electrodes. The lower hydrogen overvoltage of the antimony electrode, however, prevented us from achieving sufficiently high negative potentials at which oxidation of hydrogen atoms could be neglected. It was not possible to detect the "hydrogen drop" in the photocurrent on other investigated solid metals. On a lead electrode, hydrogen ionization is apparently noticeable; that is, its rate is comparable with the electrochemical desorption rate at more positive potentials than on bismuth or antimony.

magnitude of $(2j_o - j)/j$ and thus construct the experimental curves in $\phi - \log\{(2j_o - j)/j\}$ coordinates.

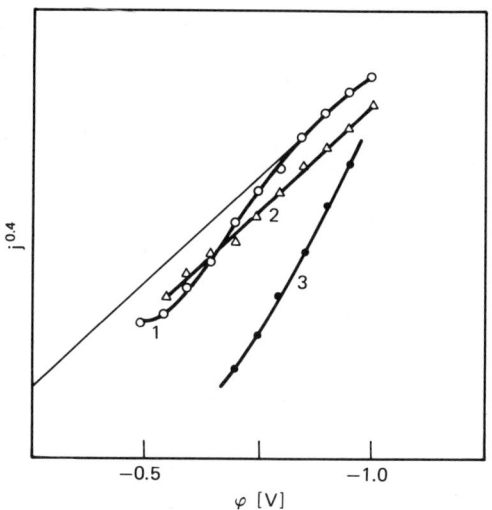

Fig. 18. "Hydrogen drop" of photocurrent on bismuth electrode (151): (1) 0.4N KCl + 0.1N HCl, (2) 0.4N KCl + 0.1N HCl + 2 × 10^{-4}M TBA, (3) 0.2N KI + 0.1N HCl.

Such curves were obtained for mercury and bismuth electrodes and are shown in Figs. 19 and 20, respectively. A linear relationship holds true between $\log\{(2j_o - j)/j\}$ and ϕ for the mercury electrode in the entire range of potential and for all the investigated solutions. On the bismuth electrode, however, although the logarithm of the ratio $(2j_o - j)/j$ likewise increases linearly with the potential in the range of − 0.8 to − 0.6 V, it is practically independent of the potential for $\phi > - 0.6$ V. For the mercury electrode the slope of the line is about 100 to 120 mV. For the bismuth electrode at potentials more negative than -0.6 V the slope is 140 to 150 mV. Using Eqs. 37 - 40 we can determine from these slopes the transfer coefficients of the respective reactions.

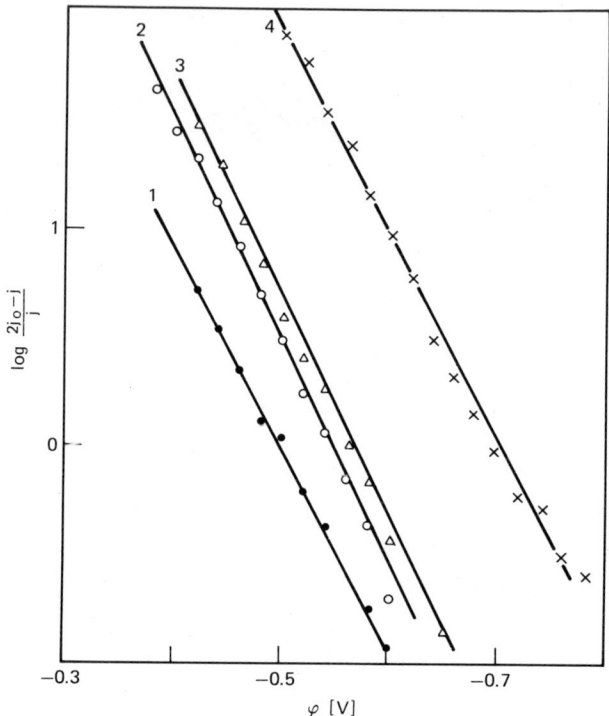

Fig. 19. Determination of transfer coefficients for atomic hydrogen reactions on mercury electrode (142). Solutions: (1) 0.1N Na_2SO_4, (2) 0.1N KCl, (3) 0.1N KBr, (4) 0.1N KI containing 0.01 mol/liter of HCl.

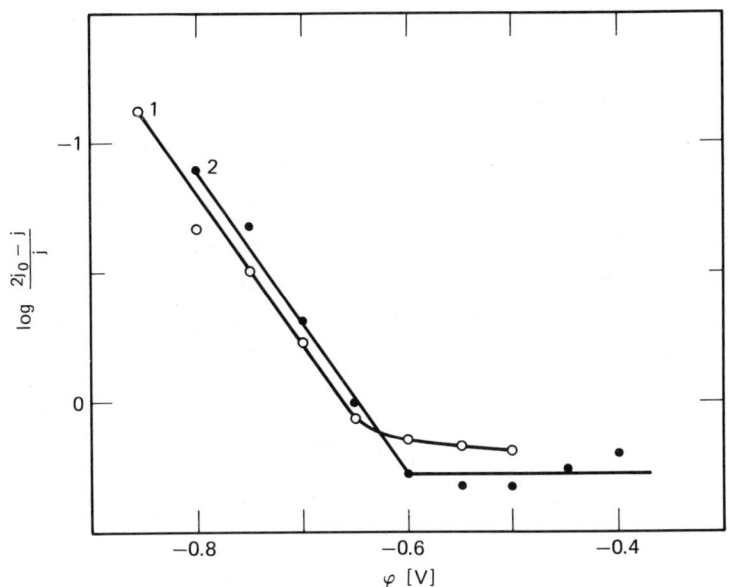

Fig. 20. Determination of transfer coefficients for atomic hydrogen reactions on bismuth electrode (151): (1) 0.5N KBr + 0.003N HCl, (2) 0.5N KCl + 0.06N HCl.

Effect of Specific Adsorption

The kinetics of the electrochemical removal of atomic hydrogen depends not only on the electrode potential, but also on the structure of the electrical double layer and the composition of the electrolyte solution. From Figs. 17 and 18 it is clear that the specific adsorption of halide ions leads to an increase in the current ratio ($\overleftarrow{j}_H / \overrightarrow{j}_H$) at any fixed potential. In the adsorption of I^- on bismuth, in contrast to the mercury electrode, it has not been possible to obtain a section corresponding to the linear relationship between $j^{0.4}$ and ϕ, as the drop in the photocurrent due to hydrogen ionization and the decrease in the photocurrent due to the "dark" discharge of hydrogen ions at the electrode take place in the same potential range.

In addition to the adsorption of halides, the adsorption of tetrabutylammonium cations has a definite influence on the atomic hydrogen removal rate. While the adsorption of Br^- and I^- leads to a shift in the potential, where the photocurrent decays, to more negative values, by adsorption of tetrabutylammonium, though the photocurrent shows a certain decrease over the entire potential range,[+] the "hydrogen drop" as such vanishes almost altogether.

Effect of pH

As Reaction B is one of the two conjugate reactions of removal of atomic hydrogen, then according to Eqs. 37, 38, and 40 the kinetics of the total process must depend on the concentration of hydrogen ions in the solution. For a quantitative investigation of the pH effect on the photocurrent it is more convenient to introduce the potential ϕ^* at which the cathodic and anodic

---

[+] The reduction in the photocurrent magnitude due to the adsorption of tetrabutylammonium is not specific to acid solution; it is also observed in N_2O solutions, as already mentioned in section 5, as a result of the effect of adsorption on the emission act proper.

currents are equal,[+] that is, $(2 j_o - j)/j = (\overleftarrow{j}_H/\overrightarrow{j}_H) = 1$. Figure 21 shows the dependence of ϕ^* on the hydrogen ion concentration ranging from 0.001 to 1 mol/liter for the mercury electrode. The potential ϕ^* monotonically increases with the acidity of the solution, and its greatest variation is observed for high $c_{H_3O^+}$. The ϕ^* potential practically does not depend on the pH value in the low acidity range (< 0.01M). Such a pH dependence of the potential ϕ^* shows that both hydrogen ions and water molecules participate in the electrochemical desorption.

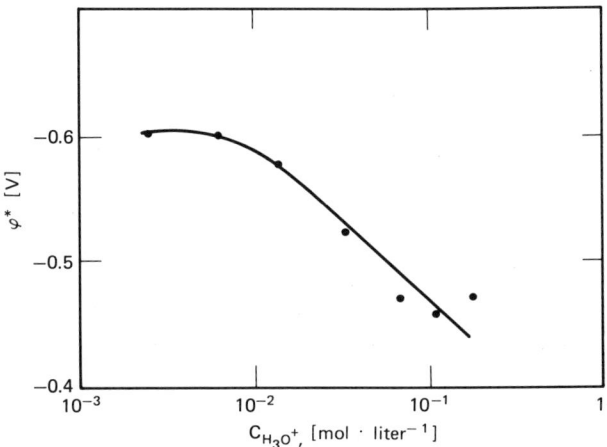

Fig. 21. ϕ^* potential on mercury electrode versus hydrogen ion concentration (142).

[+] The potential ϕ^* can be found by the point of intersection of the experimental $j^{0.4} - \phi$ curve with the straight line drawn from the threshold point with a slope of the experimental line in the negative potential range divided by $2^{2/5}$.

The pH dependence of the potential ϕ^* is weakly pronounced at the bismuth electrode. Within the limits of experimental error the potential ϕ^* remains almost constant in the hydrogen ion concentration range from 10^{-3} to 1 mol/liter. With the help of Eqs. 37, 38, and 40, and the plots shown in Fig. 21, it is not difficult to determine the relationship between the rate constants of the cathodic removal of atomic hydrogen involving hydrogen ions and water molecules. For this purpose we rewrite Eq. 38 as follows:

$$\phi^* = b \log\left(\frac{\vec{k}'_{2o}}{\overleftarrow{k}_{2o}} + X \frac{\vec{k}''_{2o}}{\overleftarrow{k}_{2o}}\right) \tag{41}$$

This gives the ratios $(\vec{k}'_{2o} / \overleftarrow{k}_{2o})$, $(\vec{k}''_{2o} / \overleftarrow{k}_{2o})$, and consequently $(\vec{k}''_{2o} / \vec{k}'_{2o})$ from the corresponding values of ϕ^* for different X values. The slope "b" is equal to the slope of the line $\phi - \log\{(2j_o-j)/j\}$ (Figs. 19 and 20). For the mercury electrode the ratio $(\vec{k}''_{2o} / \vec{k}'_{2o})$ is equal to 1600, but for the bismuth electrode it is less (~ 50). In 1M acid solution (X = 1/55) the electrochemical desorption on the mercury electrode proceeds mainly involving hydrogen ions. This is also evidenced by the slope of the line $\phi^* - \log c_{H_3O^+}$ in the high $c_{H_3O^+}$ range, which is close to 100 mV. Indeed, from a comparison of Eqs. 37 and 41, it is obvious that if the electrochemical removal of hydrogen via Reaction B takes place totally through hydrogen ions, the slope of the line $\phi^* - \log c_{H_3O^+}$ must be the same as that of the line[+] $\phi - \log\{(2j_o-j)/j\}$. Only in low acidic solutions ($c_{H_3O^+} < 0.01$ mol/liter) do we have $X (\vec{k}''_{2o} / \vec{k}'_{2o}) \ll 1$, and there is practically no participation of hydrogen ions in the process.

[+] This was pointed out to us by Dr. D. J. Schiffrin (private communication).

The role of hydrogen ions in the electrochemical desorption of hydrogen atoms on bismuth electrode is weakly expressed. Reaction B in 1M HCl takes place with about equal rates involving hydrogen ions or water molecules. At still lower $c_{H_3O^+}$ it proceeds mainly involving water molecules.

6.1.3. MECHANISM OF ELECTRODE REACTIONS INVOLVING ATOMIC HYDROGEN

Role of the Adsorption Stage in the Total Photoprocess

Reactions B and C of electrochemical removal of atomic hydrogen from solutions are extremely exothermic processes. Their heat effects when dissolved hydrogen directly enters into the reactions are equal to

$$q_1 = \Delta H + q \quad \text{(for Reaction B)}$$
$$q_2 = \Delta H - q \quad \text{(for Reaction C)}$$

where ΔH, the enthalpy of atomic hydrogen formation, is 52 kcal/mol (100), and q is the equilibrium latent heat of the electrode process $H_3O^+ + \bar{e}\ (M) \longrightarrow 1/2\ H_2 + H_2O$. Here it has been assumed that the H_3O^+ ions act as the proton source in Reaction B. The value of q can be determined from the experimental data on the barrierless H_3O^+ discharge (108). This latent heat is equal to -3.6 kcal/mol at the potential of normal hydrogen electrode (151). Hence we find that q_1 = 48.4 and q_2 = 55.6 kcal/mol. When both of these reactions occur via the adsorption step, their heat effects are diminished by an amount equal to the atomic hydrogen adsorption energy, which is 29 kcal/mol (or 1.26 eV) for the mercury electrode (106).

The high values of the heat effects of both reactions suggest that they are activationless processes with rate costants independent of the potential (i.e., at potentials more negative than that of the normal hydrogen electrode). Such a conclusion can be drawn for the barrierless discharge of H_3O^+ on mercury. In particular, the electrochemical desorption of hydrogen, in which atomic hydrogen exists as H_{ads} in the initial state, is an activationless process over the entire investigated potential range. Therefore, there is little doubt that Reaction B, in which free hydrogen atoms participate (with an energy greater than that of H_{ads} by 29 kcal), is an activationless process.

The hydrogen ionization is a reverse reaction with respect to the discharge of H_3O^+ ions provided that hydrogen atoms are adsorbed in the initial stage of ionization. The transition from the ordinary (activated) ionization to the activationless one, in this case, corresponds to the transition from the ordinary discharge of H_3O^+ to the barrierless discharge. In the absence of any specific adsorption, the transition to barrierless discharge on a mercury electrode in 0.25N H_2SO_4 + 1N Na_2SO_4 occurs at an overvoltage of 0.177 V (108), which corresponds to a potential of - 0.5 V (SCE). Taking into account the hydrogen adsorption energy we can conclude that the ionization of dissolved hydrogen atoms must be an activationless process right up to - 1.7 V.

The activationless mechanism inferred for Reactions B and C in which atomic hydrogen enters directly from the solution is valid for any metal as the initial and final states in such a process do not depend on the adsorption energy. Thus if both of the conjugate reactions would proceed directly from the solution, the ratio of their rate constants should not depend on the potential over the entire investigated potential range. But this is in contradiction with the experimental results (Figs. 19 and 20). Also inconsistent with the experiment are two other mechanisms for hydrogen removal with one of the conjugate reactions taking

Mechanism of Hydrogen Atom Removal

place directly from the solution, with the other proceeding via the adsorption step (Eqs. 39 and 40). The preceding arguments suggest that the transfer coefficients to be inserted into Eqs. 39 and 40 should be zero.

Thus we arrive at the conclusion that the removal of atomic hydrogen from the solution proceeds via the following steps:

$$H \text{ (solution)} \longrightarrow H_{ads}$$

$$H_{ads} + HS + e^- \longrightarrow H_2 + S^- \qquad (B')$$

$$H_{ads} + H_2O \longrightarrow H_3O^+ + e^- \qquad (C')$$

The kinetic equation (Eq. 38) gives a quantitative description of these steps.

Barker and McKeown (16,17,30) reached the same conclusion in their studies on the photocurrents in acidic solutions in the presence of ethanol.

Electrochemical Transformation of Atomic Hydrogen on Mercury, Bismuth, and Other Metals

The slope of the $\phi - \log\{(2j_o-j)/j\}$ lines is close to 100 - 120 mV on mercury electrodes for different solutions (see Fig. 19). It suggests that the sum of the transfer coefficients for Reactions B and C is close to $1/2$.[+] This can be explained either

---

[+] In interpreting the slope one must note the distortion caused in the transfer coefficients α_i and β_i by the dependence of ψ' on the electrode potential. Parsons (122) asserts that such a change of the transfer coefficients in halide solutions does not exceed 10%. Apparently, the lower slope of the line $\phi - \log\{(2j_o-j)/j\}$ in KI solution (100 mV) (Fig. 19) can be attributed to this effect.

by the fact that the value of the transfer coefficient is less than 1/2 for each single reaction (e.g., 0.2 and 0.3) or by the fact that one of these reactions is activationless, that is, that its transfer coefficient is close to zero and the transfer coefficient of the other reaction is 1/2. The coeffcient α_2 is not known for the electrochemical desorption of hydrogen on mercury, but it can be easily determined for ionization if we know the transfer coefficient for the reverse reaction - the hydrogen ion discharge. As already mentioned, the discharge of hydrogen ions on mercury has been investigated in thorough detail (75), and thus we can assume that $\alpha = 1/2$ is an established fact for sufficiently negative potentials (i.e., ordinary discharge range). In the more positive potential range where the discharge of H_3O^+ is a barrierless process, $\alpha = 1$ (106). Thus only two values are possible for the transfer coefficient β_2, namely, $\beta_2 = 0$ or $\beta_2 = 1/2$. Since it is established by experiment that the transfer coefficient of electrochemical desorption α_2 is linked with the coefficient β_2 by the relationship $\alpha_2 + \beta_2 = 1/2$, it follows that α_2 might also be equal to 0 or 1/2. Thus one of the two reactions - electrochemical desorption of hydrogen or its ionization - is activationless. The experimental data that we have cited for the mercury electrode, however, are not adequate for drawing an unequivocal conclusion regarding which of these reactions is the activationless one.

It is on bismuth that definite results have been obtained. Before we proceed to a discussion of these results, we examine the general character of the $(\overleftarrow{j}_H/\overrightarrow{j}_H) - \phi$ curve. For the two conjugate reactions considered three slopes may be tenable for the lines expressing the potential dependence of the current ratio $(\overleftarrow{j}_H/\overrightarrow{j}_h)$ depicted schematically in Fig. 22. At high negative potentials, the electrochemical desorption is an activationless reaction ($\alpha_2 = 0$). However, the ionization of atomic hydrogen, as its rate decreases with the shift of the potential toward negative values, proceeds via the ordinary mechanism. Yet if such conditions are attained that the ionization activation energy is equal to the heat effect, this reaction becomes a barrierless process ($\beta_2 = 1$). Therefore, in the limiting case the potential dependence of ($\overleftarrow{j}_H/\overrightarrow{j}_H$) has a slope equal to 60 mV, that is, $\alpha_2 + \beta_2 = 1$ (region 1 in Fig. 22). When the electrode potential is highly positive the slope once again tends to 60 mV. In this case, the hydrogen ionization constitutes the activationless reaction, and the electrochemical desorption will be the barrierless process (region 5). Segments 2 and 4 shown on these lines correspond to the intermediate cases where one of the reactions is activationless, and the other proceeds via the ordinary mechanism with a transfer coefficient close to 1/2. For example, on segment 2 the ionization is activated while the electrochemical desorption is activationless, and the reverse applies on segment 4. On segment 3 both reactions are activationless and the ratio ($\overleftarrow{j}_H/\overrightarrow{j}_H$) is independent of the potential ($\alpha_2 + \beta_2 = 0$).[+] A comparison of the experimental curve (Fig. 19) and the theoretical scheme (Fig. 22) clearly shows that a mere knowledge of only one slope (120 mV) of a smooth line is insufficient for deciding whether a case pertains to segments 2 or 4 represented in Fig. 22. This choice will be unique if the experimental curve simultaneously covers any two of the segments shown in Fig. 22. It is precisely this case that is realized for the bismuth electrode.

The experimental data given in Fig. 20 show that a "limiting current" is observed at positive potentials ($\phi > -0.6$ V); that

---

[+] This case, in principle, might not be realized in practice.

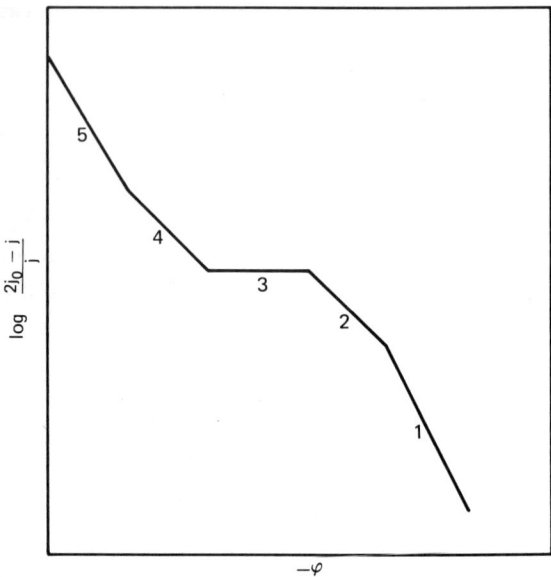

Fig. 22. Diagram for the determination of transfer coefficients of atomic hydrogen reactions.

is, the ratio of currents practically does not depend on the potential. This means, that both the reactions at potentials more positive than − 0.6 V are activationless (segment 3 in Fig. 22). As the electrochemical desorption at more negative potentials must apparently be activationless, the inflection observed at − 0.6 V is associated with the transition of hydrogen ionization from the activationless mechanism (ϕ > − 0.6 V) to the ordinary (activated) mechanism where $\beta_2 \neq 0$ (segment 2 in Fig. 22). Thus the slope of the line in the potential range ϕ < − 0.6 V characterizes the transfer coefficient for the ionization of adsorbed hydrogen. The slope of the line ϕ − log $\{(2j_o-j)/j\}$ (140 to 150 mV) shows that

Results for Mercury and Bismuth

β_2 = 0.4, which is noticeably less than on mercury. If we compare the value of β_2 with the transfer coefficient α for the H_3O^+ discharge reaction on bismuth, which is close to 0.6 (121,164), we find that $\beta + \alpha = 1$. This relationship, which is consistent with the thermodynamics, unambiguously shows that β_2 = 0.4 corresponds precisely to hydrogen ionization.[†] Thus the electrochemical desorption on the bismuth electrode is an activationless process in the entire investigated potential range and ionization at potentials $\phi < -0.6$ V proceeds with a finite activation energy which, however, becomes zero for $\phi > -0.6$ V.

In the region of limiting values of $(\overleftarrow{j}_H/\overrightarrow{j}_H)$ on bismuth the ioniziation current is not more than double that of the electrochemical desorption current. Hence it follows that in the electrochemical desorption of hydrogen atoms mainly water molecules, and not hydrogen ions, participate. Otherwise, the ratio of the desorption current to the ionization current, as on mercury, would have been proportional to the mole fraction of H_3O^+ ions. Such a conclusion is in agreement with the very weak $c_{H_3O^+}$ dependence of the potential ϕ^* obtained for bismuth. The activationless ionization observed on the bismuth electrode at potentials more positive than -0.6 V corresponds to the barrierless mechanism of H_3O^+ discharge if we assume that the energy states of H_{ads} arising in photoemission and in H_3O^+ discharge are identical. No data, however, are reported in the literature regarding the

---

[†] Here too, as in the case of mercury (see footnote on p. 89), we must bear in mind the variations in the transfer coefficients caused by the ϕ-dependence of the Ψ'-potential. However, in the potential range more negative than -0.6 to -0.8 V, the slope of Tafel line for H_3O^+ discharge from chloride solutions corresponds to a transfer coefficient close to 0.6 (164).

hydrogen overvoltage on bismuth in this potential range. Activationless ionization and electrochemical desorption of atomic hydrogen were also observed on an antimony electrode (149). Here the transition from the ordinary to activationless ionization of H_{ads} takes place at the same potential (- 0.7 V) where the ordinary discharge of hydrogen ions becomes a barrierless process (134).

It has not been possible to detect photoemission current on the platinum electrode in acid solutions (6). The cause for this, perhaps, is the rapid oxidation of atomic hydrogen that approaches the electrode surface. Indeed, if an excess H atom is planted onto the platinum electrode surface already covered with hydrogen, it will most likely be oxidized due to the equilibrium existing between H_{ads} and H_3O^+. The recombination of H_{ads} atoms on platinum proceeds at much slower rate than the discharge-ionization.

Structure of Electrical Double Layer and Mechanism of Removal of Atomic Hydrogen

The experimental results quoted in the previous pages clearly demonstrate that the specific adsorption has considerable influence on the rate of the removal of atomic hydrogen. Whereas adsorption of Br^- and I^- anions leads to an increase in the relative rate of oxidation of atomic hydrogen $[(\overleftarrow{j}_H/\overrightarrow{j}_H)$ increases; see Figs. 17 and 18], the "hydrogen drop" of the photocurrent vanishes almost completely in the course of adsorption of the TBA cation. Thus it is logical to attribute these effects of specific adsorption to the variations in the structure of the electrical double layer.

With respect to the Ψ'-potential, the kinetic equation for $(\overleftarrow{j}_H/\overrightarrow{j}_H)$ can be rewritten as

$$\frac{\overleftarrow{j}_H}{\overrightarrow{j}_H} = \frac{\overleftarrow{k}_{2o} \exp\left[\beta_2(\phi-\Psi')F/RT\right]}{\overrightarrow{k}'_{2o}\exp\left[-\alpha'_2(\phi-\Psi')F/RT\right]+X\overrightarrow{k}''_{2o}\exp(-F\Psi'/RT)\exp\left[-\alpha''_2(\phi-\Psi')F/RT\right]} \quad (42)$$

where α'_2 and α''_2 denote the transfer coefficients for the electrochemical desorption involving water molecules and hydrogen ions, respectively.

We assume here that the electrochemical desorption is activationless, and furthermore that it involves mainly water molecules (the case corresponding to the bismuth electrode). Thus we have

$$\frac{\overleftarrow{j}_H}{\overrightarrow{j}_H} = \frac{\overleftarrow{k}_{2o}}{\overrightarrow{k}'_{2o}} \exp \frac{\beta_2(\phi-\Psi')F}{RT} \quad (43)$$

The shift in Ψ'-potential toward negative values should give rise to an increase in the ratio $(\overleftarrow{j}_H/\overrightarrow{j}_H)$. On the other hand, the shift in the Ψ'-potential toward positive values results in an opposite effect. The experimental results obtained for bismuth are consistent with the above predictions.

In the specific acsorption of TBA cations the concentration of hydrogen ions participating in the electrochemical desorption shows a strong diminuition near the electrode surface, and thus their reactions can be discarded. From Eq. 42 it follows that

$$\frac{\overleftarrow{j}_H}{\overrightarrow{j}_H} = \frac{\overleftarrow{k}_{2o}}{\overrightarrow{k}'_{2o}} \exp \frac{(\alpha'_2 + \beta_2)(\phi-\Psi')F}{RT}$$

The shift in Ψ'-potential toward the positive side, irrespective of α'_2, is accompanied by a diminuition in $(\overleftarrow{j}_H/\overrightarrow{j}_H)$; in other words, the "hydrogen drop" might vanish at this potential, in agreement with the experimental results.

It must be emphasized here that during the adsorption of TBA, a screening of the electrode surface takes place together with the variation in the Ψ'-potential. This screening prevents hydrogen atoms from approaching the metal surface. But it influences both the cathodic and anodic processes equally and thus has no effect on the ratio $(\overleftarrow{j}_H/\overrightarrow{j}_H)$.

In concluding we wish to mention that the idea of activationless processes in electrochemistry was suggested quite early (97,99). Nonetheless, because of diffusion limitations, no one has thus far succeeded in observing limiting currents of activationless discharge using direct methods. With the help of photoemission we can avoid these hindrances by measuring not the rate of individual processes but the ratio of rates of two simultaneous processes in which the same substance (atomic hydrogen) takes part. In this way, the diffusion distortions are mutually canceled out.

6.2. Electrochemical Reactions Following Chemical Transformations in the Bulk of the Solution

6.2.1. KINETIC EQUATIONS. Despite the diversity of the processes initiated in the electrolyte solution by photoemission of electrons, they can be represented by a common and sufficiently general scheme and thus described by one set of kinetic equations. As a rule, immediately after the capture of a solvated electron by a scavenger, free radicals are generated:

$$e^-_{aq} + A \longrightarrow R_1^{\cdot} \tag{D}$$

Depending on the nature of the scavenger (H_3O^+, N_2O, NO_3^-, NO_2^-, CO_2, etc.), these radicals might be, say, $H^{\cdot}$, $OH^{\cdot}$, NO_3^{2-}, NO_2^{2-}, or CO_2^-. The radical $R_1^{\cdot}$ in turn, might enter into electrochemical and chemical reactions:

$$R_1^{\cdot} + e^- \longrightarrow M \qquad (E)$$

$$R_1^{\cdot} \xrightarrow{\longleftarrow} A + e^- \qquad (E')$$

$$R_1^{\cdot} + P \longrightarrow N + R_2^{\cdot} \qquad (F)$$

$$R_2^{\cdot} + e^- \longrightarrow M_2 \qquad (G)$$

$$R_2^{\cdot} \longrightarrow B + e^- \qquad (G')$$

From Reactions D through G it is obvious that the electro-oxidation of $R_1^{\cdot}$ (Reaction E') and $R_2^{\cdot}$ (Reaction G') do not contribute to the toal photocurrent, as this current compensates for those photoemitted electrons that lead to the formation of the radical $R_1^{\cdot}$ (Reaction D). On the other hand, the resultant photocurrent will be doubled due to electroreduction (Reactions E and G). The resulting photocurrent can be determined by solving the system of diffusion equations for the solvated electrons (Eq. 22) and for the radicals $R_1^{\cdot}$:

$$D' \frac{d^2 c'}{dx^2} - k_v c' + k_A c_A = 0 \qquad (44)$$

under the boundary conditions:

$$k_{eA} c'(o) = D' \left.\frac{dc'}{dx}\right|_{x=0} \qquad \text{and} \qquad c'(\infty) = 0 \qquad (45)$$

where D' is the diffusion coefficient for the radical $R_1^{\cdot}$, $c'(x)$ is its concentration in the plane x, k_v is the rate constant of the homogeneous Reaction F, and k_{eA} is the rate constant of the heterogeneous oxidation (reduction) of $R_1^{\cdot}$.

In the following pages we show that if the radical undergoes oxidation at the electrode, the product of its homogeneous transformation is usually reducible, and vice versa. Therefore, our problem consists in finding the diffusion fluxes of $R_1^{\cdot}$ (if $R_2^{\cdot}$ is oxidized) or of $R_2^{\cdot}$ (if $R_1^{\cdot}$ is oxidized), which in the steady-state are interrelated by the equation:

$$eD' \left.\frac{dc'}{dx}\right|_{x=0} = I - I_e - eD'' \left.\frac{dc''}{dx}\right|_{x=0}$$

where D'' and c'' are the diffusion coefficient and the concentration, respectively, of the radical $R_2^{\cdot}$. With regard to Eqs. 22, 24, 44, and the boundary conditions (Eq. 45), we obtain the following expressions for the photocurrent (142):

$$j = \frac{2\,ek_{eA}}{Q_v D' + k_{eA}} \cdot \frac{Q^2}{Q^2 - Q_v^2} \int_0^\infty \Phi(x) \left(e^{-Q_v x} - e^{-Qx}\right) dx \qquad (46)$$

(if $R_1^{\cdot}$ is reduced) and

$$j = 2\left[I - e\int_0^\infty \Phi(x) e^{-Qx}\,dx - \frac{ek_{eA}}{Q_v D' + k_{eA}} \cdot \frac{Q^2}{Q^2 - Q_v^2} \int_0^\infty \Phi(x)\left(e^{-Q_v x} - e^{-Qx}\right) dx\right] \quad (47)$$

(if $R_1^{\cdot}$ is oxidized). Here Q_v stands for $(k_v/D')^{1/2}$.

6.2.2. ELECTROOXIDATION AND HOMOGENEOUS DECOMPOSITION OF THE ANION RADICAL NO_3^{2-}. The anion radical (= $R_1^{\cdot}$) formed after the capture of a hydrated electron by NO_3^- entered into the following reaction:

$$NO_3^{2-} + H_2O \longrightarrow NO_2^- + OH^{\cdot} + OH^-$$

$$NO_3^{2-} \longrightarrow NO_3^- + e^-$$

$$OH^{\cdot} + e^- \longrightarrow OH^-$$

In our scheme this process corresponds to Reactions E', F, and G' and is described by Eq. 47.[+]

At high negative potentials, when $k_{eA} \ll Q_v D'$ (oxidation of NO_3^{2-} can be neglected), we find that $j = 2(I - I_e)$. For the inverse limiting case ($k_{eA} \gg Q_v D'$), if the inequalities $Qx_o \gg 1$ and $Q_v x_o \ll 1$ also hold true, we have

$$j = 2 x_o Q_v I \qquad (48)$$

Experimental investigations of this reaction on electrodes of mercury (17,28,67,104,142) and solid metals (151) have shown that both limiting cases are indeed realized on the photocurrent curve (Fig. 23). The rectilinear segments in $j^{0.4} - \phi$ coordinates at high and low potentials are separated by a region of monotonic increase in the photocurrent (on the mercury electrode between -0.7 V and -1.2 V). Addition of small amounts (10^{-4} g equiv/l) of TBA bromide into the solution changes the shape of the curve; it becomes practically a straight line in the whole potential range of TBA adsorption. From this curve it is obvious that the influence of the adsorbed TBA consists of two effects: (a) reduction in the emission current (see preceding text) and (b) inhibition of NO_3^{2-} oxidation. Both (a) and (b) are responsible for the fact that there is no photocurrent drop at positive potentials.

Adsorption of the iodide anion (151) increases the photocurrent at positive potentials as a result of inhibition of the

---

[+] Another possible mechanism is (17): $NO_3^{2-} + H_2O \longrightarrow NO_2 + 2\ OH^-$; $NO_2 + e^- \longrightarrow NO_2^-$; $NO_3^{2-} \longrightarrow NO_3^- + e^-$. This change in mechanism would not affect the validity of the following discussion and, in particular, that of Eq. 47.

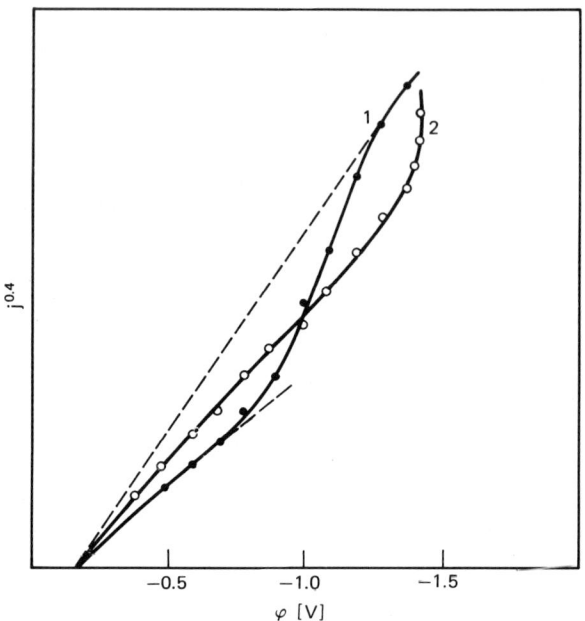

Fig. 23. Photocurrent on mercury electrode in 1N KNO_3 (1) and in the same solution with the addition of TBA (2) (67,142).

NO_3^{2-} oxidation. Iodide has practically no effect on the photocurrent outside the potential range of its adsorption on the electrode surface.

From Fig. 24 it is evident that the nature of the metal has considerable bearing on the kinetics of the NO_3^{2-} oxidation. On mercury, lead, and bismuth it begins at approximately the same potential. There is practically no photocurrent drop on indium due to the NO_3^{2-} oxidation in the investigated potential range. This is probably due to the decomposition, prior to oxidation, of NO_3^{2-}, catalyzed by the unreduced oxides present on the electrode surface.

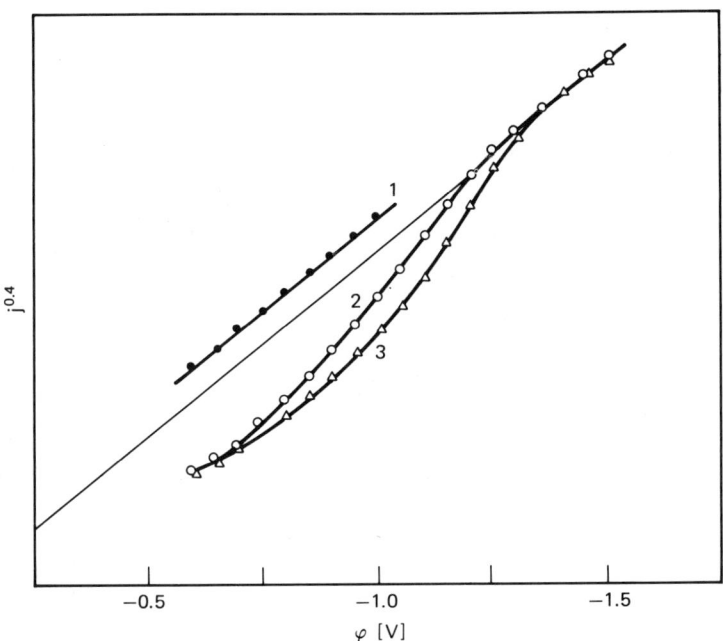

Fig. 24. Photocurrent in 1N KNO_3 on indium (1), bismuth (2), and lead (3) electrodes (151).

The role of surface oxides in the catalytic decomposition of NO_3^{2-} is very pronounced, especially on platinum electrodes. A noticeable photocurrent is observed in neutral or weakly alkaline solutions of NO_3^- at positive potentials (at which platinum is usually covered with an oxide layer), and it completely vanishes in weakly acidic solutions. Probably, here, just as in the case of an indium electrode, the surface oxides catalyze the NO_3^{2-} decomposition via Reaction F.

The parameter Q can be determined from a comparison of the actual photocurrent (Eq. 48) with the emission current (which, in the potential range corresponding to the photocurrent drop, is found by extrapolating the linear segment from the negative potential range); with the help of this parametric value we can estimate the rate constant k_v. Equation 48, however, is valid only if diffusion controls the NO_3^{2-} oxidation, and in the absence of catalytic decomposition of NO_3^{2-} on the electrode surface. The values of k_v found under these assumptions using three investigated metals (mercury, bismuth, and lead) as electrodes proved to be of the same order ($10^6 - 10^7$ s^{-1}).

The flash method presents another possibility for estimating the rate constant k_v from the slow part of the photoresponse transient indicating the formation of OH· radicals and their reduction on the electrode. The values of k_v found by this method were $4 \cdot 10^5$ s^{-1} (169) and 10^5 s^{-1} (17); thus they are somewhat lower than those obtained in stationary measurements.

This discrepancy in the values of k_v can be explained in two ways. First, it is possible that the homogeneous decomposition of NO_3^{2-} is more complex than that described by Eqs. 46 and 47. In the steady illumination method, the role of the first stage of the process (vanishing of NO_3^{2-}) is measured, but in the flash method it is the last stage that is obtained from the measurements (formation of products reduceable on the electrode). We cannot rule out the possibility that these stages exhibit considerable differences in their kinetic constants, nor that the rate of NO_3^{2-} oxidation on the investigated electrodes is not diffusion-controlled, with the observed photocurrent possibly depending on the rate constant of the electrooxidation. Similarly, we cannot preclude the possibility of catalytic decomposition of NO_3^{2-} on the electrode surface, a fact not accounted for in Eq. 48.

The potential dependence of k_{eA} and hence the transfer coefficient β for the electrooxidation of NO_3^{2-} on different metals were determined form the experimental data with the help of Eq. 47 for the limiting case: $Q \gg Q_v$ and $Qx_o \gg 1$. Thus it was found that $\beta = 0.25$ for mercury electrode, and 0.35 for lead and bismuth electrodes (67,104,151).

6.2.3. CHEMICAL AND ELECTROCHEMICAL REACTIONS INVOLVING CO_2^- AND $CO_3^{\cdot -}$ RADICALS. The redox transformation of CO_2^- in solution resembles the reactions with the participation of NO_3^{2-} that we considered in preceding sections. These processes, which are developed during the photoemission of electrons into solutions saturated with CO_2, have been investigated in detail by Schiffrin (153,154). Practically all CO_2^- radicals are oxidized at the electrode at positive potentials. Therefore, a steady photocurrent is observed only at potentials more negative than -1.4 V, when the oxidation of CO_2^- ceases, and this radical is reduced to the formiate ion in the presence of proton donors: $CO_2^- + H_3O^+ + e^- \longrightarrow HCOO^- + H_2O$.

Figure 25 shows the potential dependence of the stoichiometric number ν_{CO_2} of the total photoprocess. The half-wave potential $\phi_{1/2}$ can be easily determined from this curve (at the half-wave potential the rates of oxidation and reduction of CO_2^- are equal and hence $\nu_{CO_2} = 1$). For pH > 6.5 we have $\phi_{1/2} = -1.48$ V (SCE). Since this value does not depend on the pH value of the solution, it has been concluded that in this pH range water molecules act as the proton donor. In acid solution the half-wave potential $\phi_{1/2}$ shows a strong dependence on the pH value, and this is a clear indication that H_3O^+ ions participate in the reduction reaction.

Schiffrin used a similar approach for studying the reactions involving $CH_3^{\cdot}$ radicals formed after the capture of hydrated electrons by dissolved CH_3Cl. Unlike CO_2^-, the $CH_3^{\cdot}$ radical does

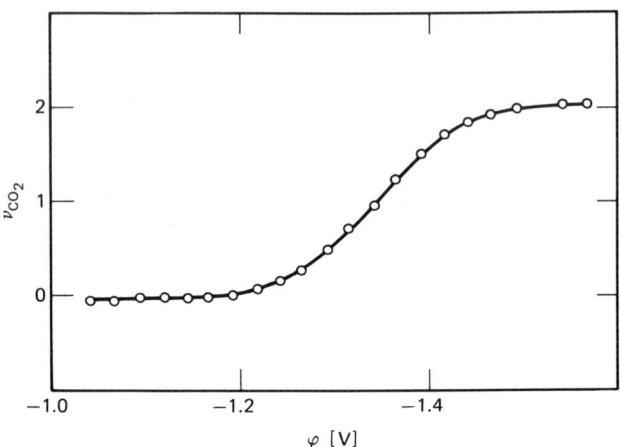

Fig. 25. Potential dependence of the stoichiometric number of CO_2^- reactions (154). Solution: 0.05N $KHCO_3$ + 0.05N KCl.

not undergo oxidation on a mercury electrode, but it dimerizes, yielding ethane. At potentials more negative than - 1.4 V it is reduced to methane. We must mention here that aliphatic monochloro derivatives cannot be directly reduced in aqueous solutions by the usual polarographic methods.

6.2.4. HOMOGENEOUS CHEMICAL REACTIONS INVOLVING H˙ AND OH˙ RADICALS. The radicals H˙ and OH˙ are the final products of the reaction of hydrated electrons with many scavengers. By the photoemission method it has been possible to qualitatively investigate the properties of these radicals, which participate in many fast reactions of radiation chemistry. Barker and colleagues have thoroughly investigated the interaction of atomic hydrogen and hydroxyl radicals with methanol and ethanol (16,20,21) and with formiate ions (22).

Radiation chemical investigations have shown that the H˙ and OH˙ radicals easily abstract α-hydrogen from alcohols. The new radicals RCHOH so formed might at the electrode surface be reduced to alcohol or oxidized to aldehyde, depending on the potential.

The scheme proposed for the conversion of hydrated electrons in the presence of alcohol when N_2O is used as a scavenger (16) is as follows:

$$N_2O + e^-_{aq} + H_2O \longrightarrow N_2 + OH^{\cdot} + OH^- \qquad (H)$$

$$OH^{\cdot} + e^- \longrightarrow OH^- \qquad (I)$$

$$OH^{\cdot} + RCH_2OH \longrightarrow R\dot{C}HOH + H_2O \qquad (J)$$

$$R\dot{C}HOH + H_2O \longrightarrow RCOH + e^- + H_3O^+ \qquad (K)$$

$$R\dot{C}HOH + H_2O + e^- \longrightarrow RCH_2OH + OH^- \qquad (L)$$

The total photocurrent is twice the current for the electroreduction of the radicals:

$$j = 2(j_1 + j_2) \qquad (49)$$

where j_1 is the current of reduction of OH˙ described by Eq. 46 over the entire investigated potential range and j_2 is the current resulting from the electroreduction of $R\dot{C}HOH$, which can be described by Eq. 47 only at negative potentials when Reaction K can be neglected. In a general case the value of j_2 constitutes only a fraction of this current (Eq. 47) and tends toward zero for positive potentials.

At low concentrations of nitrous oxide and of alcohol, when $Qx_o \ll 1$ and $Q_v x_o \ll 1$, and at high rates of OH˙ reduction ($k_{eA} \gg Q_v D'$), the expression for the resultant photocurrent (accounting for Eqs. 46, 47, and 49) takes the form:

$$\frac{j}{j_a} = 1 - y \left(1 - \frac{Q}{Q + Q_v}\right) \qquad (50)$$

where j_a is the photocurrent in the absence of alcohol in the solution, y is the fraction of RCHOH radicals formed that are oxidized at the electrode. If we neglect the reduction reaction (Reaction L) (positive potentials), then

$$\frac{j}{j_a} = \frac{Q}{Q + Q_v} \qquad (51)$$

In the other limiting case (high negative potentials) we have $(j/j_a) = 1$ and the addition of alcohol does not alter the photocurrent.

Equation 51 was extensively used by Barker (16) for analyzing the experimental data on photoemission into solutions containing alcohols (ethanol and methanol) and formiate ions.[+]

According to Eq. 51, the photocurrent must decrease with increasing alcohol concentration. Such a reduction in photocurrent is observed in the presence of methanol at potentials more positive than -1.4 V (Reactions I and K proceed on the electrode). At potentials more negative than -1.5 V the photocurrent shows a steep rise and attains the same value as in a solution devoid of alcohol. Such a picture is observed also for ethanol, however, the photocurrent increases only at -1.7 V.

[+] The CO_2^- anion radical is formed when $OH^{\cdot}$ reacts with $HCCO^-$ and is oxidized on the electrode at positive potentials. Therefore, Eqs. 50 and 51 are suitable for a quantitative description of the process and are also valid for the electron photoemission into acid solutions containing alcohol where a hydrogen atom is abstracted from alcohol molecules in the course of the reaction with atomic hydrogen.

The experimental curves plotted in $(1/j) - (cp)^{1/2}$ coordinates ($p \equiv C_2H_5OH$) are shown in Fig. 26. With the help of Eq. 51 it is possible to estimate the rate constant of $OH^\cdot$-alcohol interaction. It was found to be $2 \cdot 10^9$ mol$^{-1} \cdot$ l $\cdot$ s^{-1} for both methanol and ethanol. The rate constant of the interaction of atomic hydrogen with alcohols is $2.4 \cdot 10^7$ mol$^{-1} \cdot$ l $\cdot$ s^{-1}. These values are in satisfactory agreement with the results obtained in pulse radiolysis of alcohol solutions.+

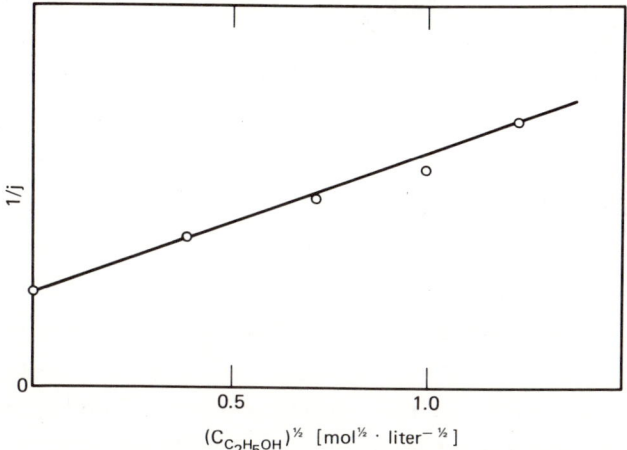

Fig. 26. Reaction of atomic hydrogen with ethanol (16). Solution: 0.2N KCl + 0.001N HCl with addition of ethanol; potential - 1.2 V.

+ In the photocurrent measurements in alcohol solutions we must also account for the effect of alcohols on the photoemission act proper as a result of alcohol adsorption on the electrode. This effect is distinctly pronounced for readily adsorbable alcohols with a long hydrocarbon chain (66). Therefore, a quantitative investigation of homogeneous $OH^\cdot$-alcohol interactions is possible only at such potentials where alcohol is desorbed from the surface.

Besides α-hydrogen other hydrogen atoms might as well be abstracted from the alcohol molecule in the interaction with OH·. A mixture of radicals is formed as a result of such an interaction, and it is very difficult to determine the "photoemission yield" for each of these radicals separately. A method has been proposed that can be effectively applied for estimating the yield of α-radicals in the total process y_α (20,21). Such data, together with the rate constants of the interaction of alcohols with OH· and H· radicals, are presented in Table III, which also shows the yields of α-radicals. The photoemission measurements are in good agreement with the radiation chemical measurements (1,5).

Separate stages of the total photoprocess have been studied in detail using flash techniques with the help of ruby lasers (second harmonics) (30). During the emission of electrons into N_2O solution containing $HCOO^-$, a reduction in photopotential is observed for a time interval of the order of 0.2 to 0.4 μs as a result of the oxidation of CO_2^-. From the kinetics of this potential drop the rate constant of CO_2^- electrooxidation was found to be $0.4 \text{ cm} \cdot \text{s}^{-1}$ at a potential of -1.3 V (SCE).

The electrooxidation of ethanol radicals proceeds at a much slower rate than that of CO_2^-, and its rate constant is 10^{-3} $\text{cm} \cdot \text{s}^{-1}$ in the potential range of -1.6 V to -1.4 V. Most probably, this slow step is the following chemical reaction

$$H_2O + \underset{\underset{Hg}{|}}{CH_3CHOH} \longrightarrow \underset{\underset{Hg}{|}}{CH_3CHO^-} + H_3O^+$$

accompanied by a fast electron transfer and desorption of the formed aldehyde. At more positive potentials the oxidation rate strongly depends on the potential. It is supposed (20) that the

TABLE III

Kinetic Characteristics of Interaction of H· and OH· with Alcohols (21,22)

Radical	Alcohol	$k_v \cdot 10^{-9}$, $mol^{-1} \cdot liter \cdot s^{-1}$		y_α	
		Photo-emission	Radiation chemistry (5)	Photo-emission	Radiation chemistry (1)
OH·	Methanol	0.63	0.43 to 0.65		
OH·	Ethanol	1.2	1.0 to 1.85	0.9	0.97
OH·	1-Propanol	0.9	1.7	0.43	0.61
OH·	2-Propanol	1.0	1.1 to 1.7	0.85	0.89
OH·	1-Butanol	1.5	2.2	0.4	0.34
OH·	tert-Butanol	0.3	0.25 to 0.42		
OH·	1-Pentanol	1.8		0.4	
H·	Methanol	0.024			
H·	Ethanol	0.024			
OH·	Formiate	2.4			

electron transfer from the adsorbed radical to the metal is a slow step.

When the product of the interaction between a hydrogen donor P (see Reaction F) and OH˙ radicals is reduced at the electrode, the abstraction of hydrogen atoms can be studied using a competing scavenger for OH˙ radicals (e.g., methanol). By investigating the competition for the capture of OH˙ between phenol and methanol the rate constant of the reaction:

$$C_6H_5OH + OH\text{˙} \longrightarrow C_6\text{˙}H_4OH + H_2O$$

was found to be equal to $2.8 \cdot 10^9$ mol^{-1} · liter · s^{-1} (45). The formed radicals, as flash measurements have shown, are reduced to phenol in the entire investigated potential range, in agreement with the previous results of stationary measurements (66).

The reactions of H˙ and OH˙ radicals with alcohols may serve as a convenient model for developing methods to investigate the mechanisms of more complicated reactions. An example of such a reaction is the homogeneous chemical conversion of the product of the reaction between hydrated electrons and sulfur hexafluoride (SF_6). Studying the photoemission into SF_6 solutions, Concialini, Tubertini, et al. (56) detected a reduction in the photocurrent in the presence of ethanol. Hence it was concluded that OH˙ radicals are formed in one of the steps of the process.

7. CONCLUSIONS

The study of photoelectron emission from metals into electrolyte solutions has become an original trend in fundamental electrochemistry. In addition to being a means for solving purely

electrochemical problems, photoemission into solutions is a
new approach for investigating the behavior of electrons with
an energy of the order of 1 eV in condensed media. Therefore,
we briefly discuss certain problems for which the photoemission
method, in our opinion will prove to be a promising investigation tool.

First, the range of objects for investigation is not limited
to metals that exhibit the property of an ideally polarizable
electrode. Of course, special interest still lies in further
investigation of the photoemission behavior of platinum group
metals where the structure of the interface is more complicated
than at the Hg-like electrodes. We expect that on Pt-type
metals with high surface coverage of hydrogen the heterogeneous
processes involving the products of photoemission induced
reactions might have special kinetic features.

Among other metals of special note are the metals covered with
oxide layers where the mechanism of generation of photoelectrons
and, consequently, the photoemission laws, might differ from those
governing the metals with clean surface. Quantitative data on the
photoemission in such systems might elucidate the mechanism of
oxide formation and its role in corrosion and passivation of
metals.

Photoemission of electrons into nonaqueous solutions is the
second promising field. The problems that arise here are associated with the energy and kinetic characteristics of excess
electrons in condensed media with different dielectric permeabilities and microstructure. Speical interest is likewise represented by the investigation of the specific features of
photoemission into solid electrolytes whose properties have
attracted increasing attention in recent years.

The third trend that has already gained sufficient weight in the photoemission investigations is the study of different mechanisms for the photoexcitation (certain mechanisms are mentioned in the Section 2.1.). Essentially new information can indeed be obtained with the use of laser sources of irradiation.

A number of problems can be solved with the help of the modern flash techniques. Equipment of high resolution (of the order of $10^{-8} - 10^{-9}$ s) and high accuracy are needed for a detailed study of diffusion processes that take place in the immediate vicinity of the elctrode surface and the kinetics of electrode reactions.

When these problems are solved, it will be possible to give a more complete and detailed description of the photoemission phenomena in electrochemical systems.

8. ACKNOWLEDGMENTS

The authors express their deep gratitude to the late Professor A. N. Frumkin, Professor L. I. Krishtalik and Dr. Yu.Ya. Gurevich for their valuable discussion of the problems outlined in this review and to Dr. Barker, who disclosed the results of some of his works prior to publication, and to Dr. K. P. Charlé for his constructive criticism.

9. LIST OF SYMBOLS

c_A Scavenger concentration
c_e Solvated electron concentration
c_{el} Electrolyte concentration
D_A Scavenger diffusion constant

List of Symbols

D_e	Solvated electron diffusion constant
e	Electron charge (absoute value)
E	Energy
F	Faraday constant
$\hbar=(h/2\pi)$	Planck constant
I	Photoemission current
j	Measured photocurrent
$\vec{j}_H, \overleftarrow{j}_H$	Electroreduction and electrooxidation current for atomic hydrogen
k	Boltzmann constant
k_A	Rate constant for scavenging of solvated electrons
k_o	Rate constant for atomic hydrogen adsorption at electrode
$\vec{k}_1, \overleftarrow{k}_1$	Rate constants for electroreduction and electrooxidation of dissolved atomic hydrogen
$\vec{k}_2, \overleftarrow{k}_2$	The same for adsorbed atomic hydrogen
p	x-Component of emitted electron momentum
$p_{\shortparallel} \equiv \{p_y, p_z\}$	Components of emitted electron momentum parallel to the surface
$Q \equiv (k_A c_A / D_e)^{1/2}$	Reciprocal mean diffusion length of solvated electrons before being scavenged
T	Temperature in degrees Kelvin
V_{ms}	Metal-solution Volta potential
w	Electron work function
X	mole fraction
x_o	Electron solvation length
α, β	Transfer coefficients
κ^{-1}	Debye Length
ν	Stoichiometric coefficient
ϕ	Electrode potential
ϕ_o	Photoemission threshold potential

$\Phi(x)$ Solvated electron source function
Ψ Electron wave function
Ψ' Outer Helmholtz plane potential
ω Light frequency
ω_o Threshold light frequency

10. REFERENCES

1. G. E. Adams and R. L. Willson, Trans. Faraday Soc., <u>65</u>, 2981 (1969).
1a. I. Adawi, Phys. Rev., <u>134A</u>, 788 (1968).
2. J. E. Aldrich, M. J. Bronskill, R. K. Wolff, and J. W. Hunt, J. Chem. Phys., <u>55</u>, 530 (1971).
3. N. M. Alpatova, L. I. Krishtalik, and M. G. Fomicheva, Elektrokhimiya, <u>8</u>, 535 (1972).
4. M. Anbar, Quart. Rev. (London), <u>22</u>, 579 (1968).
5. M. Anbar and P. Neta, Int. J. Appl. Radiat. Isot., <u>18</u>, 493 (1967).
6. I. K. Ansone, Z. A. Rotenberg, G. Slaidins, and Yu. V. Pleskov, Elektrokhimiya, <u>12</u>, 1552 (1976).
7. L. I. Antropov, Itogi Nauki, Elektrokhimiya, Vol. 6, VINITI, Moscow, 1971, p. 5.
8. A. N. Arsen'eva-Geil, Vneshnii Fotoeffekt iz Poluprovodnikov i Dielektrikov, Gostekhteoretizdat, Moscow, 1957.
9. R. Audubert, C. R. Acad. Sci., <u>117</u>, 818, 1110 (1923).
10. R. Audubert, C. R. Acad. Sci., <u>189</u>, 1265 (1925).
11. R. Audubert, J. Chim. Phys., <u>27</u>, 169 (1930).
12. S. D. Babenko, V. A. Benderskii, A. G. Krivenko, and T. S. Rudenko, Elektrokhimiya, <u>10</u>, 793 (1974).

13. S. D. Babenko, V. A. Benderskii, A. G. Krivenko, and T. S. Rudenko, Fiz. Tverd. Tela, 16, 1337 (1974).
14. S. D. Babenko, V. A. Benderskii, and T. S. Rudenko, ZhETF, pis'ma, 17, 71 (1973).
15. I. A. Bagotskaya and A. I. Oshe, Trudy IV Soveshchaniya po Elektrokhimii, Moscow, 1959, p. 82.
16. G. C. Barker, Electrochim. Acta, 13, 1221 (1968).
17. G. C. Barker, Ber. Bunsenges. Phys. Chem., 75, 728 (1971).
18. G. C. Barker, J. Electroanal. Chem., 45, 320 (1973).
19. G. C. Barker, J. Electroanal. Chem., 46, 35 (1973).
20. G. C. Barker and J. A. Bolzan, J. Electroanal. Chem., 49, 227 (1974).
21. G. C. Barker and J. A. Bolzan, J. Electroanal. Chem., 49, 239 (1974).
22. G. C. Barker and G. Bottura, J. Electroanal. Chem., 46, 35 (1973).
23. G. C. Barker, G. Bottura, G. Cloke, A. W. Gardner, and M. J. Williams, J. Electroanal. Chem., 50, 323 (1974).
24. G. C. Barker and V. Concialini, J. Electroanal. Chem., 45, 320 (1973).
25. G.C. Barker and A.W. Gardner, Osnovnye voprosy sovremennoi teoreticheskoi Elektrokhimii, "Mir", Moscow, 1965, p. 118.
26. G.C. Barker and A.W. Gardner, Elektrokhimiya, 9, 1684 (1973).
27. G. C. Barker, A. W. Gardner, and G. Bottura, J. Electroanal. Chem., 45, 21 (1973).
28. G. C. Barker, A. W. Gardner, and D. C. Sammon, J. Electrochem. Soc., 113, 1182 (1966).
29. G. C. Barker and D. McKeown, J. Electroanal. Chem., 62, 341 (1975).

30. G. C. Barker, D. McKeown, M. J. Williams, G. Bottura, and V. Concialini, Faraday Disc. Chem. Soc., 56, 41 (1973).
31. G. C. Barker, B. Stringer, and M. J. Williams, J. Electroanal. Chem, 51, 305 (1974).
32. A. P. Baz', Ya. B. Zeldovich, and A. M. Perelomov, Rasseyanie, Reaktsiī i Raspady v Nerelativistskoi Kvantovoi Mekhanike, "Nauka", Moscow, 1971.
33. E. Becquerel, C. R. Acad. Sci., 9, 145 (1839).
34. V. A. Benderskii, S. D. Babenko, Ya. M. Zolotovitskii, A. G. Krivenko, and T. S. Rudenko, J. Electroanal. Chem., 56, 325 (1974).
35. H. Berg, Naturwiss., 47, 320 (1960).
36. H. Berg, Coll. Czech. Chem. Commun., 25, 3404 (1960).
37. H. Berg, Ber. Bunsenges. Phys. Chem., 65, 710 (1961).
38. H. Berg, in Modern Aspects of Polarography, T. Kambara, ed., Plenum, New York, 1966, p. 29.
39. H. Berg, Electrochim. Acta, 13, 1249 (1968).
40. H. Berg and P. Reissmann, J. Electroanal. Chem, 24, 427 (1970).
41. H. Berg and H. Schweiss, Osnovnye voprosy sovremennoi teoreticheskoi Elektrokhimii, "Mir", Moscow, 1965, p. 130.
42. H. Berg, H. Schweiss, E. Stutter, and K. Weller, J. Electroanal. Chem., 15, 415 (1967).
43. S. S. Bhatnagar, H. L. Anand, and A. N. Gupta, J. Ind. Chem. Soc., 5, 49 (1928).
44. G. Bomchil, D. J. Schiffrin, and J. T. D'Alessio, J. Electroanal. Chem., 25, 107 (1970).
45. G. Bottura, B. Bubani, and G. C. Barker, J. Electroanal. Chem., 62, 259 (1975).
46. F. P. Bowden, Trans. Faraday Soc., 27, 505 (1931).

47. A. M. Brodsky and A. N. Frumkin, Elektrokhimiya, 6, 658 (1970).
48. A. M. Brodsky and Yu. Ya. Gurevich, Zh. Eksp. Teor. Fiz., 54, 213 (1968).
49. A. M. Brodsky and Yu. Ya. Gurevich, Dokl. Akad. Nauk SSSR, 178, 868 (1968).
50. A. M. Brodsky and Yu. Ya. Gurevich, Izv. Akad. Nauk SSSR, Ser. Fiz., 33, 388 (1969).
51. A. M. Brodsky and Yu. Ya. Gurevich, Teoriya elektronnoi emissii iz metallov, "Nauka", Moscow, 1973.
51a. A. M. Brodsky, Yu. Ya. Gurevich, and V. G. Levich, Phys. Stat. Sol., 40, 139 (1970).
52. A. M. Brodsky, Yu. Ya. Gurevich, Yu. V. Pleskov, and Z. A. Rotenberg, Sovremennaya fotoelektrokhimiya. Fotoemissionnye yavleniya, "Nauka", Moscow, 1974.
53. A. M. Brodsky, Yu. Ya. Gurevich, and S. V. Sheberstov, J. Electroanal. Chem., 32, 353 (1971).
54. A. M. Brodsky, Yu. Ya. Gurevich, and S. V. Sheberstov, Adsorbtsiya i dvoinoi sloi v elektrokhimii, "Nauka", Moscow, 1972, p. 25.
55. A. M. Brodsky, Yu. V. Pleskov, in Progress in Surface Science, Vol. 2, part 1, S. G. Davidson, ed., Pergamon, New York, 1972.
56. V. Concialini, O. Tubertini, and G. C. Barker, J. Electroanal. Chem., 57, 413 (1974).
57. B. E. Conway, Modern Aspects of Electrochemistry, Vol. 7, Plenum, New York, 1972, p. 83.
58. A. W. Copeland, O. D. Black, and A. B. Garrett, Chem. Rev., 31, 177 (1942).
59. P. Delahay, Double Layer and Electrode Kinetics, Interscience, New York, 1965.

60. P. Delahay and V. S. Srinivasan, J. Phys. Chem., $\underline{70}$, 420 (1966).
61. R. De Levie and A. Husovsky, J. Electroanal. Chem., $\underline{20}$, 181 (1969).
62. R. De Levie and J. C. Kreuser, J. Electroanal. Chem., $\underline{21}$, 221 (1969).
63. L. N. Dobretsov and M. V. Gomoyunova, Emissionnaya elektronika, "Nauka", Moscow, 1966.
64. R. R. Dogonadze, L. I. Krishtalik and Yu. V. Pleskov, Elektrokhimiya, $\underline{10}$, 507 (1974).
65. A. Einstein, Ann. Phys., $\underline{17}$, 132 (1905).
66. V. V. Eletsky and Yu. V. Pleskov, Elektrokhimiya, $\underline{10}$, 179 (1974).
67. V. V. Eletsky, Z. A. Rotenberg, and Yu. V. Pleskov, Khimiya Vysokikh Energii, $\underline{5}$, 325 (1971).
68. R. H. Fowler, Phys. Rev., $\underline{38}$, 45 (1931).
69. A. N. Frumkin, Z. Phys. Chem., $\underline{164}$, 121 (1933).
70. A. N. Frumkin, quoted in J. Horiuti and M. Polanyi, Acta physicochim. USSR, $\underline{2}$, 505 (1935).
71. A. N. Frumkin, Zh. Fiz. Khim., $\underline{31}$, 1875 (1957).
72. A. N. Frumkin, Elektrokhimiya, $\underline{1}$, 394 (1965).
73. A. N. Frumkin, Svensk Kem. Tidskrift, $\underline{77}$, 300 (1965).
74. A. N. Frumkin, Dvoinoi Sloi i Adsorbtsiya na Tverdykh Elektrodakh., Izdatel'stvo Tartuskogo Universiteta, Vol. III, Tartu, 1972, p. 5.
75. A. N. Frumkin, V. S. Bagotzky, Z. A. Jofa, and B. N. Kabanov, Kinetika Elektrodnykh Protsessov, University Press, Moscow, 1952.
76. A. N. Frumkin, O. A. Petry, and N. V. Nikolaeva-Fedorovich, Dokl. Akad. Nauk SSSR, $\underline{128}$, 1006 (1959).

References

77. A. N. Frumkin, O. A. Petry, and N. V. Nikolaeva-Fedorovich, Dokl. Akad. Nauk SSSR, 147, 878 (1962).
78. H. Gerischer, J. Electrochem. Soc., 113, 1174 (1966).
79. H. Gerischer, Ber. Bunsenges. Phys. Chem., 77, 771 (1973).
80. H. Gerischer, E. Meyer, and J. K. Sass, Ber. Bunsenges. Phys. Chem., 76, 1191 (1972).
81. P. Görlich, Photoeffekte, Vols. 1 - 3, Leipzig, 1963 - 1966.
82. D. E. Gray et al., eds., Handbook of Physics and Chemistry, Chemical Rubber Co., Cleveland, Ohio, 1967, p. 47.
83. Yu. Ya. Gurevich, Elektrokhimiya, 8, 1564 (1972).
84. Yu. Ya. Gurevich and Z. A. Rotenberg, Elektrokhimiya, 4, 529 (1968).
85. E. J. Hart and M. Anbar, The Hydrated Electron, Wiley - Interscience, New York, 1970.
86. H. Hertz, Ann. Phys., 31, 983 (1887).
87. M. Heyrovský, Nature, 200, 1356 (1965).
88. M. Heyrovský, Nature, 209, 708 (1966).
89. M. Heyrovský, Z. Phys. Chem., N. F., 52, 1 (1967).
90. M. Heyrovský, Proc. Roy. Soc. (London), A301, 411 (1967).
91. M. Heyrovský, Croat. Chem. Acta, 45, 247 (1973).
92. M. Heyrovský, R. G. W. Norrish, Nature, 200, 880 (1963).
93. P. J. Hillson and E. K. Rideal, Proc. Roy. Soc. (London), A199, 295 (1949).
94. P. J. Hillson and E. K. Rideal, Proc. Roy. Soc. (London), A216, 458 (1953).
95. K. Honda, Koge Kagaku Dzassi, 72, 63 (1969).
96. K. Honda, S. Kikuchi, and T. Yura, Seisan kenkyu, Monthly J. Inst. Industr. Sci. Univ. Tokyo, 18, 165 (1966).
97. Z. A. Jofa and K. P. Mikulin, Zh. Fiz. Khim., 18, 137 (1944).

98. F. Itskovich, Zh. Eksp. Teor. Fiz., 51, 301 (1966).
99. B. N. Kabanov, Zh. Fiz. Khim., 8, 486 (1936).
100. M. Kh. Karapet'yanz, and M. L. Karapet'yanz, Osnovnye termodinamicheskie konstanty neorganicheskikh i organicheskikh veshchestv, "Khimiya", Moscow, 1968.
101. G. A. Kenney and D. C. Walker, Electroanalytical Chemistry, Vol. 5, A. J. Bard, ed., Dekker, New York, 1971, p. 1.
102. L. I. Korshunov, V. A. Benderskii, V. I. Goldanskii, and Ya. M. Zolotovitskii, ZhETF, pis'ma, 7, 55, (1968).
103. L. I. Korshunov, Ya. M. Zolotovitskii, and V. A. Benderskii, Uspekhi Khimii, 40, 1511 (1971).
104. L. I. Korshunov, Ya. M. Zolotovitskii, V. A. Benderskii, and V. I. Goldanskii, Khimiya Vysokikh Energii, 4, 346 (1970).
105. L. I. Korshunov, Ya. M. Zolotovitskii, V. A. Benderskii, and V. I. Goldanskii, Khimiya Vysokikh Energii, 4, 461 (1970).
106. L. I. Krishtalik, Uspekhi Khimii, 34, 1831 (1965).
107. L. I. Krishtalik, Elektrokhimiya, 2, 229 (1966).
108. L. I. Krishtalik, Elektrokhimiya, 5, 3 (1969).
109. L. I. Krishtalik, J. Electroanal. Chem., 35, 157 (1972).
110. L. I. Krishtalik, N. M. Alpatova, and M. G. Fomicheva, Elektrokhimiya, 7, 1393 (1971).
111. L. I. Krishtalik, N. M. Alpatova, and M. G. Fomicheva, Croat. Chem. Acta, 44, 1 (1972).
112. V. I. Lakomov, Z. A. Rotenberg, and Yu. V. Pleskov, Elektrokhimiya 9, 152 (1973).
113. D. I. Leikis, K. V. Rybalka, E. S. Sevastyanov, and A. N. Frumkin, J. Electroanal. Chem., 46, 161 (1973).

References

114. S. D. Levina and T. V. Kalish, Dokl. Akad. Nauk SSSR, 109, 97 (1956).
115. D. B. Matthews, Austral. J. Chem., 24, 1 (1971); 25, 2061 (1972).
116. W. Mehl and J. M. Hale, in Advances in Electrochemistry and Electrochemical Engeneering, Vol. 6, P. Delahay, ed., Interscience, New York, 1967, p. 399.
117. K. Mitchell, Proc. Roy. Soc. (London), A146, 442 (1934).
118. K. Mitchell, Proc. Camb. Phil. Soc., 31, 416 (1935).
119. N. F. Mott and E. A. Davis, Electronic Processes in Non-crystalline Materials, Clarendon, Oxford, 1971.
120. V. A. Myamlin and Yu. V. Pleskov, Electrochemistry of Semiconductors, Plenum, New York, 1967.
121. U. Palm and T. Tenno, J. Electroanal. Chem., 42, 457 (1973).
122. R. Parsons, in Advances in Electrochemistry and Electrochemical Engineering, Vol. 1, P. Delahay, ed., Interscience, New York, 1961, p. 1.
123. R. Parsons, Surface Sci., 2, 418 (1964).
124. A. K. Pikaev, Solvatirovannyi elektron v radiatsionnoi khimii, "Nauka", Moscow, 1969.
125. R. L. Platzman, in Physical and Chemical Aspects of Basic Mechanisms in Radiobiology, J. L. Magee et al., eds., U. S. Natl. Acad. Sci., Publ. No. 305, Washington, D. C., 1953, p. 34.
126. R. L. Platzman and J. Frank, Z. Phys., 138, 411 (1951).
127. Yu. V. Pleskov and Z. A. Rotenberg, Uspekhi Khimii, 41, 40 (1972); Khimiya Vysokikh Energii, 8, 99 (1974).
128. Yu. V. Pleskov, Z. A. Rotenberg and V. I. Lakomov, Elektrokhimiya, 6, 1787 (1970).

129. Yu. V. Pleskov and Z. A. Rotenberg, J. Electroanal. Chem., 20, 1 (1969).
130. G. Porter, Photochemistry and Reaction Kinetics, Cambridge University Press, London, 1967.
131. D. Postl and U. Schindewolf, Ber. Bunsenges. Phys. Chem., 75, 662 (1971).
132. Yu. A. Prishchepa, Z. A. Rotenberg, and Yu. V. Pleskov, Elektrokhimiya, 10, 1824 (1974).
133. T. Pyle and C. Roberts, J. Electrochem. Soc., 115, 247 (1968).
134. K. Punning and V. Past, Uchenye zapiski Tartuskogo gosudarstvennogo universiteta, No. 235, 35 (1969).
135. Z. A. Rotenberg, Elektrokhimiya, 8, 1198 (1972).
136. Z. A. Rotenberg, Elektrokhimiya, 9, 511 (1973).
137. Z. A. Rotenberg, Elektrokhimiya, 10, 682 (1974).
138. Z. A. Rotenberg, Elektrokhimiya, 10, 1031 (1974).
139. Z. A. Rotenberg and Yu. Ya. Gurevich, Elektrokhimiya, 9, 159 (1973).
140. Z. A. Rotenberg and Yu. Ya. Gurevich, J. Electroanal. Chem., 66, 165 (1975).
141. Z. A. Rotenberg, V. I. Lakomov, A. M. Brodsky, and Yu. V. Pleskov, Elektrokhimiya, 6, 1387 (1970).
142. Z. A. Rotenberg, V. I. Lakomov, and Yu. V. Pleskov, J. Electroanal. Chem., 27, 403 (1970).
143. Z. A. Rotenberg, V. I. Lakomov, and Yu. V. Pleskov, Elektrokhimiya, 8, 313 (1972).
144. Z. A. Rotenberg and Yu. V. Pleskov, Elektrokhimiya, 4, 826 (1968).
145. Z. A. Rotenberg and Yu. V. Pleskov, Elektrokhimiya, 5, 982 (1969); 6, 418 (1970).

146. Z. A. Rotenberg and Yu. V. Pleskov, Elektrokhimiya 9, 1419 (1973).
147. Z. A. Rotenberg, Yu. V. Pleskov, and V. I. Lakomov, Elektrokhimiya, 4, 1022 (1968).
148. Z. A. Rotenberg, Yu. A. Prishchepa, and T. K. Ansone, Elektrokhimiya, 11, 651 (1975).
149. Z. A. Rotenberg, Yu. A. Prishchepa, I. K. Ansone, G. Slaidins, and Yu. V. Pleskov, Dvoinoi Sloi i Adsorptiya na Tverdykh Elektrodakh, Izdatel'stvo Tartuskogo Universiteta, Vol. IV, Tartu, 1975.
150. Z. A. Rotenberg, Yu. A. Prishchepa, and Yu. V. Pleskov, J. Elektroanal. Chem., 66, 3 (1975).
151. Z. A. Rotenberg, Yu. A. Prishchepa, and Yu. V. Pleskov, J. Electroanal. Chem., 56, 345 (1974).
152. J. K. Sass, R. K. Sen, E. Meyer, and H. Gerischer, Surface Sci., 44, 515 (1974).
153. D. J. Schiffrin, Croat. Chem. Acta, 44, 139 (1972).
154. D. J. Schiffrin, Faraday Disc. Chem. Soc., 56, 41 (1974).
155. V. P. Sharma, P. Delahay, G. C. Susbielles, and G. C. Tessari, J. Electroanal. Chem., 16, 285 (1968).
156. S. V. Sheberstov and A. M. Brodsky, Elektrokhimiya, 6, 1762 (1970).
157. S. V. Sheberstov, A. M. Brodsky, and Yu. Ya. Gurevich, Elektrokhimiya, 6, 1182 (1970).
158. V. Sihvonen, Ann. Acad. Sci. Fennicae, A26, 3 (1926).
159. A. V. Sokolov, Opticheskie Svoistva Metallov, Fizmatgiz, Moscow, 1961.
160. A. Sommerfeld, Z. Phys., 47, 1 (1928).
161. A. Stoletov, C. R. Acad. Sci., 106, 1149, 1593 (1888).
162. I. Tamm and S. Shubin, Z. Phys., 68, 97 (1931).

163. M. I. Temkin and A. N. Frumkin, Zh. Fiz. Khim., 29, 1513 (1955).
164. T. T. Tenno and U. V. Palm, Elektrokhimiya, 9, 1545 (1973).
165. V. I. Veselovsky, Zh. Fiz. Khim., 20, 1493 (1946).
166. D. C. Walker, Can. J. Chem., 45, 807 (1967).
167. D. C. Walker, Quart. Rev. (London), 21, 79 (1967).
168. Ya. M. Zolotovitskii, L. I. Korshunov, and V. A. Benderskii, Izv. Akad. Nauk SSSR, Ser. Khim., 802 (1972).
169. Ya. M. Zolotovitskii, V. A. Benderskii, S. D. Babenko, and A. T. Krivenko, in Dvoinoi Sloi i Adsorbtsiya na Tverdykh Elektrodakh, Vol. IV, Izdatel'stvo Tertuskogo Universiteta, Tartu, 1975, p. 100.

Physical and Electrochemical Properties of Metal Monolayers on Metallic Substrates

DIETER M. KOLB
Fritz-Haber-Institut der Max-Planck-Gesellschaft
Berlin, Germany

1. Introduction 127
2. Electrochemical Studies of Metal Monolayer Formation in the Underpotential Region 129
   2.1. Historical Survey 129
   2.2. Experimental Techniques 132
   2.3. Underpotential Deposition on Polycrystalline Electrodes in Aqueous Solutions 138
   2.4. The Thermodynamics of Predeposition 152
   2.5. The Electrosorption Valency 160
   2.6. Monolayer Adsorption Isotherms and Kinetics 165
   2.7. Measurements on Single-crystal Surfaces 172
   2.8. Anion Effects 177
   2.9. Measurements in Nonaqueous Solutions 184
   2.10. Alloy Formation in the Underpotential Range 187
3. Studies of Metal Monolayers in a Nonelectrochemical Environment 191
   3.1. XPS Studies on Monolayers Deposited at Underpotentials 191

	3.2.	Metal Monolayer Formation on Metal Substrates in Vacuum	194
4.	Optical Studies of the Electronic Properties of Metal Monolayers		201
	4.1.	Instrumentation	201
	4.2.	Evaluation of the Monolayer Dielectric Function from Reflectance Measurements	205
		4.2.1. Inversion of Linear Approximation Equations	207
		4.2.2. Kramers-Kronig Analysis	208
		4.2.3. Combination of Ellipsometry and Differential Reflectance Spectroscopy	210
		4.2.4. Iteration Methods	211
	4.3.	Differential Reflectance Spectroscopy	214
	4.4.	Electroreflectance	228
	4.5.	Related Studies	231
		4.5.1. Surface Conductance Measurements	231
		4.5.2. Photoemission into Electrolytes	233
		4.5.3. Mössbauer Spectroscopy	233
5.	Catalytic Effects of Metal Monolayers on Electrochemical Reactions		234
6.	Models Explaining Predeposition		239
	6.1.	The Pauling Model	239
	6.2.	Correlation between Underpotential Shifts and Work-function Differences	242
	6.3.	The Modified Levine and Gyftopoulos Model	250
7.	References		258

1. INTRODUCTION

The aim of this chapter is to survey the experimental investigation of the physical and electrochemical properties of submonolayer and monolayer metal deposits on foreign metallic substrates, where the so-called underpotential deposition is observed. This effect describes the formation of a metal monolayer at potentials positive from the reversible Nernst potential, that is, before bulk deposition can occur. Underpotential deposition offers not only a unique possibility to form metal monolayer deposits at equilibrium conditions and to vary their coverage by the potential, thus yielding exact thermodynamic data on relative binding energies, but also allows the electronic and structural properties of the deposit to be studied conveniently by various electrochemical and optical techniques. A thorough understanding of underpotential deposition has a strong bearing on galvanic processes, since it has often been found in metal deposition reactions that predeposition of one monolayer of metal occurs prior to bulk deposition. While in metal deposition processes nucleation has generally been considered to be the initial step of crystallization, even for deposition onto foreign metal substrates, this statement has now to be reconsidered in the light of our knowledge on underpotential deposition. In many cases formation of a monolayer is the very first step of metal deposition (even in the overpotential regime), and nucleation occurs on top of this first monolayer, that is, on a surface of a nearly like metal. Since the properties of the first layer are influenced by the substrate, the further growth of the deposit may also be affected. The effect of certain trace metal impurities on the nucleation behavior of metal coatings, which is phenomenologically well known for at least several decades, may also be explained by the underpotential behavior of metals.

Since the investigations on the metal monolayers in electrochemical systems are far from complete, the purpose of this article can only be to reveal the as yet unresolved problems and to summarize the present state of the art. During the preparation of this chapter many groups focused attention increasingly on the monolayer formation on single crystal surfaces, which promises to be a wealth of new interesting information in the near future. Data available up to the end of 1976 have been included. New experimental techniques that combine pure electrochemical methods with other non-electrochemical ones, such as in situ reflectance spectroscopy, are found to be powerful tools that have only very recently been used and are now beginning to produce interesting results. First attempts are described in the literature in which even the very sensitive electron spectroscopy (ECSA) is used after transferring the electrode from the electrolytic cell into a UHV system for a more specific characterization of electrode surfaces.

Emphasis is also placed on bridging the gap between electrochemistry and surface physics, where similar problems are investigated but usually without much dialogue between the two groups. The physical properties of metal monolayers and their influence on the substrate surface properties have been of considerable interest in many respects to fundamental and theoretical as well as to applied physics. Ever since it was shown in Taylor and Langmuir's classic work in the mid-1930s the extent to which metal adatoms can change the surface properties of metal substrates, many investigators have studied these phenomena in vacuum by surface analytical methods as work-function measurements, flash desorption, LEED, and many more. Initially the change in substrate surface

properties was of main interest, but more recently the properties
of the adsorbed atoms or aggregates themselves have become increasingly attractive, especially to theoreticians, for study of:
(a) general chemisorption and chemisorptive bond and, (b) the
change in properties during the transition from a single atom to
the solid state. It is shown here that the results obtained at the
electrode-electrolyte interface can be extremely useful in elucidating problems in surface physics, and vice versa. Some typical
results obtained at the solid-vacuum interface are also discussed
to the extent that the knowledge gained there is relevant to the
electrochemical studies. There is little doubt that the knowledge
of this topic is very important for a thorough understanding of
problems dealing with electrocatalysis, heterogeneous catalysis
or chemisorption in general.

2. ELECTROCHEMICAL STUDIES OF METAL MONOLAYER FORMATION IN THE UNDERPOTENTIAL REGION

2.1. Historical Survey

The effect of underpotential deposition, that is, the deposition of a submonolayer amount ($\leq$ ca. one monolayer) of a metal
onto a foreign metallic substrate at potentials positive from the
reversible Nernst potential, has now been known for a very long
time. One of the first might have been Hevesy (1), who reported
in 1912 a deviation from Nernst's law when spurious amounts of
radioactive metals were deposited onto Cu. His curves, corresponding to coverage-potential-isotherms, showed an unsymmetrical tailing off for positive potentials, indicating metal adsorption to
take place up to several tenths of a volt positive from the Nernst

potential. Shortly after this finding Herzfeld (2) offered a formalistic explanation for this apparent violation of Nernst's law by assuming that the activity of the solid phase is a function of coverage Θ, as long as the deposit does not cover the total substrate surface. He concluded that the dissolution rate for an incompletely covered electrode was smaller than that for a completely covered one and dependent on Θ, whereas the deposition rate remained the same for both electrodes, thus yielding a coverage-dependent equilibrium potential as found in the experiments.

In the late 1940s several groups investigated the deposition and stripping behavior of samll amounts of metal deposits on foreign metal electrodes. Rogers and colleagues studied in detail the deposition of trace amounts of radioactive Ag and Cu on different substrates such as Pt, Au, Pd, Rh, and W as a function of various parameters, such as metal ion concentration, supporting electrolyte, and complexing agents and so on (3,4). Their main concern was, as is true for most of the other groups, the applicability of the stripping analysis for analytical purposes. They usually decided between macro- and trace behavior as being due to bulk or multilayer and monolayer deposits (5) and were able to analyze solutions at concentrations as low as 5×10^{-8} M (12). Their results agree reasonably well with more recent data as far as the amount of underpotential is concerned (see, e.g., Table II in ref. 4). To explain their results, Rogers and Stehney picked up Herzfeld's concept and extended it somewhat (6). Besides the assumption that the activity of the solid phase in the submonolayer region is no longer constant and equal to unity but

$$a_{ML} = \begin{cases} \Theta & \text{for } 0 \leqslant \Theta \leqslant 1 \\ 1 & \text{for } \Theta \geqslant 1 \end{cases} \quad (1)$$

an additional term, E_a, was introduced to account for any specific interaction between reduced species (deposit) and substrate surface such as alloying: E_a is then dependent on the substrate material. The final expression for the modified Nernst equation in the submonolayer range has the form:

$$E_r = E_o + E_a + \frac{RT}{zF} \ln \frac{c_{ox}}{\Theta} \qquad (2)$$

Haissinsky (7-9) essentially followed this treatment in his studies of underpotential deposition. Later Nicholson investigated the systems Ag, Pb, and Cu on Pt (10) and Ni on Pt and Au (11), again with the main emphasis placed on analytical applications by stripping of minute quantities. Two of the first to pioneer the investigation of the electrolytic deposition of the first atomic layer of a metal on a foreign metallic substrate and its special properties were Mills and Willis, studying monolayer formation of Pb, Tl, Sb, and Ni on Au and Pb on Ag (13).

Despite their different scope, all of these early investigations led to a number of remarkable conclusions that were proved in recent years to be quite reasonable. It was clearly seen that the monolayer deposition at underpotentials results from a strong interaction between monolayer atoms and substrate, which can be described in a formal way by an activity less than unity for the deposit in the submonolayer range (for more details, see Section 2.4.). Various experimental results led to the conclusion that the monolayer is uniformly distributed and does not grow on active sites of the substrate (10,13), and its structure might predominantly be determined by that of the substrate (4). It was assumed that the monolayer deposition as the initial step of electroplating would be of the upmost importance for the further growth of the film and hence the physical properties of the final deposit. This was supported by the observations that not only the deposition and stripping behavior of the first monolayer depends on the substrate (which up to then was assumed as an "inert" electrode), but also to some extent that of a multilayer (4). It was found that the second layer would not be deposited until the first monolayer was completed, which required an amount of charge of roughly 250 µC/cm^2 for a monovalent metal ion. It was also demonstrated in galvanostatic charging curves for Pb on Au that the monolayer adsorption isotherm may have a rather complicated structure with more than just one adsorption step (13). Finally, the point has been raised that there should be a correlation between underpotential deposition and mismatch in lattice parameters between substrate and deposit materials (4). It was observed that the underpotential is inversely proportional to the degree of mismatch of lattice dimensions. This again seemed to support the idea of an epitaxial growth of the first monolayer onto the substrate electrode (14).

About 10 years later the monolayer formation of metals at underpotentials again received the interest of several groups, mainly that around E. Schmidt, who applied much more sophisticated techniques and focused exclusively on the phenomenon of underpotential deposition. The results of this group instigated vast and intensive investigation in this field during the last 10 years, which is reviewed and discussed critically in detail in the following sections. The results obtained prior to 1965 are considered as interesting and relevant only from a historical point of view, with a few exceptions that are stated as such.

2.2. Experimental Techniques

The experimental techniques usually applied in these studies are described only briefly to the extent necessary to interpret

the results. A short survey is also given by Lorenz (15). With
the exception of Bowles's work (16-18), where radiotracer technique was involved, classical electrochemical methods were applied.

One of the most simple and yet powerful techniques often used
for underpotential studies is cyclic voltammetry. Here the potential of the working electrode controlled by a potentiostat is
changed continuously with a constant scan rate (dE/dt) by a ramp
potential from a sweep generator. Any reaction at the electrode
surface, faradayic or nonfaradayic, will usually be detected in
the cyclic current-potential curves as a current superimposed to
the base current due to double-layer charging. A typical example
is shown in Fig. 3. The amount of metal deposited at a certain
potential can be obtained as charge equivalent by scanning the
potential positive and integrating the current caused by oxidative
desorption of the deposit. However, double-layer charging is also
incorporated and has to be subtracted. The latter one is usually
determined by integrating the current when applying the same
potential scan in the blank electrolyte. For reversible adsorption
the anodic and cathodic scans (i.e., desorption and deposition)
are usually identical, if the scan rate is low enough to avoid
concentration polarisation. This yields information on the adsorption isotherms, even at low coverages, because of the high
sensitivity of current measurements. Changing the scan rate over
a large range can supply a rough kinetic information on the
adsorption-desorption behavior and rate-determining reactions
involved. Therefore, this method will rapidly furnish thermodynamic and to some extent kinetic data and is well suited for detecting minute amounts of reactive species. The main disadvantage
is that the surface concentration of the adsorbed species, $\Gamma(E)$,
cannot be determined directly but only from the measured charge
$q(E)$ which entails additional assumptions on the electrosorption

valency (see Section 2.5.). Some typical examples of applications of this method are given elsewhere (19-21).

A more sophisticated technique originating from the above described one is the use of a <u>rotating ring-disk electrode</u>, where the potentials of disk and ring can be controlled independently. The geometry is such that the convective transport of ions is rather simple as far as the mathematical treatment of the problem is concerned. Usually, the potential of the disk electrode is scanned at a constant rate (dE/dt) while the potential of the ring remains fixed at certain values. By choosing appropriate potentials at the ring electrode, reaction products and intermediates can be detected there as they are generated at the disk electrode during the potential scan. The power of this technique, which was improved to a large extent by Bruckenstein and is used throughout his work on underpotential deposition, is demonstrated nicely in a study of the codeposition of two different metals on Pt (22). This method allows in principle to determine the surface concentration of a deposited metal, $\Gamma(E)$, on the disk electrode by measuring the charge due to redeposition at the ring during stripping from the disk, provided the ring electrode is made out of the same metal as that to be deposited. A disadvantage might be that the ring potential is usually far from the equilibrium potential of the reaction to be studied. This increases the chance of side reactions to contribute to the ring current.

Another powerful tool for studying underpotential deposition of metals is the use of <u>twin electrode thin layer cells</u> as developed and applied mainly by Schmidt and his group and based on a concept of Anderson and Reilley (23,24). The main part of such a cell (Fig. 1) consists in two plane-parallel electrodes closely spaced at a distance of about 50 μ with independent potential regulation. One electrode, the so-called generator electrode, is composed of the same metal as that to be deposited at the second (working electrode) and represents a reversible metal/metal ion electrode. The metal ion concentration in the thin-layer cell is given by the potential of the generator electrode according to Nernst's law. At an appropriate potential the metal is deposited onto the working electrode, causing a current I there due to the reaction:

$$Me^{z+} + ze^- \rightleftharpoons Me \qquad (3)$$

Integration of the current I at the working electrode yields the surface charge corresponding to the metal-deposition process when the amount due to the double layer charging current I_b as determined in the blank solution is subtracted. A blank solution in this system is readily obtained just by polarizing the generator electrode sufficiently negative to yield $[Me^{z+}] < 10^{-9}$ M.

The charge Δq due to deposition of Γ_{Me} onto the working electrode is then given by (61):

$$\Delta q(\Gamma_{Me}, E_d) = q(\Gamma_{Me}, E_d) - q(\Gamma_{Me}=0, E_d) = A^{-1} \int_{E_i}^{E_d} (I-I_b) \left(\frac{dE}{dt}\right)^{-1} dE \qquad (4)$$

with $E_i \gg E_d$ and $\mu_{Me^{z+}}$ = const and A = electrode surface area. The starting point, E_i, of the potential scan has to be so positive, that $\Gamma_{Me} = 0$ at E_i. Since the constant (and reversible) potential E_r at the generator electrode will maintain a constant chemical potential $\mu_{Me^{z+}}$ for the metal ion, any change in the metal ion concentration due to deposition (or stripping) onto the working electrode will immediately be compensated by a respective current I_g at the generator electrode. Therefore, the charge corresponding to I_g is directly related to Γ_{Me}, the surface concentration of Me at the working electrode:

$$\Gamma_{Me}(\mu_{Me^{z+}}, E_d) = (zFA)^{-1} \int_{E_i}^{E_d} I_g \left(\frac{dE}{dt}\right)^{-1} dE \qquad (5)$$

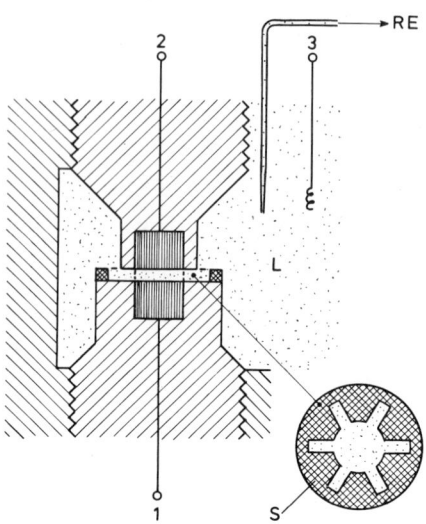

Fig. 1. Schematic view of a twin electrode thin-layer cell: (1 and 2) working and generator electrodes (metal rods in Teflon holders); (3) counter electrode; S, insulating spacer of 50 µm thickness; RE, reference electrode; L, electrolyte (15).

Since $(\partial \Delta q/\partial \Gamma_{Me})_{E_d} = \gamma z F$, any nonstoichiometry in Eq. 3 of the type

$$Me^{z+} + \gamma z e^- \rightleftharpoons Me^{(1-\gamma)z+} \qquad (6)$$

can be determined from differences in Δq and $zF\Gamma_{Me}$. In Fig. 2 generator-, working-electrode-, and base-currents are shown for the adsorption and desorption of Pb on Au as a function of the potential of the working electrode. Generator- and working-electrode currents respond practically identical during the potential sweep, indicating a complete (stoichiometric, $\gamma \approx 1$) electron transfer.

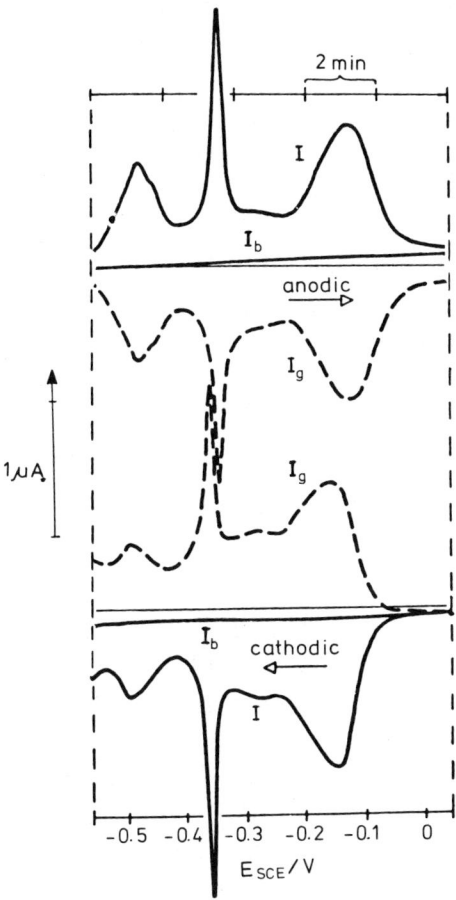

Fig. 2. Cyclic current-potential curves for Pb on Au, obtained in 0.5M $NaClO_4$ (pH 4.7 acetate-buffered) with a thin layer cell. Scan rate, 1 mVs^{-1}. $E_g = -0.565$ V_{SCE} ≙ 1.8×10^{-4}M Pb^{2+}. I, indicator current; I_g, generator current; I_b, base current (for $c_{Pb^{2+}} = 0$) (15).

The thin-layer technique yields thermodynamic data such as adsorption isotherms $\Gamma_{Me} = \Gamma_{Me}(E)$ or $\Delta q = \Delta q(\Gamma_{Me})$ (see Section 2.6) with a high degree of accuracy. However, kinetic data cannot be obtained by this method because of the high impedance of the cell. In some cases <u>potentiostatic or galvanostatic pulse measurements</u> have been used for studying the kinetics of metal deposition (15, 25,26,93) (see Section 2.6).

All of these measurements have one feature in common, in that to establish the charge due to metal deposition or stripping, double-layer charging has to be taken into account. This is usually done by simply subtracting the charge obtained in a blank solution from that measured with metal deposition when identical potential steps or cycles are applied. This is a first-order approximation, since it assumes that $(\partial q/\partial E)_{\Gamma_{Me}}$ is the same for all Γ_{Me}. However, it has been generally overlooked that the potential of zero charge (PZC) will change during deposition from that of the substrate metal at $\Gamma_{Me} = 0$ to that of the deposited metal at approximately a monolayer coverage. Since in nearly all of the studied systems the PZC of the deposit is more negative than that of the substrate, the procedure described in this paragraph will yield an adsorbate charge that is too small by an amount of approximately $C_{DL} \cdot \Delta E_{PZC}$ (19). This has direct consequences for some of the results presented in the sections that follow.

2.3. Underpotential Deposition on Polycrystalline Electrodes in Aqueous Solutions

Numerous metal pairs have been investigated by pure electrochemical methods for obtaining information on monolayer formation at underpotentials. In this chapter we refer only to those studied in aqueous solutions and on polycrystalline electrode materials. To supply the reader with a quick survey on what has been done so far, all studied systems are summarized as follows, with the corresponding references: (a) on platinum: Ag (19,27-31), Cu (16,19,25,28,30-37), Hg (19,38), Au (28,39), Pb (28,30,39-41, 219), Bi (18,41-44,219), Cd (41,45,219), Sn (46), Tl (17,41,219), Ge (44), As (44), Sb (44), (b) on palladium: Ag (47), Cu (47), Bi (48,49), Cd (48), Pb (48,220), Sn (48), Tl (48), (c) on gold: Pd (212), Ag (19,20,29,50), Cu (19,20,51), Hg (19), Pb (19,52-55), Cd (19,56), Bi (57,58), Tl (19,20,59), Sb (58), Sn (60), Zn (47), (d) on silver: Cu (19,51), Pb (19,59,61-63), Tl (19,59,63,64), Bi (59), Cd (19,65), Sn (66), Zn (47), (e) on copper: Cd (19), Tl (19,59), Pb (59), (f) on lead: Cd (59,63), (g) on bismuth: Cd (59), Sn (59), (h) on tin: Cd (59), Tl (59), and (i) on rhodium and iridium: Pb (220).

Some very characteristic features of underpotential deposition can be seen very clearly in simple cyclic current-potential curves as is demonstrated for Tl on Ag in Fig. 3.

When starting the potential scan into cathodic direction from an anodic value (e.g., 0 V vs. SCE), where the substrate surface is bare, a current peak is seen around -0.6 V due to predeposition of Tl onto Ag, which is stripped at the same potential on the reversed scan, provided the scan rate is low enough not to cause concentration polarisation. When the potential scan is stopped in the underpotential range the current immediately decreases to zero, indicating a reversible potential-dependent adsorption process. Scanning into the region negative of the reversible Nernst potential E_r for (Tl/Tl^+), bulk Tl will be deposited at a diffusion limited rate. On the anodic scan this bulk Tl will be stripped around E_r within a rather small potential range, which is mainly defined by the scan rate, the amount of Tl and hence the height of the bulk stripping peak depending strongly on the deposition time and to some extent on the potential. The more positive stripping peak around −0.55 V is referred to as the "monolayer peak". It is noted that this monolayer peak represents only an incomplete monolayer (usually less than 50% of a close-packed monolayer), a considerable fraction still being deposited underpotentially closer to E_r. The maximum amount of Tl deposited at underpotentials (i.e. ⩽ ca. 10 mV positive of E_r), which obviously is a constant proportional only to the substrate surface area, can be determined most precisely by a twin electrode thin layer cell and is about 2×10^{-9} mol/cm^2. This is what one would expect for a monolayer coverage since it corresponds approximately to the number of surface atoms per cm^2 for a metal. Nearly all of the systems studied thus far yielded about the same value. In addition, one finds that the charge required to deposit or remove this maximum amount of underpotential deposit is around $z \times 200$ µC/cm^2, which means that the charge transfer is close to the value expected for a total discharge according to Eq. (reaction) 3. Only a small (usually positive) charge, which is in the order of a few percent and dependent on the coverage, remains at the adatoms. This is discussed in detail later. This, however, implies that one has to talk about metal deposition rather than "cation adsorption", as this effect is still frequently referred to. The latter expression is not only misleading but also confusing since often metal ion adsorption may also take place at positive potentials (55,60).

Since the maximum amount deposited at underpotentials corresponds in most cases very closely to a monolayer coverage, it suggests very strongly monolayer formation. A direct proof has been obtained for platinum substrates by measuring the suppression of the hydrogen adsorption due to metal deposition in the underpotential region (16-18, 28, 35, 38, 42, 67). Hydrogen is known to adsorb on platinum prior to hydrogen evolution in two distinct

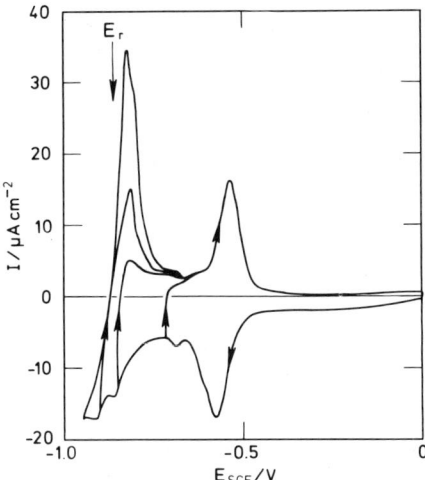

Fig. 3. Cyclic current-potential curves for Ag electrode in 0.5M Na_2SO_4 (pH 3) + 2 x 10^{-4}M $TlNO_3$. Scan rate, 20 mVs^{-1}.

potential regions, which correspond to strongly and weakly bound atoms (68). The total amount of H adsorbed on Pt corresponds to the charge equivalent of about 210 $\mu C/cm^2$, although this value differs for different crystal faces of Pt (69). This is considered as a monolayer coverage corresponding to a 1:1 ratio of H_{ads} and Pt surface sites. On the other hand, it is known that H does not adsorb on those metals which are usually deposited in these studies. For instance, when Cu is adsorbed in small amounts at underpotentials the hydrogen adsorption is suppressed to a certain extent (Fig. 4). It is noted that for small coverages of Cu the weakly bound hydrogen is affected more by the Cu atoms than the strongly bound, as one would expect from simple thermodynamic considerations. When one monolayer of Cu is deposited (i.e., when about the maximum amount of Cu atoms is adsorbed at underpotentials) the hydrogen adsorption is completely suppressed, proving that the Cu is distibuted uniformly on the Pt surface and is not piling up at a few places (due to preferential deposition and nucleation at active sites) leaving some Pt surface sites uncovered.

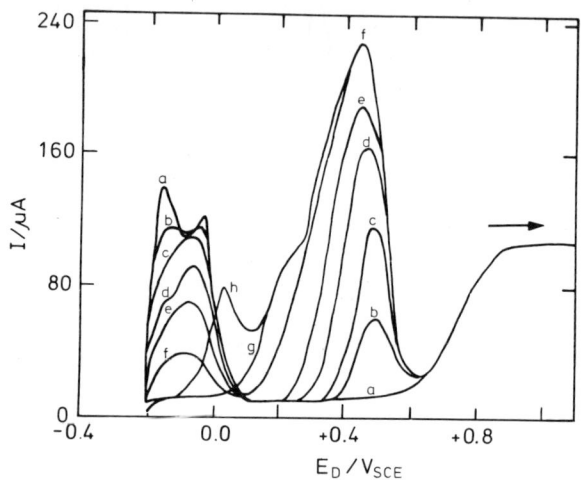

Fig. 4. Stripping current-potential curves for a Pt electrode in 0.2M H_2SO_4 + 5 x 10^{-6}M $CuSO_4$, with different amounts of Cu, ranging between zero (a) and approximately full monolayer (h) coverage, deposited at -0.25 V. Scan rate, 200 mVs^{-1} (28).

There are generally two different ways of defining a monolayer. The first one, mainly used in surface physics, refers to a monolayer coverage as the number of adsorbed atoms being equal to the number of substrate surface atoms. The second one, often used in electrochemistry, defines a monolayer coverage as that amount per unit surface area deposited on the substrate before the second layer grows. Here the monolayer term signals an important change in the adsorbate bond because it changes from the substrate-deposit bond to a pure deposit-deposit bond. We usually refer to the latter definition when talking about monolayers for obvious reasons, since the electrochemical experiments yield directly the amount of underpotentially deposited metal as contrasted by the second and next layers deposited at the Nernst potential or even at overpotentials. When the first definition of a monolayer coverage is implied, it is stated as such.

In Fig. 5 the amount of deposited metal is plotted against the amount of H_{ads} still adsorbed at the platinum surface for Cu (28), Ag (28), Bi (42), and Tl (67).[+] There is a simple linear relation between both quantities, again proving quantitatively the formation of a uniform monolayer. For the smaller metal atoms such as Cu and Ag the slope is unity corresponding to a 1:1 displacement of H_{ads} whereas for the larger metal atoms such as Bi and Tl the efficiency of the H_{ads} displacement increases, being about 2:1 for Bi. This implies a much smaller surface concentration of Bi as necessary to form a complete, close-packed monolayer. Bowles has demonstrated that for Cu, Cd, Sn, Tl, and Bi (67) the inhibition of hydrogen adsorption expressed by the ratio $\Gamma_{Me}^{\Theta=1} / \Gamma(H_{ads}^{max})$ is directly proportional to the square of the ratio of the atomic radii of platinum and the adsorbed metal atom. Although the exact linearity with unity slope must be considered as somewhat fortuitous when comparing it with the results of other authors, the trend is clearly seen, indicating that a completely covered Pt surface results from an underpotential deposit. An interesting observation has been reported (28,42) that large atoms such as Bi,

---

[+] The various scales were normalized for $H_{ads}^{max} = 2.1$ nmol/cm^2 for better comparison, since different H_{ads}^{max} values should reflect only different surface-roughness factors.

Pb, and Au inhibit adsorption of weakly and strongly bound H_{ads} to the same extent at all coverages, whereas smaller atoms such as Cu and Ag show a preferential suppression of the weakly adsorbed H at low coverages. Mikuni and Takamura (41) determined the amount of various metals adsorbed on Pt in the underpotential region as a function of potential by measuring the resulting displacement of adsorbed hydrogen. However, they extended these investigations on to Pd (48), where hydrogen dissolves into the metal rather than being adsorbed only at the surface as in the case of Pt. Therefore, these results have to be questioned.

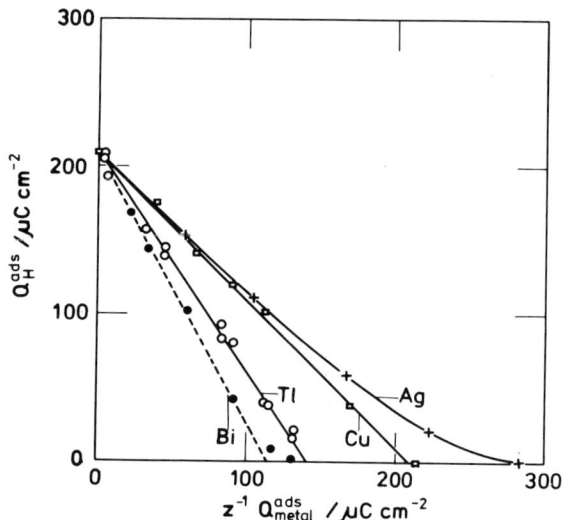

Fig. 5. Total amount of hydrogen that can be adsorbed on a Pt electrode, as a function of the surface concentration of adsorbed metal atoms for Bi, Tl, Cu, and Ag (28,42,67).

Although this review is focused on metal electrodes only, some results obtained with <u>graphite</u> electrodes should be mentioned here because of their wide spread use in analytical and technical chemistry. It has been found quite early that graphite is also able to adsorb metal monolayers at underpotentials. Some typical examples studied up to now are Ag (70-72), Cu (72,73), Bi (71), Pb (71,72), and Hg (72) on graphite. However, the data published by different authors for the same systems differ markedly, indicating the need for much more work to obtain reliable information. Whereas in earlier work the emphasis was placed mainly on stripping analysis for analytical purpose, more recently the studies were directed toward the question of monolayer deposition. Vassos and Mark (73) tried to present direct proof of whether the monolayer amount of Cu deposited at underpotentials represents massive (i. e., multilayer) deposition of Cu on active sites or a uniform monolayer caused by the strong interaction between deposit and substrate. Monitoring the X-ray fluorescence of the 1.542-Å emission line of Cu during electron bombardment after the graphite electrode had been removed from the electrolyte, they demonstrated that the surface gave a uniform distribution of counts when Cu was deposited at underpotentials in a monolayer amount. Deposition of bulk Cu (at potentials negative of the reversible Nernst potential) corresponding to approximately five or six layers (if the deposit were uniformly distributed), however, yielded clustering of the deposit several microns in thickness, whereas the vast areas between the Cu clusters were apparently covered only with a very thin Cu film (monolayer). These results also sustain the assumption of a uniform distribution of the metal atoms deposited at an underpotential, although direct proof for a monoatomic distribution is not given because of the limitation in instrumental sensitivity. They also indicate that the bulk Cu starts growing as three-dimensional nuclei at specific sites on top of the Cu monolayer on graphite following completion of monolayer deposition.

Having shown now that the metal atoms deposited at underpotentials form a uniform monolayer on the substrate, further details on the electrochemical behavior of these monolayers are reported. According to Nernst's law a shift in the reversible equilibrium potential E_r with the respective metal ion concentration in solution of about 60 mV/z per decade is observed for the bulk phase. The same behavior was found for the submonolayer metal deposit.

$$\left(\frac{\partial E}{\partial \log c_{Me^{z+}}}\right)_{\Gamma_{Me}} = \frac{60 \text{ mV}}{z} \tag{7}$$

It has been observed for Ag on Au (20) and for Ag on Pt (74) that the monolayer peak potential determined by cyclic voltammetry shifts toward positive values with 60 mV/decade when the Ag^+-concentration is raised from 10^{-5} to 10^{-3}M. The same effect was observed by Schmidt and colleagues for various systems investigated by thin-layer technique. They obtained identical isotherms for various concentrations in the range between 1×10^{-5} and 2×10^{-3}M, when they plotted the surface concentration Γ_{Me} versus the potential difference $E-E_r$ (52). This of course implies also that the charge transfer is almost identical with z.

In the experiments usually described in the literature, the metal monolayer is deposited at underpotentials, that is, under thermodynamic equilibrium conditions, where the coverage is changed reversibly with the potential applied. An important question for various applications is, however, whether the monolayer is also formed at much more cathodic potentials in a non-equilibrium situation, where bulk deposition and hence three-dimensional nucleation is allowed to take place. Furthermore, it is interesting to know whether the monolayer can be maintained on the electrode surface when no corresponding metal ions are in solution. This is important, for example, when studying the electrochemical, optical, and catalytic properties of the monolayer deposit proper in a wider potential range without changing the coverage with potential (see Sections 4 and 5). Earlier measurements have indicated that the monolayer should be stable at nonequilibrium conditions (17,28). To study this in detail, bulk deposition has to be avoided. This is done most conveniently in a twin electrode thin-layer cell, where the metal ion concentration in solution can be dropped to zero simply by driving the generator electrode potential negatively after monolayer deposition onto the working electrode has been achieved. It has been shown that Ag and Cu deposited in submonolayer or monolayer amounts onto Pt at a potential negative from E_r again forms a uniformly distributed adsorbate (31,75). This was done by generating a certain, well-defined amount of Ag^+ or Cu^{2+} at the generator electrode, which was subsequently deposited at the Pt working electrode, the potential of which was held sufficiently negative to maintain nonequilibrium conditions. The displacement of adsorbed hydrogen on Pt was then used to verify the monolayer formation.

Hydrogen adsorption was also used to study the monolayer formation of Au on Pt (39). The findings seem to be in contradiction to Bruckenstein's finding (28) that Au is not deposited onto Pt as monolayer. This, however, is a matter of <u>surface treatment</u>. It has been shown (39) that indeed Au does not form a uniform monolayer on Pt when potential cycles are applied that reach into the oxygen adsorption and oxid formation region (28). The corresponding inhibition of hydrogen adsorption is incomplete, even when the amount of deposited Au adds up to the equivalent of several monolayers. Here, because of oxide formation and reduction, interchange processes will most likely favor alloying between Pt and Au, always allowing Pt atoms to reach the surface to some extent. However, when the oxide region is carefully avoided after Au has been deposited in submonolayer amounts, the hydrogen displacement measurements indicate that a uniform monolayer of Au is built up. It may also be interesting to note that the presence of an oxide layer on Pt modifies underpotential deposition. It has been shown (27) that a silver monolayer cannot be deposited onto an oxidized Pt surface. Hence the monolayer adsorption of Ag at underpotentials is mainly governed by the oxygen reduction. On the other hand, oxygen will not adsorb onto the silver monolayer, and formation of Pt-O will be inhibited until vacant Pt sites become available after Ag desorption. This is demonstrated clearly in a current-potential curve shown in Fig. 6. The I-E curve due to Pt-O formation alone but in the presence of Ag^+ in solution has been calculated from the corresponding ring-disk data, which allow the separation of Ag monolayer desorption and Pt-O formation. The inhibition of oxygen adsorption on Pt due to Ag adatoms as compared to oxygen adsorption in the blank electrolyte (Fig. 6, dashed line) is clearly seen on the anodic scan.

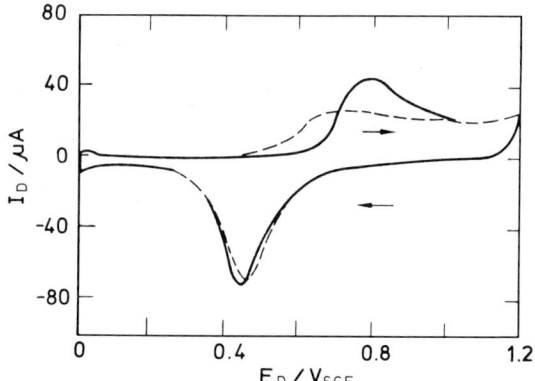

Fig. 6. Current-potential curve for a Pt electrode related to oxide formation and reduction in the presence of Ag^+ in solution as calculated from ring-disk data (solid line). For comparison, the current-potential curve in the blank solution ($c_{Ag^+} = 0$) is also shown (dashed line) (27).

The results on underpotential deposition obtained in recent years throw new light onto the question as to whether monolayer formation or nucleation is the primary step in metal deposition reactions on foreign metallic substrate. In those systems where underpotential deposition is observed the formation of a uniform monolayer will most likely occur first, even for $E < E_r$, before nucleation takes place. In some cases the distinction between monolayer and bulk formation (i.e., first and second layers) is exceptionally clear, since a small nucleation overpotential is observed for the deposition of the second layer (25). In most cases, however, the second layer starts as soon as the first is completed or even somewhat earlier, due to surface imperfection. It has sometimes been postulated (27,93) that two monolayers are deposited at underpotentials before the bulk deposition occurs, indicating that the substrate influence reaches beyond the first layer. Although it is often difficult to measure the amount of underpotentially deposited atoms close to E_r because of side effects (alloying, penetration into microcracks in the substrate surface, etc.; see Section 2.10), it is not unreasonable to assume that the underpotential deposition influences the further growth of the film. Optical measurements on Pt electrodes support this assumption (Section 4.3.). It is interesting to note that nucleation experiments in the gas phase observed by field-emission microscopy revealed similar results for deposition of various metals onto tungsten (76). It was found that a continuous and uniform film of even a few monolayers was deposited first onto the substrate before nucleation took place. Hence not only will nucleation not occur on the foreign substrate directly, but in some cases even a second or third layer may be deposited uniformly before nucleation commences.

Finally, we discuss shortly the coadsorption respectively competitive adsorption of two different metals in the underpotential region and the monolayer formation on top of another metal monolayer. The deposition of small amounts of Bi and Pb on Ag and Pb and Tl on Ag has been studied by Schmidt and Gygax (77), who found a drastic change in the adsorption and desorption behavior of the more electronegative metal caused by the underpotential deposition of the more electropositive entity. It was clearly demonstrated that the presence of predeposited Bi on Ag suppresses the underpotential deposition of Pb that obviously will not go on top of Bi. The same behavior is reported for Tl deposition on Ag when Pb has been predeposited in submonolayer amounts on the surface. Increasing the surface concentration of Pb from zero to monolayer coverage decreases the monolayer peak for Tl on Ag from full size to zero without observable changes in peak potential. Here again the total Tl is deposited as bulk at the respective Nernst potential when the Ag electrode is completely covered with a lead monolayer, although underpotential deposition of Tl on Pb is expected to take place (and has been observed in nonaqueous solutions) with a monolayer peak potential about 70 mV positive of E_r. However, halide ion adsorption may shift the monolayer peak too far negative (see Section 2.8.) to be observed before bulk deposition begins. The codeposition of Cu and Ag onto platinum as studied by rotating ring-disk technique (22) yields about the same finding. Here, however, it is well known that Cu on Ag does not show any appreciable underpotential effect (51). Underpotential deposition of Cu onto Pt is suppressed by the presence of adsorbed Ag that will have been already deposited at more positive potentials.

The underpotential behavior of a metal can also be used to study the structural and adsorptive properties of composite metal surfaces. Stucki (75) has investigated the deposition of an "indicator" metal such as Tl or Pb on metal electrodes, which are already covered with a submonolayer or monolayer amount of another metal.

In Fig. 7a the anodic stripping curves for Tl from an Au electrode, which is covered with various amounts of Ag, are shown together with the Tl stripping curve from a bulk Ag electrode for comparison. Desorption of Tl from Au as well as from Ag sites are clearly seen at medium coverages. However, the presence of Ag adatoms on the surface obviously alters the bond strength between the indicator metal Tl and the gold surface as indicated by the shift in

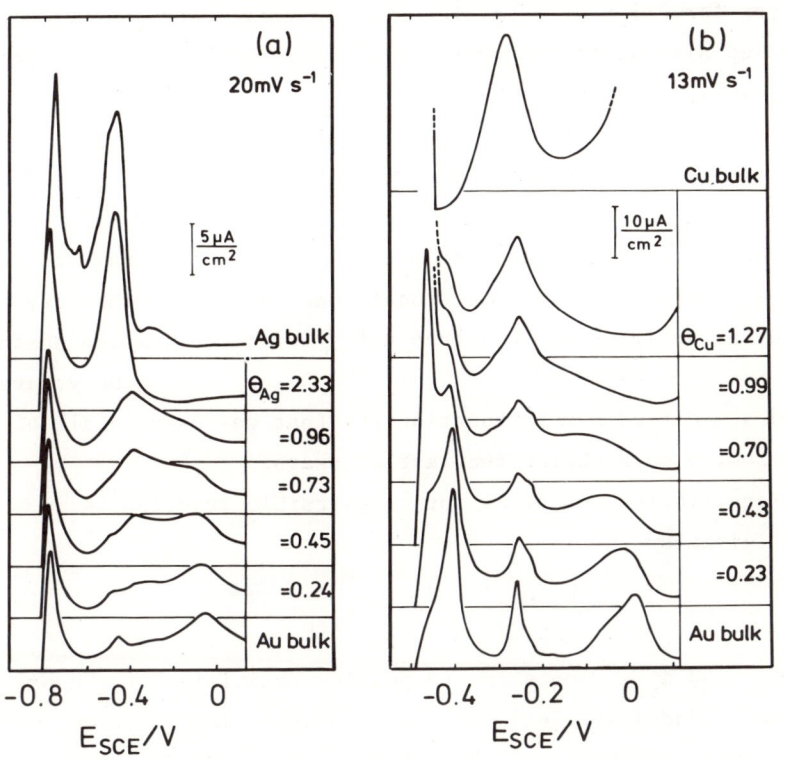

Fig. 7. Current-potential curves due to Tl (a) and Pb (b) stripping from Au electrode, covered with various amounts of Ag (a) and Cu (b), respectively, in 0.5M NaClO$_4$ (pH 2). [Me^{z+}] = 5 x 10^{-4}M (75).

potential of the monolayer peak for Tl on Au. The same behavior is found for Pb deposition on Au surfaces covered with various amounts of Cu (Fig. 7b). Here also a weakening of the Pb-Au bond is observed as the coverage of Cu on Au is increased (reduction of underpotential shift for Pb on Au). This method provides a surface specific tool in a true sense because only the surface atoms are involved in the adsorption process. Thus valuable information on the surface composition of alloys may be gained. In addition, monolayer formation of metal deposits on nearly any substrate can be checked similar to the hydrogen displacement method, the latter being restricted to Pt.

2.4. The Thermodynamics of Predeposition

The deposition of a metal at potentials positive from its reversible Nernst potential has usually been described thermodynamically by an activity of less than unity for the deposited phase. On the basis of Herzfeld's concept (2), Rogers and Stehney have derived a modified Nernst equation (6) that was used in all of the following work during the last 20 years.

The equilibrium potential for a reversible reaction is in general given by

$$E_r = E_o + \frac{RT}{zF} \ln \frac{a_{ox}}{a_{red}} \qquad (8)$$

where a_{ox} and a_{red} denote the respective activities of the oxidized and reduced species; for a metal deposition reaction, where a solid phase is involved, the activity of the deposit is assumed to be constant and set to unity. However, when the electrode surface is only incompletely covered by the trace amount of a foreign deposit, the activity of the deposit (in the following termed as "monolayer") should be less than unity and vary with surface coverage (2). The most simple and straightforward assumption will be a linear relation between deposit activity a_{ML} and coverage Θ, $a_{ML} = f_{Me} \cdot \Theta$ for $0 \leq \Theta \leq 1$, and $a_{ML} = f_{Me}$ for $\Theta > 1$, with $f_{Me} = 1$. This modifies Eq. 8 to:

$$_{ML}E_r(\Theta) = E_o + \frac{RT}{zF} \ln \frac{c_{Me^{z+}} \cdot f_{Me^{z+}}}{\Theta \cdot f_{Me}} > E_r \qquad (9)$$

with $f_{Me^{z+}}$ and f_{Me} being the activity coefficients of the metal ions in solution and the bulk deposit, $c_{Me^{z+}}$ the metal ion concentration. Equation 9 yields equilibrium potentials that are always more positive than the corresponding Nernst potential for the bulk phase. However, one still has to take into account the specific interaction between deposit and substrate resulting in different underpotential ranges for the same deposit on various substrates. This is done either by allowing f_{Me} to become a parameter, usually also less than unity and depending on the substrate material as well as on the deposit, or by adding a system-specific quantity, E_a, to E_o (6). Equation 9 also reveals that the monolayer equilibrium potential will shift with metal ion concentration by 60 mV/z per decade for constant coverage Θ.

A slightly different approach was given by Brainina, Zakharchuk, et al. (71), who chose an expression for the monolayer activity that does not have the clearcut break at $\Theta = 1$ as that discussed in the preceding paragraph:

$$a_{ML} = a_{bulk} \left[1 - \exp(-\xi Q)\right] \qquad (10)$$

where ξ is a constant depending on the nature of substrate and deposit, Q is the charge equivalent of the amount deposited, and $a_{bulk} \equiv 1$ is the activity of the bulk phase. It is readily seen that for small coverages ($\xi Q \ll 1$) the result is identical to that of Rogers and Stehney (6), whereas for large amounts deposited ($\xi Q \gg 1$) Nernst behavior is found.

The assumptions on the monolayer activity coefficients are purely formalistic and hardly advance our understanding of the underpotential effect. The most important and direct thermodynamic information one obtains from the experiments is that the chemical potential of the first monolayer, μ_{ML}, is markedly different from that of the bulk metal, μ_{metal}. The chemical potentials of bulk phase and monolayer can be obtained from the equilibrium conditions for the reactions (19):

$$Me^{z+} + ze^-_{metal} \rightleftharpoons Me_{metal} \qquad (11a)$$

and

$$Me^{z+} + ze^-_{substrate} \rightleftharpoons Me_{ML} \qquad (12a)$$

The respective equations for the electrochemical potentials, $\tilde{\mu} = \mu + ze_o\varphi$, are

$$\tilde{\mu}_{Me^{z+}} + \tilde{\mu}_{e^-}(metal) = \mu_{metal} \qquad (11b)$$

and

$$\tilde{\mu}_{Me^{z+}} + \tilde{\mu}_{e^-}(substrate) = \mu_{ML} \qquad (12b)$$

Comparing the two equilibria in the same solution and for the same composition that is equal $\tilde{\mu}_{Me^{z+}}$, yields the difference between the chemical potentials of bulk deposit, μ_{metal}, and monolayer, μ_{ML}, at a certain underpotential $_{ML}E_r$. This is simply the difference in electron energies at the respective electrode potentials necessary to establish the equilibrium conditions for Eqs. 11a and 12a:

$$ze_o \left(_{ML}E_r(\Theta) - E_r\right) = \mu_{ML}^{\Theta} - \mu_{metal} \qquad (13)$$

(chemical potential μ in atomic units). The expressions on both sides in Eq. 13 represents a direct measure for the difference in binding energies between a metal adatom (at a certain coverage) on the substrate and this metal atom on a like surface. This difference is independent from the metal ion concentration, since both values are shifted with concentration by the same amount, and it is independent from the nature of the electrolyte (even for complexing agents), as long as no different specific interaction takes place between the electrolyte and the two solid phases (see Section 2.8.).

To establish a quantitative description of the underpotential deposition Kolb, Przasnyski, et al. (19) have chosen the peak

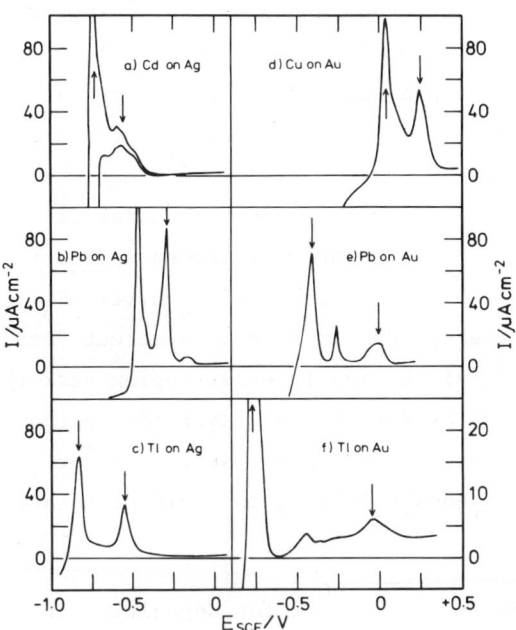

Fig. 8. Anodic stripping curves for various metal deposits on Ag and Au in 1M Na_2SO_4 (pH 3) (a and c) and in 1M $NaClO_4$ (pH 3) (b, d, e, and f). $[Me^{z+}] = 2 \times 10^{-4}$M. Scan rate, 20 mVs^{-1}. The arrows indicate the peak potentials for bulk and monolayer stripping (19).

potentials of monolayer and bulk stipping peaks as most suitable values for the difference in chemical potentials of monolayer and bulk deposit. In Fig. 8 the anodic scans of cyclic voltammograms are shown for six different systems, bulk and monolayer peak potentials being marked by arrows. While the bulk stripping peak is within some 10 mV the same as the reversible Nernst potential E_r in these solutions (the small shift usually being caused by kinetic effects[+]), the monolayer peak potential E_p corresponds fairly well to about the same surface concentration Γ_{peak} for all systems. Hence $E_p = {_{ML}}E_r(\Theta_p)$. When the monolayer isotherm exhibits more than one peak in the corresponding I-E curve as in the case of Pb on Au, the position of the more anodic (low coverage) peak was chosen as E_p. The pronounced current maximum at the peak potential E_p shows that most of the adsorbed atoms are deposited around this potential with the least variation in adsorption energy as a function of coverage. The potential difference between bulk and monolayer stripping peak is termed as <u>underpotential shift</u> ΔE_p. In Table I all underpotential shifts ΔE_p from (19) are summarized and are expanded for some more recent data. In addition, the half width $\delta_{1/2}$ of the monolayer stripping peak that is a characteristic quantity for the monolayer adsorption isotherm is given in Table I whenever possible. The values for underpotential deposition in nonaqueous solutions and salt melts are also included,

---

[+] Cyclic current-potential curves for deposition of Tl onto Ag from a solution containing 2×10^{-4} M Tl^+-ions have shown that within a sweep range from 1 mV/s up to 1 V/s the potential difference between bulk and monolayer stripping peaks remains constant when the peaks are about equal in height, although both peak potentials are shifted toward positive values with increasing scan rate resulting from kinetic effects (47).

but are discussed in greater detail in Section 2.9. Only data from experiments in halide ion-free electrolytes were taken, since it was shown that specific adsorption of halide ions may render the ΔE_p values incomparable (see Section 2.8.).

TABLE I

Data for Underpotential Shift ΔE_p and Half Width $\delta_{1/2}$ of the Monolayer Stripping Peak

Substrate/ metal ions	Supporting electrolyte	(ΔE_p/V)	($\delta_{1/2}$/V)	References
\multicolumn{5}{c}{AQUEOUS SOLUTIONS}				
Pt/Ag^+	0.5M H_2SO_4	0.44	0.25	19 (27)
Pt/Cu^{2+}	0.5M H_2SO_4	0.41	0.31	19 (32,33,37)
Pt/Hg^{2+}	0.5M H_2SO_4	0.47	0.14	19 (38)
Pt/Pb^{2+}	1M $HClO_4$	0.87	–	41 (39,219)
Pt/Bi^{3+}	0.12M $HClO_4$	0.59	–	42 (41,44,219)
Pt/Cd^{2+}	1M $HClO_4$	0.96	–	219 (41)
Pt/Tl^+	1M $HClO_4$	≈ 1.3	–	41 (219)
Pt/Sb^{3+}	1N H_2SO_4	0.64	–	44
Pt/Ge^{4+}	1N H_2SO_4	≈ 0.40	–	44
Pt/As^{3+}	1N H_2SO_4	≈ 0.75	–	44
Pd/Ag^+	0.5M $NaClO_4$ (pH 2)	≈ 0.30	–	47
Pd/Cu^{2+}	0.5M $NaClO_4$ (pH 2)	0.23	–	47

(TABLE I, cont.)

Au/Ag^+	$0.5M\ H_2SO_4$	0.51	–	19	(50)
Au/Cu^{2+}	$0.5M\ Na_2SO_4$ (pH 3)	0.22	0.11	19	(20)
Au/Cd^{2+}	$0.5M\ Na_2SO_4$ (pH 3)	0.51	0.17	19	
Au/Tl^+	$0.5M\ Na_2SO_4$ (pH 3)	0.69	0.19	19	
Au/Pb^{2+}	$1M\ NaClO_4$ (pH 3)	0.40	0.12	19	(52,55)
Au/Hg^{2+}	$0.5M\ H_2SO_4$	0.43	0.16	19	
Au/Bi^{3+}	$0.12M\ HClO_4$	0.25	–	57	
Au/Zn^{2+}	$1M\ Na_2SO_4$	0.59	–	47	
Ag/Cu^{2+}	$0.5M\ H_2SO_4$	→ 0	–	19	(51)
Ag/Cd^{2+}	$1M\ Na_2SO_4$ (pH 3)	0.16	0.16	19	
Ag/Tl^+	$1M\ Na_2SO_4$ (pH 3)	0.28	0.08	19	
Ag/Pb^{2+}	$1M\ NaClO_4$ (pH 3)	0.16	0.05	19	
Ag/Bi^{3+}	$0.5M\ HClO_4$	0.06	–	59	
Ag/Sn^{2+}	$0.5M\ NaClO_4$	0.21	–	47	
Ag/Zn^{2+}	$1M\ Na_2SO_4$	0.13	–	47	
Cu/Cd^{2+}	$1M\ Na_2SO_4$	0.23	0.14	19	
Cu/Tl^+	$1M\ Na_2SO_4$	0.34	0.08	19	
Cu/Pb^{2+}	$0.5M\ NaClO_4$ (pH 2)	0.20	0.08	39	
Bi/Sn^{2+}	$0.5M\ NH_4ClO_4$ + $0.5M\ HClO_4$	→ 0	–	59	
Bi/Cd^{2+}	$0.5M\ NH_4ClO_4$ + $0.5M\ HClO_4$	→ 0	–	59	

Underpotential Shifts 159

(TABLE I, cont.)

Sn/Cd^{2+}	0.5M NH_4ClO_4 + 0.5M $HClO_4$	→ 0	–	59
Sn/Tl^+	0.5M NH_4ClO_4 + 0.5M $HClO_4$	≈ 0.08	–	59
Pb/Cd^{2+}	0.5M NH_4ClO_4 + 0.5M $HClO_4$	→ 0	–	59

NONAQUEOUS SOLUTIONS

Pt/Ag^+	1M $LiClO_4$ (acetonitrile)	0.45	0.86	19
Au/Ag^+	1M $LiClO_4$ (acetonitrile)	0.55	0.52	19
Ag/Tl^+	1M $LiClO_4$ (acetonitrile)	0.20	–	78
Cd/Tl^+	1M $LiClO_4$ (acetonitrile)	0.08	–	78
Pb/Tl^+	1M $LiClO_4$ (acetonitrile)	0.1	–	78
Cd/Li^+	1M $LiClO_4$ (acetonitrile)	0.8	–	78
Tl/Li^+	1M $LiClO_4$ (acetonitrile)	0.6	–	78 (79)
Au/Li^+	1M $LiClO_4$ (acetonitrile)	1.23	0.53	19
Ag/Li^+	1M $LiClO_4$ (acetonitrile)	0.96	–	19
Ag/Li^+	1M $LiClO_4$ (propylene carbonate)	1.00	0.30	19
Cu/Li^+	1M $LiClO_4$ (propylene carbonate)	1.05	0.35	19

(TABLE I, cont.)

Pt/Li$^+$	2×10^{-2} M Li$^+$ (propylene carbonate)	1.56	0.50	80
Pt/Na$^+$	1×10^{-2} M Na$^+$ (propylene carbonate)	1.30	0.27	80
Pt/K$^+$	2×10^{-2} M K$^+$ (propylene carbonate)	1.50	-	80
Pt/Cs$^+$	2×10^{-3} M Cs$^+$ (propylene carbonate)	2.00	≈ 0.26	80
Pt/Ni^{2+}	salt melt	0.4	-	81

2.5. The Electrosorption Valency γ

When a metal atom is adsorbed onto a foreign metal substrate transfer of a partial charge will usually occur due to the difference in electronegativities or, in other words, due to the difference in their Fermi energies. As a consequence, a dipole layer will be built up to establish electronic equilibrium between substrate and adsorbate changing the electronic properties of the (composite) surface. At the metal-vacuum interface the dipole of the adatom-substrate bond can be determined by work function measurements, since (in cgs units) (82,83)

$$\Delta\phi(\Theta) = 4\pi n \mu(\Theta) \tag{14}$$

where $\Delta\phi(\Theta)$ is the change in work function on adsorption, $n = N_o \Theta$ is the absolute number of adatoms per cm^2 and $\mu = \delta e_o r$ is the half-dipole consisting of charge δe_o separated by the distance r from the surface or 2r from its image charge in the metal. $\mu(\Theta)$ and $\delta(\Theta)$ are important quantities to describe the adatom's properties. For most of the metal pairs studied the partial transfer is from the adatom to the substrate leaving a partial positive charge at the adatom, the amount of which depends on the constituents and the coverage. Equation 14 can be used to estimate an upper limit for the partial charge $\delta(\Theta)$, since too large values (as for ionic layers) result in unreasonably high work function changes. The value $\delta(\Theta)$ decreases with increasing Θ due to depolarization effects of the equally charged adatoms; for example, for K on W $\delta(\Theta)$ has been reported to be 0.27 e_o and 0.15 e_o for $\Theta \to 0$ and 1, respectively (83).

When dealing with the metal-electrolyte interface, the situation is somewhat more complex due to the finite width of the Helmholtz layer. There the adatom or adion can penetrate into the double layer and will be held in position at a certain point (see Fig. 9). Therefore, the adsorbed species feel only a fraction of the total potential drop across the Helmholtz layer. The penetration of the adsorbate is described by the geometric factor g, which is given by[+] (84-86):

$$g = \frac{\varphi_{ad} - \varphi_e}{\varphi_m - \varphi_e} \quad (15)$$

[+] At the metal-vacuum interface similar problems are encountered, as has been recently recognized. Since the free-electron density of the metal decreases gradually from the bulk value inside to zero outside the surface within a range of approximately 5 Å, different adsorbates may encounter different surface potentials depending on their size.

In praxi the potential difference, $\varphi_h - \varphi_e$, between the outer Helmholtz layer and the electrolyte is neglected, since excess supporting electrolyte is commonly used. The charge-transfer coefficient, δ', is given by the difference between the actual charge of the adsorbate, z_{ad}, and that of the respective ion in solution, z,

$$\delta' = z - z_{ad} \qquad (16)$$

with $0 \leqslant \delta' \leqslant z$.[+] The values δ' and g, which cannot be measured separately, are correlated with the electrosorption valency γ by (84, 86-88):

$$z\gamma = gz + \delta'(1-g) + \kappa_{ad} - \nu\kappa_{solv} - \frac{1}{F} \int_{E_{PZC}}^{E} \left(\frac{\partial C_{DL}}{\partial \Gamma_{Me}}\right)_E dE \qquad (17)$$

where κ_{ad} and κ_{solv} are dipole terms of the adsorbate and the solvent molecule, and ν is the number of solvent molecules displaced by adsorption of one adsorbate atom. For metal adatoms $\nu = 0$ if one assumes that the desorbed solvent molecules will adsorb again onto the monolayer in the same way as onto the substrate. The integral term in Eq. 17 takes into account the influence of the electrode potential on γ, which is mainly determined by the change in double-layer capacity C_{DL} with increasing surface concentration Γ_{Me}.

Since the dipole terms in Eq. 17 are usually small compared to the charge terms, the electrosorption valency γ, when determined at the potential of zero charge, can be approximated by (84):

[+] For comparison with the original papers by Vetter and Schultze (87,88) note that λ has been replaced in this work by $-\delta'$, and γ by $\gamma \cdot z$.

$$\gamma_{PZC} \approx g + \frac{\delta'}{z}(1-g) \tag{18}$$

The value γ can be determined experimentally, since it describes the potential dependence of the electrosorption equilibrium as well as that of the charge flow. Schultze and Vetter have shown that in the case of excess supporting electrolyte (87,88)

$$\gamma = \frac{1}{zF}\left(\frac{\partial \mu}{\partial E}\right)_{\Gamma_{Me}} = -\frac{1}{zF}\left(\frac{\partial q_m}{\partial \Gamma_{Me}}\right)_E \tag{19}$$

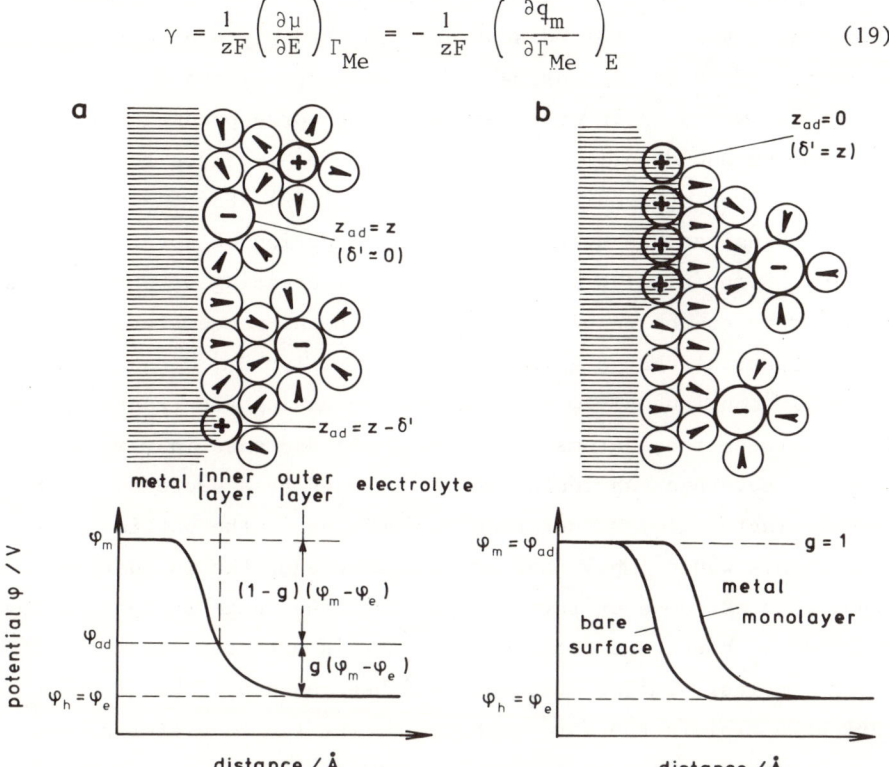

Fig. 9. Schematic diagram of the double-layer and corresponding potential distribution in the presence of a supporting electrolyte: (a) partially discharged cation and electrostatically adsorbed anion; (b) completely discharged cations (84).

where μ is the chemical potential of the adsorbate species in solution (e.g., metal ion) and q_m is the total electrode surface charge density. In addition, the relation between the electrochemical transfer coefficients α and β and the electrosorption valency γ ($\alpha + \beta = \gamma$) can be used to obtain γ. When γ is known, the partial charge tranfer δ' can be estimated by making reasonable assumptions about the geometric factor. For metal adatoms that are nearly completely discharged and strongly bonded to the substrate surface it is reasonable to assume g ≈ 1 and $\delta' \approx z$; hence $\gamma \approx 1$, which is in agreement with the corresponding data derived from vacuum experiments. A collection of γ-values for various systems is given by Schultze and Koppitz (84,85). For the system Cu on Pt, Schultze has determined γ as a function of coverage Θ (Fig. 10). Surprisingly, γ decreases with increasing Θ, which seems contradictory to the results found at the metal-vacuum interface. There the partial charge δ is always largest for $\Theta \to 0$ and decreases usually continuously with Θ due to increasing electrostatic repulsion between the adatoms. Therefore, the decrease in γ with increasing coverage was tentatively explained by a decrease in the geometric factor $g(\Theta)$ rather than an increase in the partial charge δ' (25). This would imply that at high coverages the Cu adatoms are now located somewhat further away from the metal surface, being less strongly bound and feeling to a certain extent already the solvent dipoles. Similar arguments had been used by Schmidt and Wüthrich to explain the more cathodic of the two monolayer peaks in the underpotential region for Pb on Au (52). However, it has been pointed out (89) that for Cu adsorption on Pt in sulfuric acid coadsorption of anions like HSO_4^- is strongly induced. When this effect is not taken into account, the calculated γ-values for Cu may be too low; therefore, further experimental data are necessary to answer these questions with certainty.

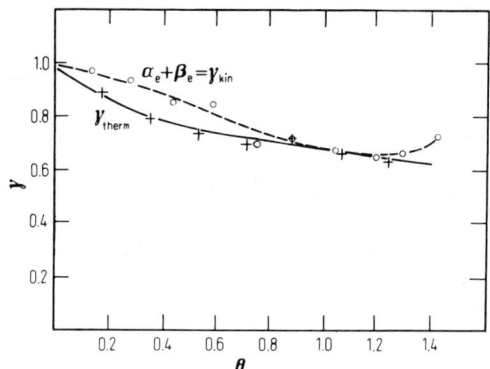

Fig. 10. Electrosorption valency γ for Cu adsorbed on Pt as a function of coverage, determined from thermodynamic (+++) and kinetic (ooo) measurements (25).

2.6. Monolayer Adsorption Isotherms and Kinetics

The amount of underpotentially adsorbed metal as a function of potential can be determined most precisely in thin-layer experiments. Usually, the monolayer adsorption on polycrystalline surfaces occurs around a single potential value, yielding one pronounced current peak in cyclic voltammetry. This means that the adsorption isotherm has a sigmoid shape and can in many cases be described by a Langmuir-type isotherm (90), or more generally by a Frumkin isotherm:

$$\ln\left(\frac{\Theta}{1-\Theta}\right) = \ln K_o + \frac{\Delta G^o}{RT} - g\Theta - \frac{zF}{RT}\Delta E \qquad (20)$$

(g: interaction parameter). It has been shown that the half width $\delta_{1/2}$ of the monolayer adsorption peak (the most anodic peak in those cases of more than one) increases with increasing underpotential shift ΔE_p, the values of $\delta_{1/2}$ for $\Delta E_p \to 0$ being around 50 - 100 mV (19). This number corresponds roughly to the half width expected from the classical Langmuir-type isotherm (g = 0). According to Frumkin the broadening is a result of repulsion between the adsorbed adatoms (91,92), which is obviously increasing with ΔE_p. This may be explained by the metal adatoms' small (positive) partial charge (see Section 2.5.), which also increases with ΔE_p (19) and leads to a progressive mutual repulsion (increasing g). This, however, should favor the formation of ordered structures to minimize the total energy of the adsorbed layer. Attraction among the adsorbed species, as in many cases with organic molecules, would cause two-dimensional island formation and result in an isotherm with a half width <u>smaller</u> than that of the Langmuir-type (-4 < g < 0). The existence of ordered structures, which is discussed in greater detail in connection with single-crystal surfaces in Section 2.7., can be seen from yet another fact. It is noted that the amount adsorbed at the monolayer peak potential E_p usually represents much less than half of the monolayer amount, as one would expect from a symmetric monolayer adsorption isotherm. In Table II the surface concentration at E_p expressed in terms of coverage Θ_p is shown for various systems. The value Θ_p was calculated from the charge stripped between E_p and $E(\Theta = 0)$ and the charge equivalent of the total number of substrate surface atoms, the latter being roughly 210 $\mu C/cm^2$ for a monovalent metal. It turns out that at the peak potential the coverage ranges between 0.1 and 0.3. This means that the monolayer peak, representing a major portion of adatoms with energetically equal adsorption sites yields a coverage of roughly 0.2 to 0.6 with respect to the total number of surface sites. As a first-order approach this can be interpreted as due to the formation of various (2 x 2) structures

TABLE II

Coverage[a] of Metal Deposit at Monolayer Peak Potential, E_p, for Various Systems

System	Au/Cu^{2+}	Au/Tl^{+}	Au/Pb^{2+}	Ag/Tl^{+}
Θ_p	0.18	0.26	0.22	0.22

System	Ag/Pb^{2+}	Ag/Cd^{2+}	Cu/Tl^{+}	Pt/Cu^{2+}
Θ_p	0.15	0.17	0.31	0.23

average: $\bar{\Theta}_p \approx 0.2$

[a] Full coverage refers here to the number of substrate surface atoms.

that correspond to $\Theta = 0.25$ and 0.5, respectively. After this structure is completed the layer may change gradually to a close-packed monolayer when the potential approaches the reversible Nernst potential or change rather abruptly to another structure before bulk deposition occurs on top of it.

An interesting example of a more complex adsorption isotherm is given in Fig. 11 for the system Pb on Au. As seen even more clearly in the corresponding current-potential curve (Fig. 2), the monolayer adsorption peak at $\Delta E_p = 0.42$ V ($\Theta_p = 0.2$) is followed by a surprisingly sharp second peak at $\Delta E_p = 0.21$ V. This may be considered as due to a change in structure of the adsorbed atoms allowing the accommodation of more Pb on the Au surface. It is interesting to note that in some cases such sudden changes in the adsorbate structure may be initiated by specifically adsorbed anions such as halides (see Section 2.8.).

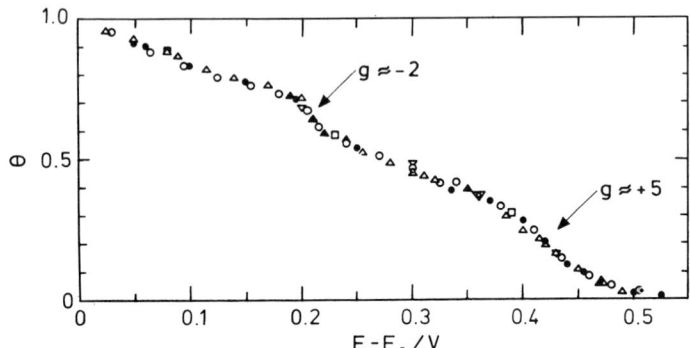

Fig. 11. Adsorption isotherm for underpotentially deposited Pb on Au. $c_{Pb^{2+}}$ = 1.8 x 10^{-5}M (△), 4.0 x 10^{-5}M (□), 1.8 x 10^{-4}M (●), 4.0 x 10^{-4}M (▽), 7.6 x 10^{-4}M (▲), 1.8 x 10^{-3}M (o) (15,52); g-values estimated from half widths of corresponding current peaks.

The adsorption isotherm of Cu on Pt exhibits a rather broad and linear range around the peak potential at low coverage. This is a result of at least two adsorption peaks that are often not resolved and which represent two different adsorption sites at the Pt surface. Schultze has analyzed the Cu monolayer formation on Pt and found the adsorption to obey a Tempkin isotherm over a large range in Θ, that is, a linear dependence of Θ on concentration and on potential, respectively (25):

$$B \cdot \Theta = RT \ln a_{Cu^{++}} - 2\gamma FE + const \qquad (21)$$

The value of B which represents the difference in adsorption energy at Θ = 0 and 1, is found to be 11.3 kcal/mol. This is more than what

one might expect from surface heterogeneity and may be explained by
the mutual repulsion of the adatoms, which increases with increasing coverage, hence decreasing the adsorption energy. The γ-value
is given as 0.73, which seems to be too small in view of a metallic
adsorbate rather than an adion. An analytical determination of Γ_{Cu}
yields $\gamma = 0.83$ (25). From the data of Tindall and Bruckenstein
for Cu on Pt (32), γ is estimated as being close to 0.9 (25),
which seems a more appropriate value. For Pb on Au a Tempkin-type
adsorption isotherm was determined from reflectance measurements
(156).

It is interesting to note that the adsorption isotherm of various systems (which exhibit a rather well-defined adsorption or
stripping peak in cyclic voltammetry), although quite different
in shape on the usual potential scale, are nearly the same when
plotting them against a normalized potential ($\Delta E/\Delta E_p$), where ΔE
is measured with respect to the reversible Nernst potential or the
bulk stripping peak. It indicates that the change in adsorption
energy, ΔH_{ads}, as a function of Θ is of about the same shape for
these systems and proportional to ΔE_p, the underpotential shift.
This means that an increase in the adsorption energy for the
adatom-substrate bond, which manifests itself in a larger ΔE_p,
also increases the change in adsorption energy between $\Theta = 0$ and 1
(see Section 6.3.). This effect (g ~ ΔE_p) again originates from
the small partial charge of the adatoms.

Recent measurements on single-crystal surfaces indicate that
the adsorption of metal atoms may not necessarily be always the
initial step in monolayer formation (93,94). It has been observed
that on the (111)-faces of Ag and Au electrodes deposition of
nearly a complete monolayer occurs within an extremely narrow
potential range, while at the (110)- and (100)-faces, as well as
at polycrystalline surfaces, rather broad adsorption and desorp-

tion peaks are found at low coverages (see Fig. 12). This observation instigated a controversy, as to whether deposition at the (111)-surface represents a first-order phase transition (g = −4) due to two-dimensional nucleation and layer growth or whether it can be described by an adsorption isotherm with strong attraction among the adatoms (−4 < g < 0) (93-95). As seen in Fig. 12b for Ag(111), formation as well as stripping of the Pb monolayer starts at a certain well-defined potential value (−356 mV$_{SCE}$) and not over an extended potential range as in the other cases, where adsorption is believed to take place. Since a clear discontinuity in the Θ − E curve was not observed (g = −3.7) (94,95), a finite width for the phase transition due to surface heterogenuities had been assumed (94). In the case of Pb on Ag(111) even small overpotentials for monolayer formation and reduction have been reported (94), which also supports the idea of a phase transition, although further experimentation is still necessary.

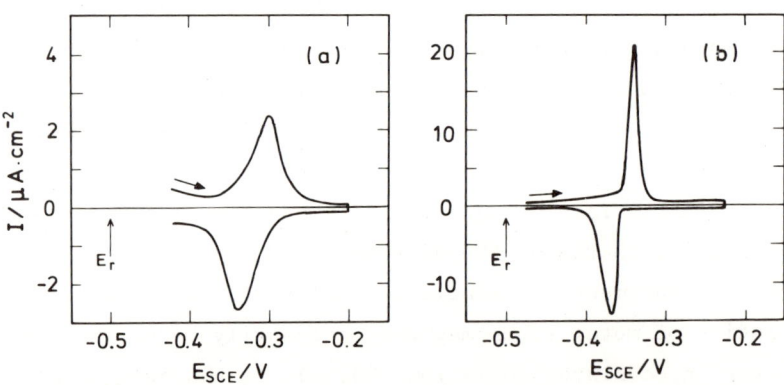

Fig. 12. Cyclic current-potential curves for Pb deposition on different Ag substrates in the underpotential range (E > E$_r$): (a) polycrystalline Ag, (b) Ag(111). [Pb^{2+}] = 2 × 10^{-4}M in 0.5M NaClO$_4$ (pH 2). Scan rate, 1 mVs^{-1}.

Up to now, relatively few papers have dealt with the kinetics of monolayer adsorption as studied by classical electrochemical means (15,25,26,93,95,156). Schultze investigated the kinetics of Cu deposition on Pt by galvanostatic pulse techniques (25). For $\Theta < 1$ the charge-transfer reaction for Cu^{++} is rate-determining, whereas for $\Theta > 1$ reactions involving Cu^+ become important. Studies of the dissolution and deposition of Cu on bulk Cu and on a Cu monolayer on Pt as a function of potential in 0.1M $CuSO_4$ show that for the Cu monolayer the exchange current density is smaller by a factor of 2 compared to that for the bulk Cu electrode. The sum of the transfer coefficients for the monolayer is 0.78, which agrees well with the γ-values determined thermodynamically (25), whereas it is about unity for the bulk Cu electrode as expected. Lorenz, Hermann, et al. have studied the kinetics of metal monolayer formation on Au substrates (15,26). They found in the case of Cu on Au that the charge corresponding to the apparent monolayer deposition or stripping is larger than that predicted theoretically and depends on the Cu^{++}-concentration in the electrolyte and on the scan rate. This supports evidence for the fact that the redox process Cu^{++}/Cu^+ is involved in the rate-determining step. For the underpotential deposition of Ag, Tl and Pb on Au charge transfer and surface diffusion of the adatom were considered as possible rate-determining steps although a distinction between both could not be made with certainty. The situation is too complicated to obtain reliable information from the present data.

The exchange currents for the monolayer formation on polycrystalline Au are calculated to be between 30 and 130 mA cm^{-2} for Ag, around 10 mA cm^{-2} for Pb and in the order of 200 mA cm^{-2} for Tl. For Ag and Pb these values are clearly smaller than those of the corresponding bulk metal electrodes, whereas for Tl it is about the same. In general, no marked dependence of the exchange currents on the (under-) potential (or coverage) was reported (26).

2.7. Measurements on Single-Crystal Surfaces

It has been pointed out that cyclic current-potential curves yield information on adsorption energies similar to flash-desorption experiments in the gas phase (96). In the first case, however, exact data for binding energies may be obtained since these measurements can be performed under thermodynamic equilibrium conditions. It is obvious that measurements on polycrystalline surfaces have the disadvantage of representing an average over all bonds formed on a rather heterogeneous surface, whereas single-crystal surfaces offer the possibility of studying underpotential deposition on (even in an atomistic sense) well-defined and uniform adsorption sites. Important questions such as those regarding the influence of the substrate structure on the metal monolayer or of the existence of ordered structures may be answered.

A pronounced influence of the substrate surface structure on the underpotential behavior of metal monolayers is shown in Fig. 13 for Tl on three different Ag substrates, a mechanically and a chemically polished polycrystalline silver surface and a (110) single-crystal surface (which will contribute most to a polycrystalline surface). An increase in fine structure is clearly discernable with improving surface homogeneity, which may indicate a substantial gain in detailed information about the adsorbate-substrate complex.

Up to now, the following systems have been investigated in detail by cyclic voltammetry and transient analysis: (a) Bi, Cu, Pb, Sb and Tl on Au (97,98), (b) Pb on Au (156), and (c) Tl and Pb on Ag (93-95, 98,99), using the three low index planes (110),

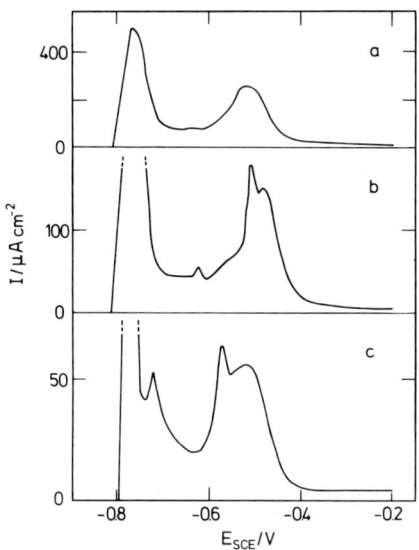

Fig. 13. Anodic stripping curves for Tl on Ag in 0.5M Na_2SO_4 (pH 3) + 0.75 × 10^{-3}M Tl_2SO_4: (a) mechanically polished polycrystalline Ag; 100 mVs^{-1}, (b) chemically polished polycrystalline Ag; 100 mVs^{-1}, (c) chemically polished single-crystal Ag(110); 30 mVs^{-1} (93).

(100), and (111) as substrate surfaces. Great care was usually taken in preparing the single-crystal planes from melt-grown crystals and avoiding electrolytic contact with other faces than that under consideration (98). In addition, metals were chosen that do not show strong alloying tendency in the underpotential region. In Fig. 14 the cyclic current-potential curves for Tl on Ag and Pb on Au are shown for three different faces. It is generally observed that these curves are much more complex than those obtained with polycrystalline electrodes where fine structure is obviously smeared out. The following information was obtained

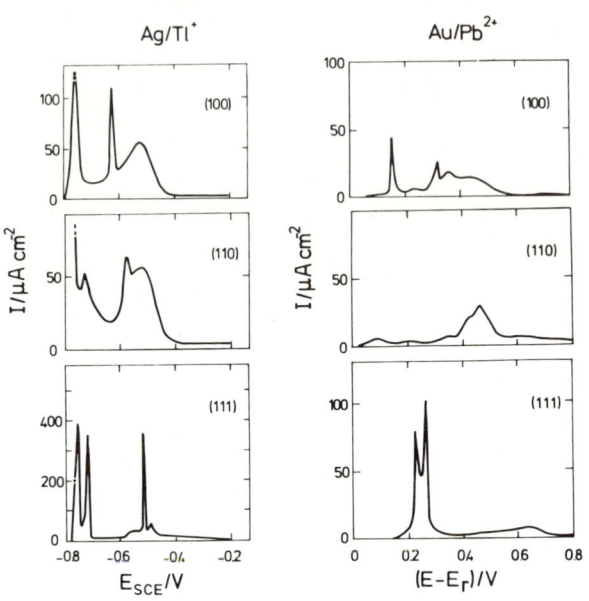

Fig. 14. Anodic stripping curves for Tl on Ag and Pb on Au single-crystal planes. Scan rates are 30 mVs^{-1} and 20 mVs^{-1}, respectively (93,97).

from the measurements on Au (97). For relatively small adatoms such as Cu the total monolayer charge is dependent on the substrate plane and agrees nicely with the number of substrate surface atoms in the corresponding crystal faces. Thus a 1:1 adsorption for the complete monolayer of Cu on Au is assumed for all planes. Contrary to this, it is found that larger atoms such as Pb obviously form a close-packed monolayer, the structure of which is independent of the substrate orientation. This is shown in Fig. 15 for Pb on Au (156). It is indeed observed that the surface concentration of adatoms in the completed monolayer depends on the adatom's radius and decreases with increasing size. In any case, the total charge for a complete monolayer is equal to or less than about z × 210 $\mu C/cm^2$, as is also found for polycrystalline surfaces. The charge values for Tl on Ag single-crystal planes suggest that two monolayers of Tl may be deposited underpotentially (93,95). It may be worthwhile to note that experiments performed in the author's laboratory with epitaxially grown Ag(111) on mica, where no surface treatment is necessary, yielded current-potential curves, which are nearly identical to those obtained with melt-grown single crystals (93).

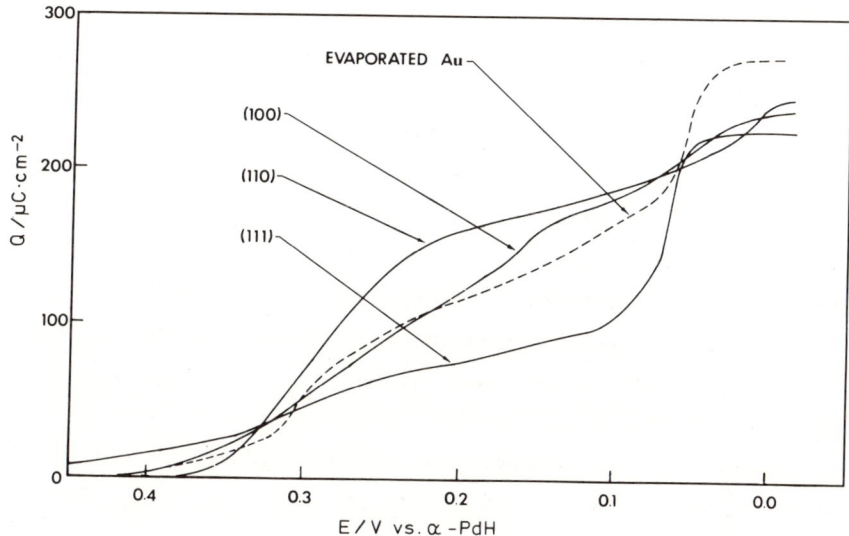

Fig. 15. Amount of underpotentially deposited Pb on different planes of an Au single crystal and on an evaporated Au electrode as a function of electrode potential (156).

In the case of Tl on Ag and Pb on Au a common feature can be seen exceptionally clearly: when driving the potential cathodically, metal deposition onto the initially bare substrate surface causes a rather broad adsorption current peak at low coverages, followed by a sharp adsorption peak at higher Θ. From the charge equivalent corresponding to the broad low coverage peak it is assumed that the initial adsorption is such that a regular superlattice is formed. For instance, for Pb on Ag the amount of charge connected with the low coverage peaks has been shown to correspond to those values obtained by assuming a superlattice with a (2 x 1) or (2 x 2) structure (see Fig. 16) (98). Similar results are reported for metal adsorbates on Au at low coverages (97) where different peaks were tentatively assigned to different ordered structures that have been observed by LEED at the metal-vacuum interface (see Section 3.2). This again emphasizes the close parallelism between the observations at the electrolyte and the vacuum interface. The formation of ordered structures is a result of the repulsion between the adatoms at low coverages. This is again seen in the shape of the corresponding adsorption or desorption peaks. Assuming a Langmuir type of isotherm (i.e., isoenergetic coverage-independent adsorption sites), a half width $\delta^L_{1/2}$ = 90 mV/z is expected. When interaction between the adatoms comes into play, the isotherm half width $\delta_{1/2}$ should be larger than $\delta^L_{1/2}$ for repulsion and smaller for attraction (see Section 2.6). Consequently, the isotherm may indicate growth by ordered structures or by island formation. The preceding examples can then be interpreted by an initial growth by ordered structures up to roughly half a monolayer [e.g., corresponding to a c(2 x 2) structure] followed by a phase transition that is indicated by a very narrow current peak [$\delta_{1/2}$ may be as low as 5 mV (94)]. This phase transition, which indicates the beginning of a lateral attraction, may be caused by the onset of a metallic-like interaction among the adatoms at high coverages. It can be interpreted as the formation of a two-dimensional electronic band due to sufficient overlap of the adatoms' electron orbitals parallel to the surface. Obviously this will be established most easily at a (111)-face, where the very sharp peaks occur at rather low coverages. Optical measurements will be more suited to reveal this transition from the adatom to the adlayer state.

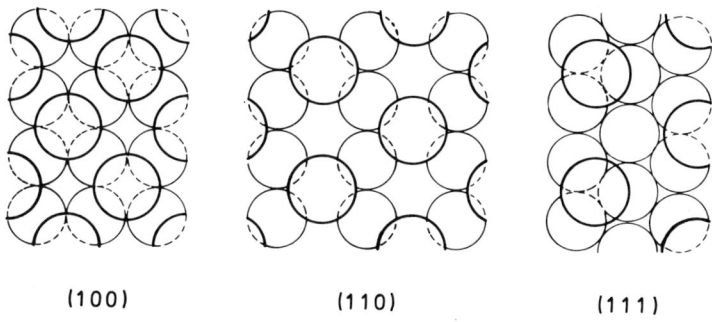

(100) (110) (111)

Fig. 16. Structure of a (2 x 2) superlattice on the three low-index crystal planes of Ag (15,100).

2.8. Anion Effects

It has been recognized for some time that anion adsorption can have a strong influence on the underpotential deposition behavior, while no such effect was reported for the cations of the supporting electrolyte (59). A typical example of the anion effect on the monolayer adsorption isotherm as demonstrated by cyclic current-potential curves is shown in Fig. 17. Two effects are noted when measuring in halide ions containing solutions, as typical examples of strongly adsorbing anions frequently used in experiments. First, the underpotential shift ΔE_p is decreased, which is equivalent to a reduction in bond strength between metal adatom and substrate surface. This change in bond energy must be caused by the interaction of the halide ion with the substrate, which is obviously stronger than that with the adsorbate. Second, the area under the monolayer peak is enlarged. This is because the metal atom adsorption is now accompanied by halide ion desorption, both processes contributing to the current in the same direction and thus producing a larger apparent surface charge (90):

$$Me^{z+} + xX^{-}_{ads} + (\gamma z + x)e^{-} \rightleftharpoons Me^{(1-\gamma)z+}_{ads} + xX^{-}_{solv} \qquad (22)$$

(X^{-} = halide). For example, for Pb adsorption on Ag in the underpotential region an effective charge transfer of 2.6 has been determined in 0.5M KCl, whereas $\gamma z \approx 1.9$ was found for the corresponding Pb bulk deposition. In the first case the experimentally determined γ can be explained by a combination of lead adsorption

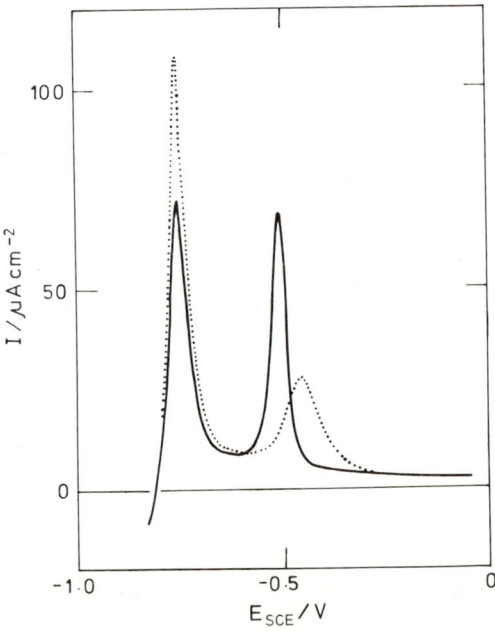

Fig. 17. Anodic stripping curves for a Tl deposit on Ag in 0.5M $NaClO_4$ (pH 3). (•••) $c_{Cl^{-}} = 0$; (———) $c_{Cl^{-}} = 3 \times 10^{-2}$M. $c_{Tl^{+}} = 2 \times 10^{-4}$M. Scan rate, 20 mVs^{-1} (19).

($z\gamma_{Pb}$=2) and Cl^- desorption (x = 0.6) (90). This effect was found to increase within the sequence ClO_4^-, F^-<SO_4^{--}<Cl^-<Br^-<SCN^-<J^-, that is, with increasing specific adsorption.

In some cases the effect of anions is more complex than described in the preceding paragraph and shown in Fig. 17. It has been reported that in J^- containing electrolytes the total adsorption isotherm for Tl on Ag is changed; a new peak arises with increasing J^- concentration at more cathodic potentials while the original monolayer peak becomes successively suppressed (101). An interesting observation is also reported by Schmidt and Stucki (50) for Ag on Au in electrolytes of various Cl^- concentrations

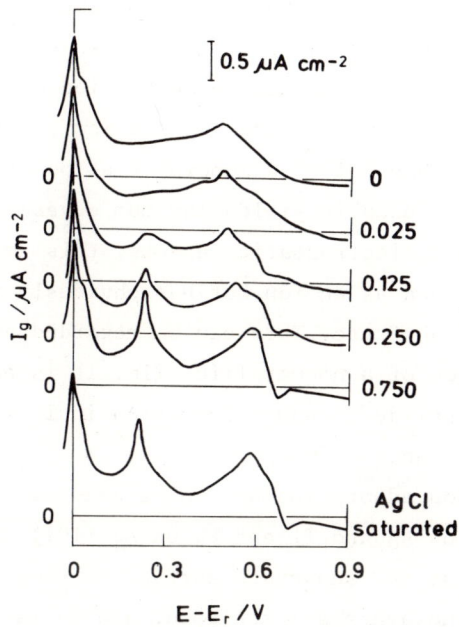

Fig. 18. Current-potential curves related to underpotential deposition of Ag on Au at various concentrations of chloride ions in the electrolyte. Numbers refer to Cl^- concentration in 10^{-9} mol/cm^2 (50).

(Fig. 18). Besides a slight retardation for the beginning of Ag adsorption at high positive potentials, most likely due to $AuCl_3$ formation, and a slight shift of E_p toward more anodic values, there is a drastic change in the total current-potential curve. At about $\Delta E=0.2$ V a rather sharp peak begins to rise with increasing Cl^- concentration, which is similar in shape to the corresponding peak found for Pb on Au and may be attributed to a change in the monolayer structure (see Section 2.6), induced by Cl^-.

Schmidt and Wüthrich (61) and Stucki (102) have discussed the halide ion effect on the underpotential deposition onto Ag and Au substrates as competitive adsorption. By measuring the adsorption isotherms of Cl^- on Ag and Au with the thin-layer technique they found that the respective surface concentration of Cl^- on Ag and Au may be as high as about 1.2 and 1.5 $nmol/cm^2$, corresponding to a coverage of more than 50% of a monolayer at positive potentials. Since ion adsorption usually yields maximum coverages of about 10 to 20% for simple electrostatic reasons, this implies that the halide is not adsorbed as an ion but has obviously formed a compound with the substrate. This can be regarded as the underpotential deposition of a nonmetallic film. It is easily understood that such a halide compound formation inhibits massively the metal monolayer deposition.

However, true ion adsorption has to be considered also. It has been shown for Tl on Ag and Tl and Pb on Au (101) that the halide ion concentration at the substrate surface at those potentials, where the metal monolayers are stripped, are by far too small (a few percent of a monolayer coverage) to account for any competitive adsorption. It is more likely that the specifically adsorbed anions change the properties of the surface as a whole (e.g., its electronegativity) thus influencing the adatom-substrate bond.

As mentioned earlier, the observed decrease in ΔE_p due to halide adsorption can only be understood, if one assumes that the interaction of the halide with the substrate is stronger than with the adsorbate. This finding seems to be in contradiction to the classical electrostatic model [at least for the cases of Pb and Tl on Ag and Au (101)], in which a stronger interaction with that metal is expected, which has the more negative potential of zero charge (e.g., Pb and Tl). However, it may not be appropriate to compare the halide-bulk metal and the respective halide-adsorbate metal interactions, since the latter occurs at such positive potentials where the bulk metal is unstable. The strong interaction between the metal adatoms and the substrate will likely diminish their ability to undergo covalent bonding with ions from solution.

In Table III the underpotential shifts $\Delta E_p^{Cl^-}$ in various chloride-containing solutions are given and compared with the respective values in Cl^--free solutions. It demonstrates that quite large differences in ΔE_p can arise due to anion adsorption that have to be taken into account when comparing results from different authors. In Fig. 19 the shift in ΔE_p as a function of halide ion concentration is given for various systems.

As already mentioned in Section 2.6., the adsorption isotherms derived from the current-potential curves in Fig. 17, which are quite different on the normal potential scale, fall nearly on top of each other when plotted versus a normalized potential ($\Delta E/\Delta E_p$). This again indicates that the anion effect should be seen in a more general way as changing the metal surface properties (PZC, work function, etc.) rather than as mere competition for bare substrate surface places.

Finally, some interesting examples of coadsorption should be mentioned. While in most cases studied thus far the adsorption of the metal monolayer in halide solutions was accompanied by halide

TABLE III

Data for Underpotential Shift $\Delta E_p^{Cl^-}$ in Cl^--containing Electrolytes and Difference $\delta E_p^{Cl^-}$ between Monolayer Stripping Peak Potentials in Cl^--containing and Cl^--free Solutions

Substrate/ metal ions	Supporting electrolyte	($\Delta E_p^{Cl^-}$/V)	($\delta E_p^{Cl^-}$/V)	References
Au/Ag$^+$	0.5M H$_2$SO$_4$ + 10^{-3}M NaCl	0.59	+0.09	50
Au/Tl$^+$	0.5M KCl	0.46	−0.22	59
Au/Pb^{2+}	1M NaClO$_4$ (pH 3) + 0.2M KCl	0.33	−0.07	19 (55)
Ag/Pb^{2+}	0.5M KCl	0.12	−0.04	19 (59,61)
Ag/Tl$^+$	0.5M KCl	0.19	−0.09	59 (100)
Ag/Cd^{2+}	0.5M Na$_2$SO$_4$ (pH 2) + 0.1M NaCl	0.14	−0.02	47 (65)
Ag/Sn^{2+}	1M KCl (pH 2)	0.18	−0.03	66
Cu/Tl$^+$	0.5M KCl	0.20	−0.13	59
Cu/Pb^{2+}	0.5M KCl	0.13	−0.07	59

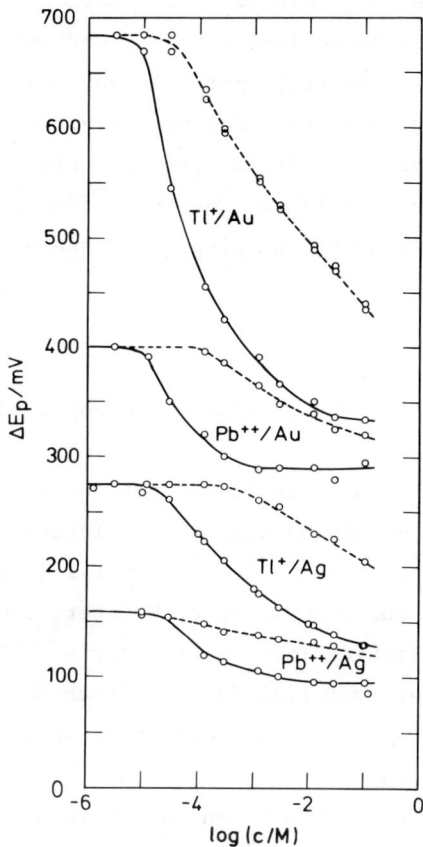

Fig. 19. Underpotential shift ΔE_p between monolayer and bulk stripping peaks for Tl and Pb on Ag and Au as a function of Cl^- (---) and Br^- (——) ion concentration in the electrolyte. $[Me^{z+}] = 2 \times 10^{-4} M$ (101).

ion desorption, it has been found that underpotential deposition of Cd (103) and Cu (104) on Pt strongly favors coadsorption of Cl^-. In these cases the apparent charge transfer is smaller than z and may thus simulate an incompletely discharged metal adatom when coadsorption is not taken into account. Horányi has shown that even HSO_4^- adsorption on Pt is markedly enhanced when Cu is deposited at underpotentials (89). Consequently, electrosorption valencies determined for Cu on Pt in sulfuric acid solutions would have to be corrected slightly for this effect. A detailed discussion of the electrosorption valency in case of coadsorption ($\gamma < 1$) and competitive adsorption ($\gamma > 1$) is given by Schultze and Vetter (105).

2.9. Measurements in Nonaqueous Solutions

The use of nonaqueous solvents in underpotential deposition studies is interesting mainly because a much larger variety of metal pairs can be investigated, including metals such as Al, Sb, Cd, and Sn which react quite strongly with water, usually forming a relatively stable hydroxide layer on the surface. This renders them useless as substrate material for monolayer deposition. Even less noble metals such as the alkali cannot be studied at all in aqueous solutions. In addition, it may be interesting to see whether the solvent molecules adsorbed in the inner Helmholtz layer have any influence on the underpotential deposition (106). The nonaqeous solvents used thus far are methanol (106), acetonitrile (19, 72, 78), and propylene carbonate (19, 79, 80, 108, 109). The systems studied in these solvents are summarized in Table I (under "Nonaqueous Solutions") together with the respective data. Unfortunately, the investigation of metal deposition in these solvents was often made impossible by the very low solubility of certain metal ions (especially polyvalent ions). Also, the cleaning procedure and dehydration of the solvents are quite cumbersome, lacking the ease of the use of water. Cyclic current-potential curves for Tl on Ag have been performed using water, methanol and acetonitrile as solvents. No difference was seen in the curves recorded in H_2O and methanol (106). A slightly broader adsorption isotherm for the Tl monolayer was found in acetonitrile, whereas the underpotential shift ΔE_p seems to be roughly the same in all three cases within experimental error. This indicates that the solvent molecules are unlikely to be involved directly in building up the metal monolayer, as we would expect if the layer consisted of partially solvated ions (52).

The underpotential deposition of alkali metals onto various substrates (mostly noble metals) were studied by Fried and Barak (80) and Kolb and colleagues (19,78). In the case of Li, alloy formation in the underpotential region interferes frequently due to the high alloying tendency of Li. Nevertheless, monolayer deposition of Li could be found on a series of metals (Pt, Au, Ag, Cu, Pb, Cd, Tl, Al and C) (19,72,78-80,109). For deposition onto Mg (108) and Sb (79) only alloy formation has been reported. In Fig. 20 the codeposition of Tl and Li on Ag in acetonitrile studied by cyclic voltammetry is shown as an example of underpotential deposition in a nonaqueous solution. The anodic scans have been recorded after keeping the potential at about -1 V for some time to allow Tl deposition in various amounts and then scanning rapidly to -2.8 V for Li deposition. Because of the very low concentration of Tl^+ (10^{-4}M) compared to Li^+ (0.5M), practically no Tl will be deposited during the potential cycle, if there is no pause for some time at the appropriate potential. The top curve (a) in Fig. 20 shows the usual stripping peak of a Li monolayer on Ag (19) at -1.9V and no Tl being detected. When allowing some Tl to be deposited for a submonolayer coverage on Ag, the presence of which can be seen in the corresponding stripping peak around -0.4 V on the anodic scan, it is found that the stripping peak due to Li adsorption on Ag at underpotentials decreases while a second peak arises around -2.4 V, which is attributed to the monolayer formation on top of the Tl layer. This is seen more clearly in curve (c), where about a complete Tl monolayer is deposited on Ag suppressing nearly totally the Li adsorption on Ag but increasing the monolayer peak due to Li on Tl. When more than one monolayer of Tl is deposited on Ag, as is indicated by the appearance of the bulk Tl stripping peak at -0.6 V on the anodic scan, Li is seen to adsorb underpotentially on a Tl surface. These experiments also substantiate earlier results from which it was

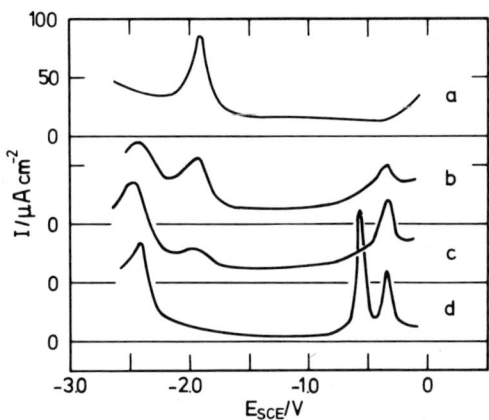

Fig. 20. Anodic stripping curves for Li and Tl deposits from Ag electrode in acetonitrile (0.5M $LiClO_4$ + 10^{-4}M $TlNO_3$). Scan rate, 100 mVs^{-1}. Tl bulk and monolayer are stripped at -0.6 and -0.4 V, respectively, from the Ag surface, and the Li monolayer is stripped at -1.9 and -2.45 V, respectively, from the Ag surface and from the Tl deposit.

concluded that: (a) one monolayer of a metal deposit has about the same thermodynamic properties as the respective bulk metal and (b) metal atoms can be deposited underpotentially on top of the monolayer of another metal.

Recently, investigations on underpotential deposition were extended even to studies in molten salts (81,110) and solid electrolytes (107). It was found that Ni, deposited from a LiCl-KCl eutectic, adsorbs onto Pt in monolayer amounts at underpotentials ($\Delta E_p \approx 0.4$ V) (81). However, the results presented thus far do not allow definite conclusions[+]. As is shown later, the Ni/Pt-result does fit into the relation established for ΔE_p in aqueous solutions and obeyed also in nonaqueous solvents.

2.10. Alloy Formation in the Underpotential Range

Whenever two different metals are in contact with each other, interchange reactions are likely to take place if a thermodynamically stable solution of both can be formed. Whereas the alloy formation has been studied frequently for two <u>bulk</u> metals, that is, in the potential region $E < E_r$ for the deposited metal or more generally speaking, for $a_{Me} = 1$, only a few investigations deal with alloying in the underpotential region. The latter process, however, is quite important because here alloying may interfere with the observation of the underpotential deposition of the simple metal monolayer. Since the reversible equilibrium potential

---

[+] These authors also studied Ag on Pt (81) in the melt and found a small peak close to the reversible Nernst potential. However, the potential region of interest, where Ag should deposit underpotentially according to measurements in aqueous solutions, has not been investigated.

of an alloy is more positive than that of the less noble component (≙ deposit), the alloy may be formed and will be stripped in the underpotential range. However, there are at least two criteria that allow a clear distinction between both phenomena:

1. While monolayer formation is a rather fast adsorption process, alloy formation is slow due to the high kinetic barrier (diffusion coefficient, $D < 10^{-16} cm^2 s^{-1}$) and hence can be distinguished by kinetic measurements (56).
2. The amount of underpotentially deposited metal atoms is confined to one monolayer yielding a limiting value of approximately 200 $\mu C/cm^2$ for a monovalent atom when stripped from the substrate. Such a limiting value of course does not exist for the alloy, the amount of which should increase with time slowly but steadily (56,66).

A typical example of a system where alloying in the underpotential range takes place at a fairly high rate is Cd on Au. In a cyclic current-potential curve (Fig. 21) all three processes occurring during metal deposition are seen exceptionally clear at the anodic scan. At the reversible Nernst potential E_r for Cd/Cd^{++} bulk Cd is stripped yielding a rather sharp peak close to E_r. Then the Cd/Au alloy formed during the cathodic scan in the region $E > E_r$ is dissolved before the Cd monolayer is desorbed in the third most anodic peak. The dependence of the Cd/Au alloy formation on temperature and potential as a function of time has been studied in detail by Schultze, Koppitz, et al. (56). While for large underpotentials and low temperatures no alloy is formed (no charge increase after monolayer formation, which occurs within 1 ms), a successive increase in charge is seen with rising temperature and with reducing the underpotential yielding alloy films up to an average of about 200 layers within a reasonably short time (100 s).

Whereas the temperature dependence of the alloy formation is explained by a temperature dependent diffusion coefficient, which was found to vary according to an activation energy of 22 kcal/mol, the pronounced potential influence may be governed by two potential dependent factors: (a) the adatom concentration (coverage) and (b) the alloy composition at the surface (56,111).

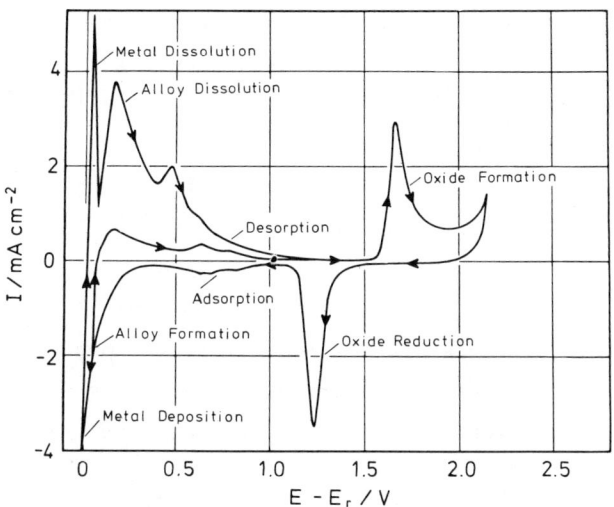

Fig. 21. Cyclic current-potential curves for Au electrode in 0.1M $CdSO_4$ + 1M CH_3COONa + 1M CH_3COOH. Scan rate, 200 mVs^{-1} (56).

Besides the study on Cd/Au mentioned in the preceding paragraph, a few more systems have been investigated. Bruckenstein, Hassan, et al. have studied the very complex system Hg on Pt by rotating ring-disk technique (38). They were able to separate the Hg monolayer from various alloy compounds being formed when more than one monolayer of Hg is deposited, although the distinction between them is unclear since the monolayer is already partly interpreted in terms on an alloy compound Pt_2Hg. To get bulk Hg onto the Pt surface, relatively large amounts ($\approx$ 50 layers) have to be deposited that will be slowly converted into an alloy with time (38). Schmidt and Wüthrich (40) have observed that lead deposited onto platinum at underpotentials yields an alloy when waiting for a relatively long time (5 x 10^4 s) at a potential very close to E_r (ΔE = 10 mV). This was identified as $PtPb_4$ by comparison of the stripping potentials of the underpotentially generated alloy and a bulk film of $PtPb_4$. On the other hand, it has been shown that Pb/Au alloys will not be formed at underpotentials, even after a 10-h polarization at a potential close to E_r (ΔE = 10 mV) (52), while for potentials negative of E_r ($E < E_r$) instantaneous formation of two alloys, $AuPb_2$ and $AuPb_3$, has been found (53). This result agrees nicely with observations in the gas phase. For Pb evaporated on the (100)-face of Au, alloying was found to take place only after completion of the first monolayer (120). Obviously, the Pb adatom is more stable at the surface than in an alloy; therefore, the monolayer is preserved. For Sn on Ag (66) and Cd on Ag (65) alloy formation was also found besides the well-known monolayer formation.

It has been noted several times that the adsorption isotherm caused by monolayer deposition indicates an increase in Γ_{Me} again near E_r after already having leveled off at more positive potentials (27,50). Bowles (17) has interpreted this extra amount of strongly and often <u>irreversibly</u> adsorbed metal atoms as due to penetration into fissures and microcracks of the substrate, Schultze, Koppitz, et al. (56) have discussed it as the filling of vacancies close to the surface by the deposited atoms and not as a real alloy formation. Since the latter will become increasingly important, even in the underpotential region with increasing diffusion coefficient, Li is expected to take part in alloying at an exceptionally rapid rate. This is demonstrated in a recent work on electrochemical behavior of metal electrodes in Li^+-containing organic electrolytes (79), where cyclic current-potential curves are shown for alloy formation of Li with Sb and Tl electrodes. In some cases, for example, for Li on Cd, Sn, and Au, alloy formation in the underpotential region takes place so rapidly that the monolayer stripping peaks are masked by alloy dissolution in cyclic voltammetry at the usual scan rates of about 100 mVs^{-1} (47).

Besides the complication alloying introduces into the monolayer studies, the combination of both effects yields an interesting aspect. Underpotential deposition allows, in principle, electrochemical alloy formation at potentials where at least one component is not expected from thermodynamic considerations to be plated onto a metallic substrate. Although it has been well known for quite some time that in electrodeposition of alloys the less noble constituent is indeed often deposited at "underpotentials" (called "anomalous codeposition"), this important effect has been explained only very recently by the underpotential behavior of the less noble component on the more noble one (112). This seems to be one of the most reasonable explanations offered thus far; then a thorough understanding of the underpotential deposition would also allow to predict the occurrence of anomalous codeposition in alloy formation.

3. STUDIES OF METAL MONOLAYERS IN A NONELECTROCHEMICAL ENVIRONMENT

3.1. XPS Studies on Monolayers Deposited at Underpotentials

Since the highly surface sensitive photoelectron spectroscopy techniques cannot be applied to <u>in situ</u> studies of electrode surfaces and electrosorbates, it is a very appealing idea to transfer the electrode in a certain reaction state from the electrochemical cell into an UHV system. There, X-ray photoelectron spectroscopy (XPS) should yield very specific molecular information about the species involved in electrode surface chemistry. However, removal of an electrode from the electrolyte will be the crucial step, since the controlled electrode potential will no longer be present, leaving the surface in a more undefined state as it drifts to its rest potential. This is especially important when the potential is necessary to stabilize unnoble reaction products on the surface. Moreover, contact with the atmosphere may readily oxidize the surface product (156), and finally, spurious amounts of salt from the supporting electrolyte will remain on the surface. One of the first to pioneer this field was the group around Winograd, who used XPS to study surface products of electrode reactions, such as electrochemically formed Pt-O on Pt, hence successfully establishing a link between electrochemistry and high-vacuum surface-science techniques (113). More recently, they extended these studies to investigations of metal monolayers formed by underpotential deposition. The electrolyte of a small electrochemical cell in which underpotential deposition was achieved was aspirated in approximately 0.1 s and the working electrode was then rapidly transferred into the UHV chamber of an ESCA spectrometer (114). The following results were obtained from the measurements for Cu on Pt and Ag on Pt (115,116). A significant shift in the electron-binding energies for the Cu $2p_{3/2}$ and the Ag $3d_{5/2}$ levels in the Cu and Ag monolayers on Pt relative to bulk Cu and Ag is observed (Fig. 22). This shift is about 0.95 eV for Cu and 0.65 eV for Ag. Surprisingly, no coverage dependence was found for the XPS shift in the measured range of $0.2 \leq \Theta \leq 1.0$, although the corresponding current-potential curves indicate several distinct adsorption states for the metal deposits in the submonolayer region (at least three states for Cu and two

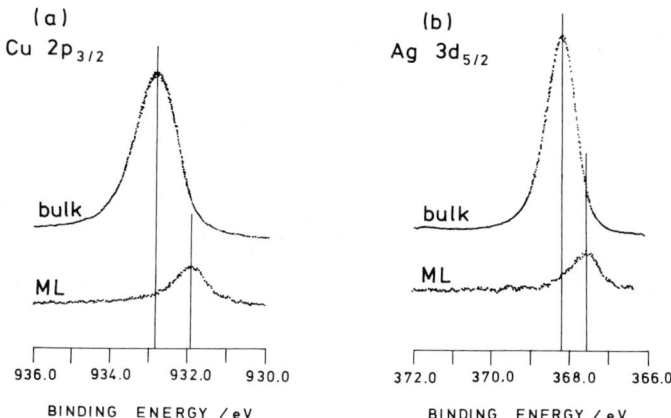

Fig. 22. XPS spectra of a monolayer (ML) on Pt and of the bulk metal for Cu(a) and Ag(b), respectively (115).

for Ag on Pt). Thickness calculations from intensity measurements show no detectable loss of Cu during the transfer. For comparison, vapor-deposited films of Cu and Ag on Pt were also investigated. Here, at low coverage ($\Theta \approx 0.1$) the shifts were identical to that observed for the underpotential deposit, which emphasizes the close similarity of both entities; however, a gradual shift with coverage to the bulk metal value was found for the vapor deposit. This was tentatively explained by different layer structures. Although the spectra indicated that some H_2O and SO_4^{--} was adsorbed on the surface, the authors could firmly rule out the presence of Cu^{2+}, proofing the metallic character of the underpotential deposit. As seen in Fig. 22, the electron-binding energies of the core levels under consideration are found in the monolayer at lower values than in the respective bulk. This looks surprising at first sight, since we know from electrochemical experiments that the adatom on the foreign substrate is more strongly bound than on a surface of its own material, and that it will usually carry a small positive charge. Therefore, the adatom's level should be located at higher binding energies. However, it must be considered that the energies in Fig. 22 are not referenced to the vacuum level and that relaxation effects have to be taken into account (116). First calculations have shown that the correct shift to higher binding energies for the adatom with respect to its bulk value may be obtained (116). Although further experimental and theoretical work seem necessary to yield the required information on the adatoms electronic properties, the experiments thus far reveal interesting data and show promise.

3.2. Metal Monolayer Formation on Metal Substrates in Vacuum

The deposition of single atoms, monolayer amounts, and thin films onto metal substrates (mostly W or other valve metals) at ultrahigh-vacuum conditions has been studied extensively by LEED, AES, field-ion microscopy, work-function measurements, and many other techniques (82,83,117-140). A review of this work is far beyond the scope of this chapter. In these studies the deposition of the first atoms up to a monolayer amount has received greatest interest because it represents the initial step of film growth. In addition, it was recognized quite early that in most cases a strong interaction between adatom and substrate metal leads to a drastic change in the substrate-surface properties, for example, inducing large work-function changes (cesiated surfaces!). Here again it has been found that the adatom-substrate bond is usually much stronger than the bond of the adatom on its own lattice. This is most convincingly demonstrated in flash-desorption experiments. The temperature of the substrate, covered with a deposit, is raised continuously at a constant rate (dT/dt), and the amount of

desorbed atoms is measured directly by a mass spectrometer. The
results are thermal desorption spectra as shown in Fig. 23 for
desorption of Cu from a W(110) surface. The similarity with the
"desorption spectra" obtained by cyclic current-potential curves
for underpotentially deposited monolayers in electrolytic solutions is evident (see Fig. 3). Whereas the bulk deposit is desorbed at a certain temperature, the temperature has to be raised
above that value to desorb those atoms which are directly bound
to the substrate. The constant amount of these extra bound adatoms,
which is of the order of roughly 10^{15} atoms/cm^2, indicates the
formation of a uniformly distributed monolayer. Although the higher
desorption temperature represents a higher bond energy, a direct

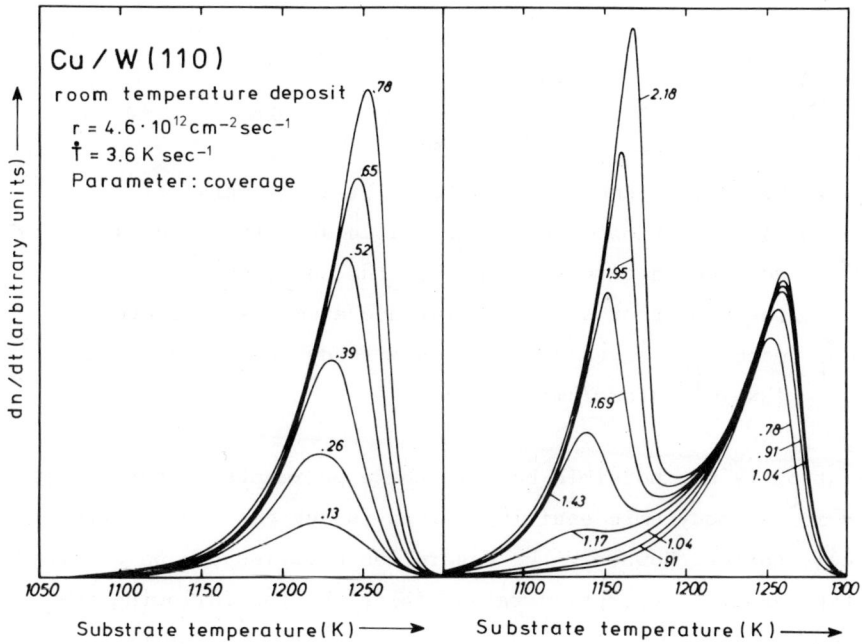

Fig. 23. Thermal desorption spectra of Cu adsorbed on W(110)
(119).

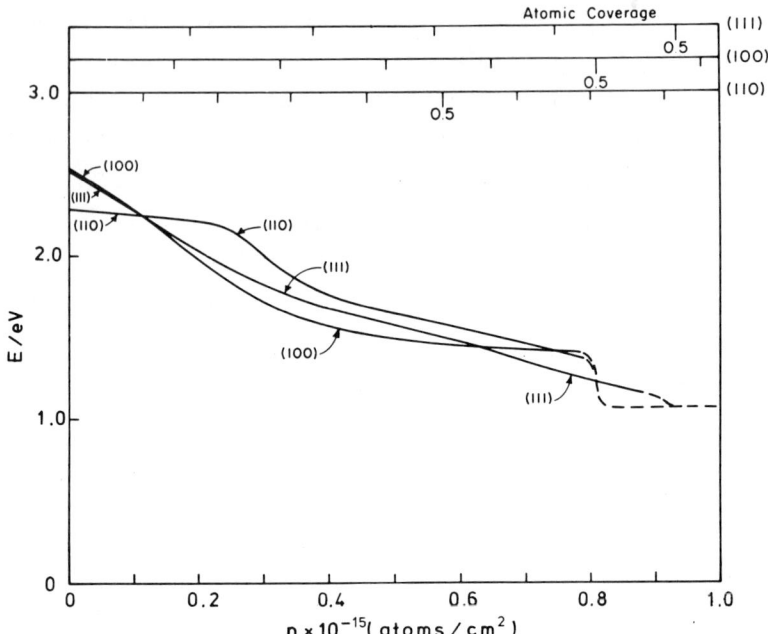

Fig. 24. Desorption energies for Na on Ni single-crystal faces as a function of coverage, calculated from thermal desorption curves. Note that the coverage Θ refers to the number of substrate surface atoms. A complete monolayer coverage of Na is already reached at about 1×10^{15} atoms/cm^2 (82).

correlation is not possible because the experiments cannot be performed in thermodynamic equilibrium, as is the case in electrochemical studies. However, the desorption energies can be obtained from the measured desorption rates (dn/dt) by the following expression:

$$\frac{1}{n} \cdot \frac{dn}{dt} = \nu(n) \, \exp\left(-\frac{E(n)}{kT}\right) \qquad (23)$$

assuming $\nu(n)$ and $E(n)$ to be temperature-independent. A typical example for the coverage dependent desorption energy is shown in Fig. 24 for Na on Ni (82). After the monolayer is completed the energy values reach those given by the heat of sublimation for the deposited metal (1.1 eV for Na). An excess binding energy for metal adatoms on a foreign metal substrate has been reported for many systems [e.g., alkali on W (83,126,127) and on Ni (82), Ga on W (125), Pb on W (118,129), Ag on W (122,139,140), and Cu on W (119,123)].

Some data obtained from measurements on W single crystals are as follows:

System	ΔE/eV	References
Au/W(110)	0.2	141
Cu/W(100)	0.8	119
Ag/W(110)	1.1	122
Pb/W(113)	1.8	129

The excess binding energy ΔE, taken at $\Theta \approx 0.5$, is only a rough estimate, indicating a similar trend with work function as found in underpotential studies (see Section 6.2). The experimentally determined correlations of $\Delta E - \Theta$ and $\Delta\phi - \Theta$ are usually rather complicated and do not allow to establish such a simple correlation between ΔE and $\Delta\phi$ as in the case of polycrystalline electrode materials.

The LEED studies of metal adatoms in the submonolayer range (118,120,121,130,132-134) indicate that at low coverages superstructures of various types are formed, where the adatoms occupy potential wells on the substrate with maximum ligancy (usually fourfold), while at high coverages close-packed layers are formed

(120,121). The various superlattice structures obtained for different coverages may be directly used to identify the structures observed in the stripping current-potential curves in underpotential studies (97). Ultraviolet photoemission spectroscopy (UPS) on metal monolayers should yield valuable data on the density of states in the valence-band region and hence information on the band structure of these monolayers. Ultraviolet photoemission spectroscopy is extremely surface-sensitive because of the small escape depth (4 to 10 Å) of the photoexcited electrons, when the photon energy is chosen around 20 to 100 eV (142). Yet UPS work on metal adsorbates on foreign metal surfaces is sparse (135-137,143, 144). Eastman and Grobman studied photoemission from thin overlayers of Cu and Pd on Ag, the average thickness ranging between half a monolayer and several layers (143). Since the escape depth of the photoelectrons is quite large for $\hbar\omega$ = 8.6 eV, which was used in these experiments, the substrate contributes considerably to the total yield. Therefore, the emission intensity from the overlayer was determined by subtracting the emission spectrum of the bare Ag surface from the total spectrum, employing scaling factors for a proper evaluation. The spectra should represent the overlayer density-of-states, which predominantly reflect the d-bands. In Fig. 25a the result for Cu on Ag is shown. A small change in the d-band density-of-states is discernible in the studied thickness range between the average of half a monolayer and bulk Cu. However, it should be noted that Cu on Ag does not form a monolayer when deposited electrolytically, nor is Pd on Ag expected to do so from general electrochemical considerations (see Section 6.2). Therefore, we can exclude the existence of a true monolayer in these two cases and have to assume some island formation. This could explain the relatively small change in d-band position in Cu and Pd when going from 0.5 monolayer to bulk.

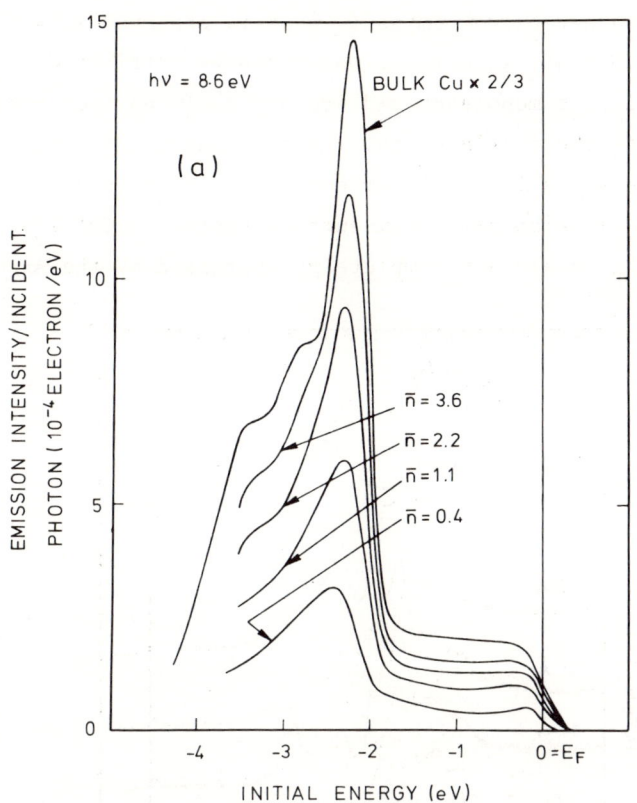

Fig. 25a. UPS spectra of thin Cu overlayers on Ag (polycrystalline). The numbers refer to the average layer thickness in monolayers (143).

On the other hand, deposition of Ag on Cu is expected to form a monolayer (see Section 6.2), although underpotential deposition cannot be observed directly in those cases where the substrate is less noble than the deposit. Recently, Neddermeyer (144) has presented UPS data for Ag overlayers on Cu(111), where a monolayer of Ag is obtained by epitaxial growth. The result is shown in

Fig. 25b. When a monolayer of Ag (d = 2.5 Å) is deposited onto Cu a strong and rather narrow emission peak is observed at 4.7 eV (beside the substrate spectrum), which can be attributed to the Ag 4d states in the monolayer and are obviously not yet broadened into a band. As the overlayer thickness is increased, these d-states broaden and shift to lower energies. It is interesting to note that the emission peak from surface states on Cu(111) about 0.4 eV below E_F is already completely suppressed by the Ag mono-

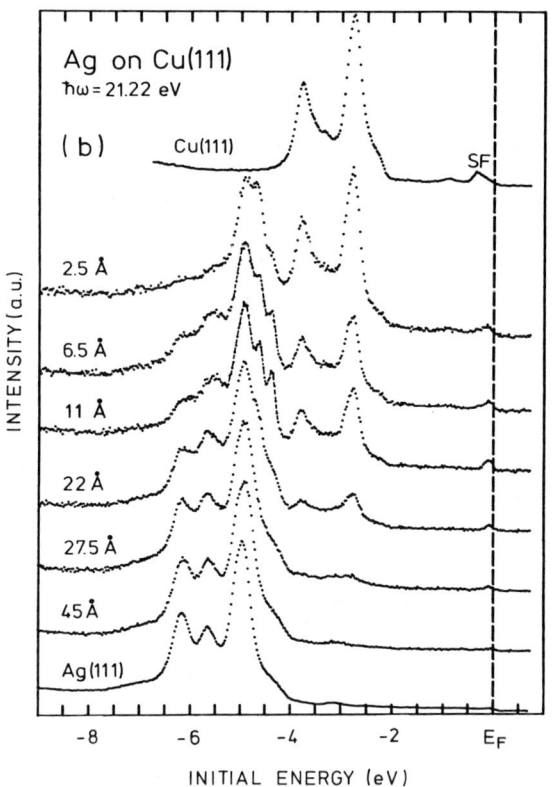

Fig. 25b. UPS spectra of thin Ag overlayers on Cu(111). The numbers refer to the average layer thickness in Ångstroms (144).

layer, while a new peak appears at an energy typical for surface states on Ag(111) (144). This is also clear evidence for the existence of a complete Ag monolayer on Cu. Experiments of this kind will have to be used to answer such interesting questions as the band structure of metal monolayers and the change in electronic properties for the transition from adatom to bulk metal.

4. OPTICAL STUDIES OF THE ELECTRONIC PROPERTIES OF METAL MONOLAYERS

4.1. Instrumentation

In optical studies, which for metallic substrates are usually confined to reflectance measurements, minute changes in absorbance or reflectance, caused by adsorption processes, film formation, or surface reactions, have to be measured as a function of photon energy over an extended VIS-UV region. The most common techniques for investigation of the properties of adsorbates on metal substrates are ellipsometry, modulation spectroscopy (e.g., electroreflectance), and differential reflectance spectroscopy. In the latter two cases relative reflectance changes, ($\Delta R/R$), are measured, thus avoiding the very difficult task of determining <u>in situ</u> absolute reflectance values. Several experimental setups have been described in the literature (145-158), two of them are reproduced as block diagram in Fig. 26.

In modulation spectroscopy the condition of the electrode surface is perturbed periodically (hence reversibly) by some external parameters, usually by a small modulation of the electrode potential, and the resulting reflectance modulation is recorded. Here a single-beam setup is most convenient, since the ratio

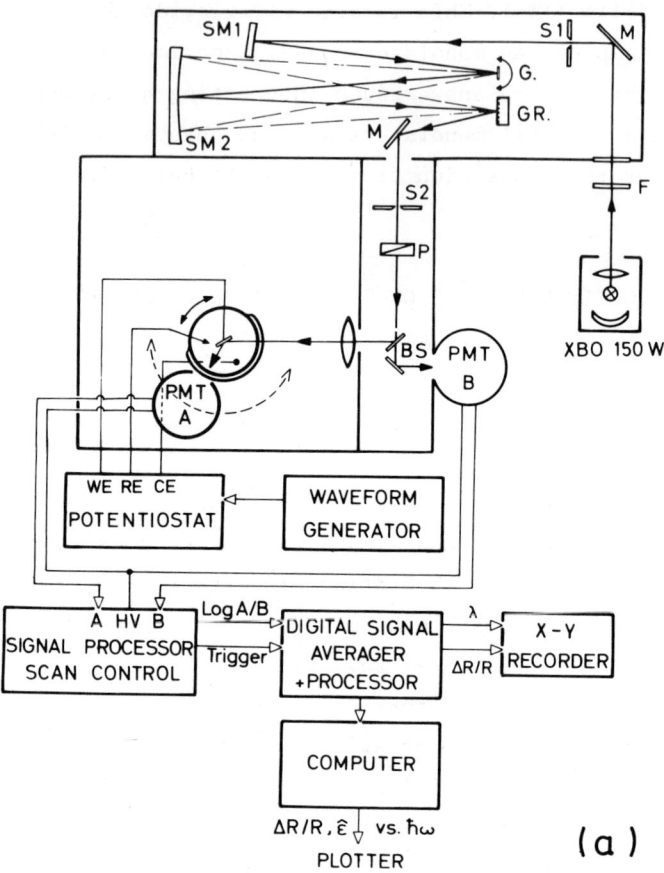

Fig. 26a. Schematic diagram of a rapid scan spectrometer. For symbols see text (158).

($\Delta R/R$) can be performed instantaneously by a ratiometer. The AC component of the signal, which is proportional to ΔR, is usually amplified by a lock-in amplifier and divided by the DC signal. In many cases a feedback of the photomultiplier signal is used to drive the high voltage supply of the photomultiplier tube (PMT)

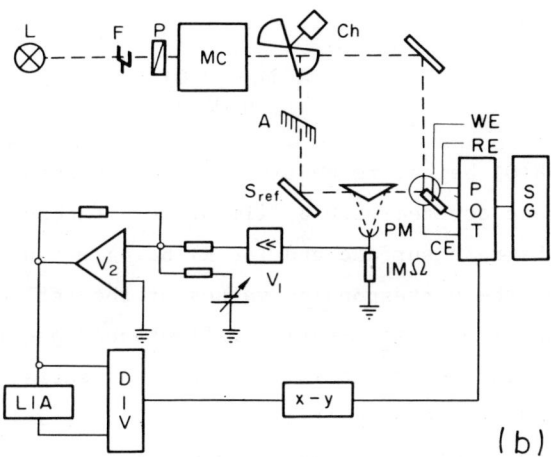

Fig. 26b. Schematic diagram of a dual-beam instrument for monitoring minute reflectance changes for slow or irreversible reactions. A, variable light attenuator; Ch, mechanical light chopper; DIV, divider; F, filter; L, light source; LIA, lock-in amplifier; MC, monochromator; P, polarizing prism; PM, photomultiplier; SG, sweep generator; S_{ref}, reference mirror; V_1, V_2, amplifier (154).

to achieve a constant DC signal output. Thus the AC signal is then directly proportional to ($\Delta R/R$), avoiding a ratiometer. This technique is especially useful for metal substrates such as Ag which exhibit strongly variing reflectance spectra where inaccuracy may be introduced by an analogue divider due to the limited dynamic range in the denominator voltage ($\sim R$).

Since the modulation spectroscopy is in general limited to reversible changes in the surface state, differential reflectance spectroscopy has a wider applicability for studies of adsorption processes or film formation. Here one measures the normalized reflectance change

$$\frac{\Delta R}{R} = \frac{R(d) - R(0)}{R(0)} \qquad (24)$$

with $R(d)$ and $R(0)$ being, respectively, the reflectances from a substrate surface covered with a film of mean thickness d and from a bare substrate surface and $(\Delta R/R)$ being measured directly as the ratio of the corresponding values of the reflected intensities, $I(d)$ and $I(0)$. Differential reflectance spectroscopy is now widely used in electrochemistry as well as surface physics to study in situ adsorbate properties (145,146,157,159-161). In Fig. 26a a rapid scan spectrometer is shown which is now beginning to be used in spectroelectrochemical surface studies (157,158) and offers a unique combination of rapid scanning and high sensitivity. A small electromagnetically driven galvanometer mirror (G) reflects the light beam via a spherical mirror (SM2) onto the grating (GR), where the angle of incidence changes continuously as the galvanometer mirror is vibrated. Light of a certain wavelength then exits at slit S2, before it is split into sample and reference beam by a beam splitter (BS). The rapid wavelength scan (typically 10 ms for a 2000-Å sweep, with about 40 repetitions/s) is ideally suited for signal averaging technique. In addition, the electrochemical cell with a semicircular quartz window is placed on a φ - 2φ goniometer table, the photomultiplier (PMT A) attached directly to the 2φ-arm, allowing easy change of the angle of incidence at the electrode surface between $10°$ and $90°$. In Fig. 26b a dual-beam setup is shown for measuring small reflectance changes as a function of potential or time, where a rotating sector mirror as chopper generates sample and reference beam. Since the reference beam can be attenuated continuously by a rotating comb or a stack of attenuation filters until its intensity equals that reflected from the bare sample electrode surface, any reflectance change of the electrode surface, such as that due to film formation, will show up as an AC signal, the reference beam representing the reflectivity of the bare substrate surface. Thus the very sensitive lock-in technique can be applied for studying very slow processes or even steady states, since the lock-in is now tuned to the chopping frequency of the rotating mirror and not to a periodic change in the electrode potential.

4.2. Evaluation of the Monolayer Dielectric Function from Reflectance Measurements

We consider briefly some calculations for obtaining information on the monolayer dielectric function from optical data (for a more thorough discussion, see refs. 145,157). Using a simple three-layer model with sharp boundaries for ambient, film, and semiinfinite substrate phases, the relative reflectance change $(\Delta R/R)$ = $[R(d) - R(0)]/R(0)$ between a substrate, covered with an adsorbate of thickness d, and a bare substrate can be calculated from Fresnel equations, when the complex dielectric constants $\hat{\varepsilon}_i$ of all three phases (ambient, film, and substrate) are known. Although $(\Delta R/R)$ can be calculated for any given set of optical data with the aid of computers very easily, the calculation is too complex to allow any direct insight as to how the film optical constants would affect the overall reflectivity. However, for very thin films (d $\ll \lambda$), McIntyre and Aspnes have expanded the exact expressions in terms of (d/λ) and neglected higher orders (162). They have shown that the range of validity is such that for relative layer thicknesses $(d/\lambda) \leq 5 \times 10^{-3}$ this linear approximation can be used. Since the monolayer thickness range is the most interesting one for studying adsorption processes and interfacial problems, this treatment can yield readily valuable information. Let $\varepsilon_1 = n_1^2$ be the real dielectric constant of the ambient phase, $\hat{\varepsilon}_2$ and $\hat{\varepsilon}_3$ the complex dielectric constants of film and substrate, respectively, with $\hat{\varepsilon}_i = \varepsilon_i' - i \cdot \varepsilon_i'' = \hat{n}_i^2$ ($\hat{n} = n - ik$), d the average film thickness, and λ the vacuum wavelength of the probing light incident at an angle φ_1. Then for d<<λ the normalized reflectance change ($\Delta R/R$) for perpendicular ($\perp$) and parallel ($\parallel$) polarized light is given by (162):

$$\left(\frac{\Delta R}{R}\right)_\perp = \frac{8\pi d n_1 \cos \varphi_1}{\lambda} \cdot \text{Im}\left(\frac{\hat{\varepsilon}_2 - \hat{\varepsilon}_3}{\varepsilon_1 - \hat{\varepsilon}_3}\right) \quad (25a)$$

and

$$\left(\frac{\Delta R}{R}\right)_\parallel = \frac{8\pi d n_1 \cos\varphi_1}{\lambda} \cdot \text{Im}\left\{\left(\frac{\hat{\varepsilon}_2 - \hat{\varepsilon}_3}{\varepsilon_1 - \hat{\varepsilon}_3}\right)\left[\frac{1-(\varepsilon_1/\hat{\varepsilon}_2\hat{\varepsilon}_3)(\hat{\varepsilon}_2+\hat{\varepsilon}_3)\sin^2\varphi_1}{1-(1/\hat{\varepsilon}_3)(\varepsilon_1+\hat{\varepsilon}_3)\sin^2\varphi_1}\right]\right\} \quad (25b)$$

As discussed in Section 4.3, in many cases the film deposition to some extent alters the optical properties of the substrate surface. Consequently, we have to deal actually with a four-phase system rather than a three-phase system. To get some idea about the contributions to ($\Delta R/R$) of any film in a multilayer system, Takamura, Takamura, et al. (163) have expanded the linear approximation theory. They showed that the total ($\Delta R/R$) of such a system is given by the sum over all ($\Delta R/R$) values due to each layer:

$$\left(\frac{\Delta R}{R}\right)_\perp = \frac{8\pi n_1 \cos \varphi_1}{\lambda} \sum_j d_j \, \text{Im}\left(\frac{\hat{\varepsilon}_{2j} - \hat{\varepsilon}_3}{\varepsilon_1 - \hat{\varepsilon}_3}\right) \quad (26a)$$

and

$$\left(\frac{\Delta R}{R}\right)_\parallel = \frac{8\pi n_1 \cos\varphi_1}{\lambda} \sum_j d_j \, \text{Im}\left\{\left(\frac{\hat{\varepsilon}_{2j}-\hat{\varepsilon}_3}{\hat{\varepsilon}_1-\hat{\varepsilon}_3}\right)\left[\frac{1-(\varepsilon_1/\hat{\varepsilon}_{2j}\hat{\varepsilon}_3)(\hat{\varepsilon}_{2j}+\hat{\varepsilon}_3)\sin^2\varphi_1}{1-(1/\hat{\varepsilon}_3)(\hat{\varepsilon}_1+\hat{\varepsilon}_3)\sin^2\varphi_1}\right]\right\}$$

(26b)

where the indices 1 and 3 refer to ambient and substrate phase, and 2j refers to j different layers of thickness d_j sandwiched in between.

At normal incidence Eqs. 25a and 25b for a three-phase system reduce to:

$$\left(\frac{\Delta R}{R}\right)_{\varphi_1=0} = \frac{8\pi n_1 d}{\lambda} \cdot \frac{\varepsilon_2''(\varepsilon_3' - \varepsilon_1) - \varepsilon_3''(\varepsilon_2' - \varepsilon_1)}{(\varepsilon_3' - \varepsilon_1)^2 + \varepsilon_3''^2}$$

(27)

From this equation it becomes evident that the $(\Delta R/R)$ spectrum is not solely determined by the optical properties of the film itself but also by those of the substrate (145). This means that the spectra will usually not look transmission-like, containing readily accessible information on absorption bands in the film material. Therefore, visual interpretation of $(\Delta R/R)$ spectra alone will be a doubtful approach to gain information about absorption processes in the adsorbate, because this information will usually be masked by the substrate response. In most cases it will be absolutely necessary to evaluate the optical constants of the film itself. This can be done in several ways.

4.2.1. INVERSION OF LINEAR APPROXIMATION EQUATIONS. By measuring the values of $(\Delta R/R)_\perp$ and $(\Delta R/R)_\parallel$ at one angle of incidence, Kolb and McIntyre (164) have shown that the film optical constants n_2 and k_2 (or real and imaginary part of the complex dielectric constant, ε_2' and ε_2'') can be obtained from Eqs. 25a and 25b by inverting and solving them for the two unknown parameters ε_2' and

ε_2''. The third unknown parameter, the layer thickness d, has to be determined independently, usually by coulometric measurements, or estimated. However, the uncertainty in d will not affect the result strongly, because in Eqs. 25a and 25b d appears only in the sensitivity factor $8\pi dn_1 \cos\varphi_1/\lambda$ and not in the shape factor $\text{Im}(\hat{\varepsilon}_i,\varphi_1)$. Thus only the absolute magnitude of ε_2' and ε_2'' will be affected to some extent, but not the shape of the spectral dependence of these quantities, which contains the most relevant information. The equation to solve finally for the film optical constants is to the third power. When the number of real roots is three instead of one, difficulties may arise in finding the physically meaningful solution, and an assignment at one wavelength may not be possible with certainty. However, when a wide spectral range in $(\Delta R/R)$ is evaluated, in most cases only one set of solutions is inherently continuous, indicating that this should correspond to the actual physical values of ε_2' and ε_2''. The other two solutions, which become identical just before turning imaginary, can then be discarded (164). When using this method one has to bear in mind that the initial Eqs. 25a and 25b have been derived for isotropic thin films. Especially in the case of monolayers, however, anisotropy of the optical constants will most likely arise. This can be tested by applying the above described procedure to $(\Delta R/R)$ values measured at different angles of incidence. In the case of anisotropy, angle-dependent optical constants will result that indicate the existence of anisotropy but do not necessarily represent realistic values. Other methods then have to be used to evaluate the optical properties of the film (see following paragraph).

4.2.2. KRAMERS-KRONIG ANALYSIS. The Kramers-Kronig analysis has frequently been used to determine the optical constants of substrates from normal incidence reflectance spectra. This causality relation is now also used to determine the film optical constants from the spectra of (ΔR/R) (165,166). In a two-phase system the complex Fresnel reflection coefficient, r_{13}, may be expressed in terms of its argument and its phase change, which then can be related to the optical constants. For normal incidence

$$r_{13} = R^{1/2} \exp(i\delta^r) = \frac{n - ik - 1}{n - ik + 1} \qquad (28)$$

Hence when δ^r is determined the optical constants can be calculated. It is well known that δ^r can be related to the reflectance R by the Kramers-Kronig-dipersion relation,

$$\delta^r(\omega_o) = \frac{\omega_o}{\pi} \int_o^\infty \frac{\ln[R(\omega)/R(\omega_o)]}{\omega^2 - \omega_o^2} d\omega \qquad (29)$$

when the reflectance $R(\omega)$ is measured over the entire frequency range. However, since $R(\omega)$ is usually determined only over a limited frequency range, assumptions have to be made about the tail fit, or additional measurements with other methods must be performed. These problems have been discussed in the literature extensively (255). In a three-phase system the same approach may be chosen, relating the corresponding ratio of the Fresnel coefficients, r_{123} and r_{13}, with the difference in the phase change for the three-phase and the two-phase systems:

$$\frac{r_{123}}{r_{13}} = \left(\frac{R_{123}}{R_{13}}\right)^{1/2} \exp\left[i(\delta^r_{123} - \delta^r_{13})\right] \qquad (30)$$

It has been shown that for $\Delta\delta^r = \delta^r_{123} - \delta^r_{13}$ an analogous expression can be derived (165):

$$\Delta\delta^r(\omega_o) = \frac{1}{\pi} \int_o^\infty \frac{\omega \ln[R_{123}/R_{13}]}{\omega_o^2 - \omega^2} d\omega \qquad (31)$$

The optical constants of the film may then be calculated from $(\Delta R/R)$ and $\Delta\delta^r$ by separating real and imaginary parts on either side of Eq. 32:

$$\frac{r_{123}}{r_{13}} = \left(\frac{R_{123}}{R_{13}}\right)^{1/2} \exp[i\Delta\delta^r] = f(\varepsilon_1, \hat{\varepsilon}_2, \hat{\varepsilon}_3, d, \lambda, \varphi_1) \quad (32)$$

and equating them. For explicit evaluation of ε_2' and ε_2'' (or n_2 and k_2) the linear approximation has to be used instead of the exact expression for f. At normal incidence Eq. 32 reduces to a simple form in the linear approximation:

$$f = \frac{r_{123}}{r_{13}} = 1 - i\frac{4\pi dn_1}{\lambda} \frac{\hat{\varepsilon}_2 - \hat{\varepsilon}_3}{\varepsilon_1 - \hat{\varepsilon}_3} \quad (33)$$

The film thickness d will still remain as parameter and has to be determined by an independent method. Problems of the tail fit are discussed to some extent by Naegele and Plieth (166). This method becomes very appealing (despite its drawback of tail-fitting problems) when studying the anisotropy of thin films. Assuming an uniaxial tensor of the film optical constants with the components $\hat{n}_{2n}$ and $\hat{n}_{2t}$ normal and tangential to the surface, all four optical constants can be determined by the Kramers-Kronig analysis when measuring the $(\Delta R/R)$ spectrum with perpendicular polarized light, preferably at normal incidence ($\rightarrow \hat{n}_{2t}$) and with parallel polarized light at higher angles of incidence ($\rightarrow \hat{n}_{2n}$).

4.2.3. COMBINATION OF ELLIPSOMETRY AND DIFFERENTIAL REFLECTANCE SPECTROSCOPY. In some cases the choice of film thickness may be a severe problem, and one might want to actually measure the optical thickness of the film directly rather than determining it indirectly; for instance, coulometric measurements yield the average number of deposited layers, but the layer thickness still has to be estimated. The usually employed ellipsometric techniques, as well as differential reflectance spectroscopy, yield only two equations; however, it has been demonstrated that the combination of both supplies three equations from which all three unknowns, namely, ε'_2, ε''_2, and d, may be calculated (167–169). Besides the azimuthal angle Ψ and the phase change difference $\Delta = \delta^r_\parallel - \delta^r_\perp$, which are determined ellipsometrically and correlated with the optical parameters of the three-phase system (157), the reflected intensity of the light is measured and the reflectance change due to film formation determined. Although in general two solutions are obtained from the ellipsometric equations as a function of film thickness d, the unique values ε'_2, ε''_2, and d may now be found by inserting these values into the appropriate equation for ($\Delta R/R$). When multiple solutions occur, the angle of incidence can be changed to avoid this problem if a physically meaningful solution cannot be found by simple reasoning.

4.2.4. ITERATION METHODS. Still another method of calculating the optical properties of thin films in differential reflectance spectroscopy is by measuring the angle dependence of $(\Delta R/R)_\parallel$ over a wide range (e.g., $5^\circ \leq \varphi_1 \leq 85^\circ$). By computational iteration those film dielectric constants ε'_2 and ε''_2 can be found which fit best the measured ($\Delta R/R$) curve. This method again is well suited to study anisotropy effects in thin films (the appropriate equations as derived from the linear approximation theory are given in the text that follows), since this fit can easily be performed also with four film dielectric constants, ε'_{2n}, ε''_{2n}, ε'_{2t}, and ε''_{2t}, as in the case of an uniaxial absorbing film. Usually more than only one minimum is found, depending on the initial values chosen for the parameters, the film optical constants, but is generally assumed that the absolute minimum is obtained only for the correct set of data. Due to the very high sensitivity of the optical constants on minute variations in ($\Delta R/R$), very accurate measurements are required. Nevertheless, one may derive different sets of optical constants, which yield about the same deviation in ($\Delta R/R$), rendering any decision very difficult or speculative. The fact is stressed that due to unavoidable errors in the measured ($\Delta R/R$) curve, the absolute minimum may even be obtained with "wrong" optical constants, whereas the physically correct set of data may yield a minimum deviation slightly larger than that of the "best" set. This might be a drawback of the explicit method proposed by Kolb and McIntyre (164), since those optical constants are calculated that fit <u>exactly</u> the measured ($\Delta R/R$) data, despite experimental errors in ($\Delta R/R$). The close vicinity of other possible solutions is not seen, whereas any ambiguity will show up in the iteration method and at least signal for cautious interpretation.

Anisotropy

Thin films, monolayers and submonolayer adsorbates, are expected to exhibit an appreciable amount of anisotropy in their optical constants. Dignam, Moskovits, et al. have derived quantitative expressions using the linear approximation for ($\Delta R/R$) arising from deposition of an uniaxial absorbing film onto an isotropic substrate (170,171). They are given by:

$$\left(\frac{\Delta R}{R}\right)_{\perp} = \frac{8\pi dn_1 \cos \varphi_1}{\lambda} \text{ Im } \left(\frac{\hat{\varepsilon}_{2t} - \hat{\varepsilon}_3}{\varepsilon_1 - \hat{\varepsilon}_3}\right) \quad (34a)$$

$$\left(\frac{\Delta R}{R}\right)_{\parallel} = \frac{8\pi dn_1 \cos \varphi_1}{\lambda} \text{ Im } \left\{\left(\frac{\hat{\varepsilon}_{2t} - \hat{\varepsilon}_3}{\varepsilon_1 - \hat{\varepsilon}_3}\right) \times \left[\frac{1-(\varepsilon_1/\hat{\varepsilon}_{2n}\hat{\varepsilon}_3)[(\hat{\varepsilon}_3^2 - \hat{\varepsilon}_{2t}\hat{\varepsilon}_{2n})/(\hat{\varepsilon}_3 - \hat{\varepsilon}_{2t})]\sin^2\varphi_1}{1-(1/\hat{\varepsilon}_3)(\varepsilon_1+\hat{\varepsilon}_3)\sin^2\varphi_1}\right]\right\} (34b)$$

where $\hat{\varepsilon}_{2n}$ and $\hat{\varepsilon}_{2t}$ represent the principal components of the thin film dielectric constant normal and tangential to the surface. These equations are very valuable to study the influence of anisotropy on $(\Delta R/R)$, since they readily provide information on the magnitude of the effect to be expected. As an interesting consequence it has been pointed out, that a thin transparent but anisotropic film on a metallic substrate yields $(\Delta R/R)$ values, which, when calculated for an isotropic film, may pretend absorbing properties (170). Therefore, film anisotropy has to be considered for quantitative analysis of monolayer optical properties. Up to now isotropic optical constants have usually been assumed as a first-order approach. Similar errors in the calculated film optical properties may be introduced by surface roughness. Anomalous plasma excitation at submicroscopic bumps and pits on the surface or at adsorbed particles may show up in the $(\Delta R/R)$ spectra, as pointed out by Dignam and Moskovits (152) and yield erroneous film optical constants when not taken into account properly. However, the true magnitude of this effect has not yet been established, and the influence has usually been neglected.

Finally, a few words should be said about the meaning of monolayer optical constants calculated by these procedures. All methods start with Fresnel equations and treat the thin-layer phase as continuum with very small d-values, neglecting the fact that d is in the order of atomic dimensions. A general microscopic theory of the optical constants is not yet available, although first approaches have been made (172-178). Short surveys are given by Bootsma and Meyer (179) and McIntyre (145,157). However, our aim is to establish general features in the properties of monolayers and adatoms as compared to their bulk values to learn something about the electronic structure of these deposits. This information still should be found in the spectra of $\hat{\varepsilon}_2$ as calculated from the classical analysis of reflectance data; for example, maxima in the imaginary part of $\hat{\varepsilon}_2$ should still correspond to absorption processes in the thin film (when all side effects have been taken care of). Thus information on the joint density of states will be obtained. Not too much emphasis should be placed at present on the accuracy of the absolute value of the optical constants, considering the very limited validity of Fresnel's equation in atomic dimensions.

4.3. Differential Reflectance Spectroscopy

Monitoring the underpotential deposition by spectroscopic techniques has been demonstrated first by Takamura, Takamura, et al. (150), who studied the reflectance change ($\Delta R/R$) of a gold electrode during monolayer formation of Cd and Pb (Fig. 27). The ($\Delta R/R-E$) curve at constant photon energy closely resembles the corresponding q-E curve for metal monolayer deposition, expressing a close relationship between ($\Delta R/R$) and coverage Θ of the deposit (180-183). Magnitude and sign of ($\Delta R/R$), however, are usually strongly dependent on $\hbar\omega$. Thus measuring reflectance changes at a fixed wavelength can furnish more specific information on certain electrode processes such as metal deposition than current measurements, and with less interference by side reactions, such as hydrogen evolution (154) (provided the hydrogen evolution is small enough not to produce gas bubbles, which will seriously distort optical measurements). Even more complex reactions like codeposition of two different metals may be studied successfully in combination with charge or current measurements (184).

It is expected that the optical properties of metal monolayers differ substantially from those of the respective bulk. In Fig. 28 the normalized reflectance changes ($\Delta R/R$) for p- and s-polarized light at λ = 4000 Å are plotted against film thickness for Ag

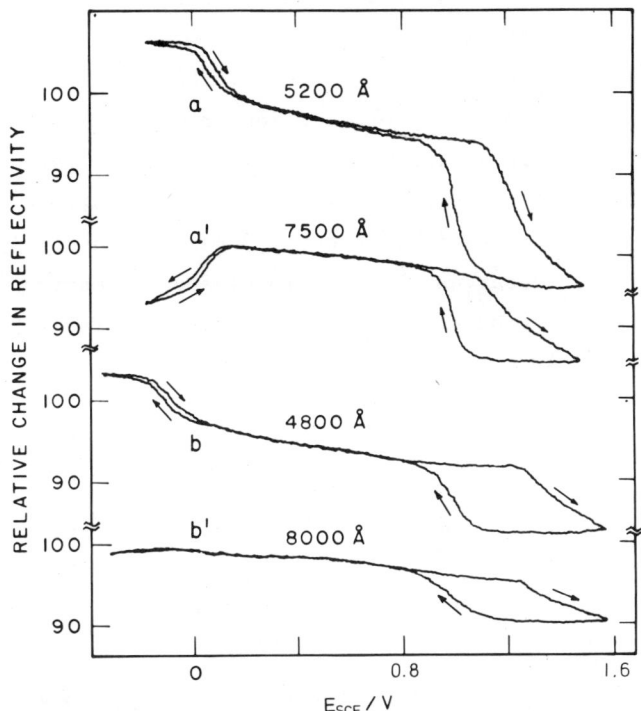

Fig. 27. Potential curves for ($\Delta R/R$) at various wavelengths for an Au electrode in 0.2M $HClO_4$. Curves a, a': 5×10^{-4}M Pb^{2+}; Curves b, b': 1×10^{-3}M Cd^{2+}; $\varphi_1 = 50°$; 18 reflections. Scan rate, 105 mVs^{-1} (150).

deposition onto a polycrystalline Pt electrode (147). The computed reflectance changes as calculated by the linear approximation equations assuming bulk properties for the Ag deposit are also shown (dashed lines). It is clearly seen that in the beginning of Ag deposition the measured ($\Delta R/R$) is actually opposite in sign to that predicted by using Ag bulk values. This is not surprising, since especially in the submonolayer region the adsorbate is more

appropriately described by an adatom-substrate complex rather
than a silverlike overlayer. It is most interesting to note that
after a layer thickness of only four to five monolayers, the film
obviously acquires bulk optical properties. This is in agreement
with recent data obtained by photoemission from Ag overlayers
(see Fig. 25) (144). This finding also contrasts that for metal
films evaporated onto semiconducting or insulating material like
quartz or glass substrates as frequently investigated for Ag. In
the latter case one observes that due to cluster formation bulk
optical properties are usually not reached at an average thickness
of under 50 to 100 Å, an order of magnitude more than in the case

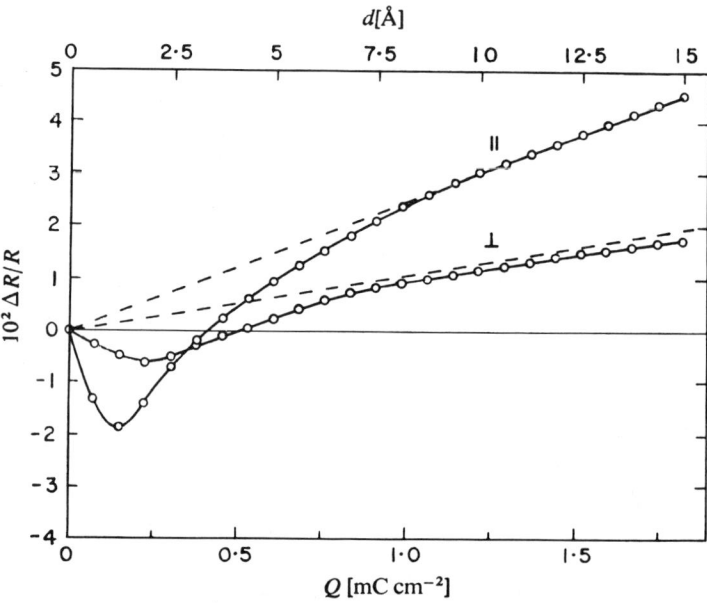

Fig. 28. The value ($\Delta R/R$) as function of film thickness during
deposition of Ag onto a Pt electrode in 1M $HClO_4$ +
5×10^{-5}M $AgNO_3$. E_d = + 0.15 V_{SCE}; λ = 4000 Å; φ_1 = 45°.
(---), computed with bulk Ag optical constants (147).

of Ag on Pt (185-188).

This confirms that the substrate's influence reaches beyond the completion of the first monolayer inducing a continuous and island-free layer growth for the next few layers. As an important result we find that underpotential deposition may initiate the growth of homogeneous, cluster-free films, which exhibit bulk optical properties at a thickness as low as four to five monolayers. When underpotential deposition is absent as in the case of Cu on Ag in electrolytic solutions, a drastic deviation from the computed ($\Delta R/R$) values is found, even at high thicknesses. Usually the reflectance decreases continuously during deposition indicating the formation and growth of three-dimensional nuclei as the very initial step. The true optical constants of the metal clusters then have to be calculated by applying the Maxwell-Garnett theory (189). In Fig. 29 the normalized reflectance changes ($\Delta R/R$) for p-polarized light are shown as a function of film thickness at various wavelengths for Ag deposition on Pt. The curve at 5000 Å shows the same behavior as that at 4000 Å (Fig. 28). Note that the assymptotic approach of the experimental curve to the computed line, which passes through the origin, implies that the deposited film in its total thickness acquires bulk values; hence the optical properties of the first monolayer, which are markedly different from those of the bulk, are also shifted toward bulk values as the deposition proceeds. This behavior is not found at $\lambda = 3250$ Å, near the plasma edge of bulk Ag. The computed line is never approached; however, its slope is reproduced at approximately five to six monolayers. This indicates that the first few layers retain their special properties at that wavelength and do not sustain volume plasma excitation as in bulk. Finally, at $\lambda = 3500$ Å, total disagreement with the calculated ($\Delta R/R$) values is seen. At this wavelength surface plasma excitation occurs in Ag and renders the reflectance strongly depend on the surface structure. Hence it is

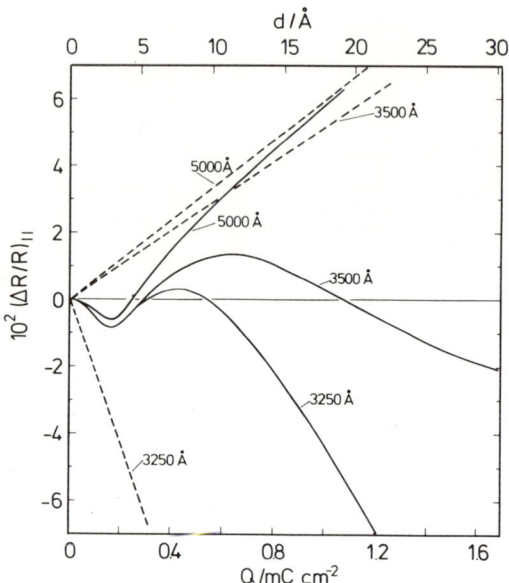

Fig. 29. The value ($\Delta R/R$) as function of film thickness for p-polarized light during deposition of Ag onto a Pt electrode in 0.5M H_2SO_4 + 4 x 10^{-5}M $AgNO_3$; E_d = - 0.15 V (SCE); φ_1 = 45°. (---), computed with bulk Ag optical constants.

not surprising that the Ag deposit exhibits strong deviation from bulk behavior, even at higher thickness. When Cu is deposited onto Pt and monitored by reflectance measurements at constant wavelengths, the ($\Delta R/R$)-thickness curves have the same characteristics (190). In the beginning the ($\Delta R/R$) decreases and reaches a minimum at about one monolayer, and then the reflectance again increases, before acquiring a slope approximately predicted by Eq. 25 from the bulk Cu values.

More interesting information is buried in the spectral dependence of the reflectance change ($\Delta R/R$) for monolayer deposition. The frequency dependence of $(\Delta R/R)_{\parallel}$ caused by Cu deposition onto Pt in the underpotential region is shown in Fig. 30 for various coverages. Since ($\Delta R/R$) is proportional to the adsorbate's average thickness (Θ) and the photon energy (λ^{-1}), the normalized reflectance change ($\Delta R/R$) per photon energy and unit coverage is plotted as a function of $\hbar\omega$ to reveal more clearly the influence of the shape factor containing the film dielectric constants (see Eq. 25). Bare and Cu-covered Pt surfaces were achieved by stepping the electrode potential from + 0.52 V(SCE) to the appropriate values in the underpotential region (158). The spectra show surprisingly sharp structures in ($\Delta R/R$), indicating absorption processes in the monolayer adsorbate. The spectral dependence

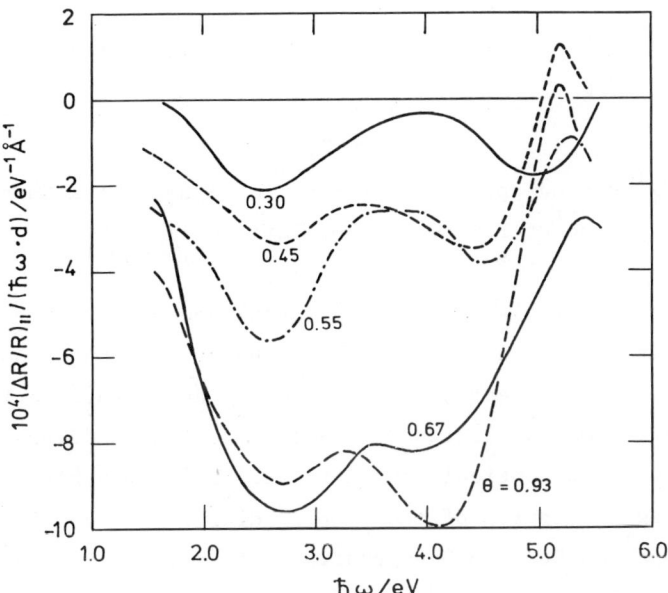

Fig. 30. Spectra of normalized reflectance change for various coverages of Cu on Pt. $\varphi_1 = 60°$. p-polarization (158).

of (ΔR/R) as calculated for a 3-Å thick Cu overlayer with bulk properties on Pt (Eq. 25) is shown in Fig. 31 for comparison to emphasize the distinct difference. It was found that the (ΔR/R) spectra for deposits of two to three Cu monolayers show already clear indication for the Cu interband transition at 2 eV [e.g., sharp decrease in (ΔR/R) around 2 eV with sign change as seen in Fig. 31].

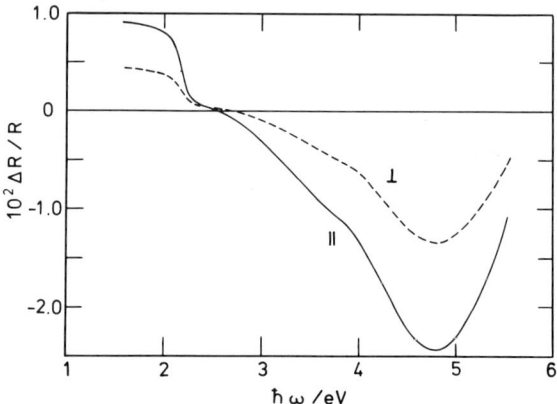

Fig. 31. Calculated spectra of (ΔR/R) for a 3-Å thick Cu overlayer on Pt for p- and s-polarized light. $\varphi_1 = 45°$.

Although the correlation between (ΔR/R) and $\hat{\varepsilon}_2$ is rather complex, even in the linear approximation theory as pointed out earlier, it is still evident that for Cu on Pt any structure in the (ΔR/R) spectrum may be assigned to absorption in the deposit, since the substrate optical constants vary only smoothly with $\hbar\omega$. Real and imaginary parts of the complex dielectric function for a Cu monolayer on Pt have been evaluated from the (ΔR/R) data by using the method of Kolb and McIntyre (164). The result, as shown in Fig. 32, bears some resemblence with ε_b of bulk Cu, the contribution of bound electrons to the total dielectric function of

Cu. A tentative assignment of the absorption bands, as seen in ε'' for the Cu monolayer, has been given by Kolb and Kötz (158). An interesting observation is reproduced as insert in Fig. 32: the optical properties (e.g., ε'') when plotted versus coverage are found to change rapidly at $\Theta = 0.6$, although the spectral shape remains nearly unchanged (except that some line broadening is noticed). This may be direct evidence for a structural change in the adsorbate occurring at about 2/3 of a monolayer.

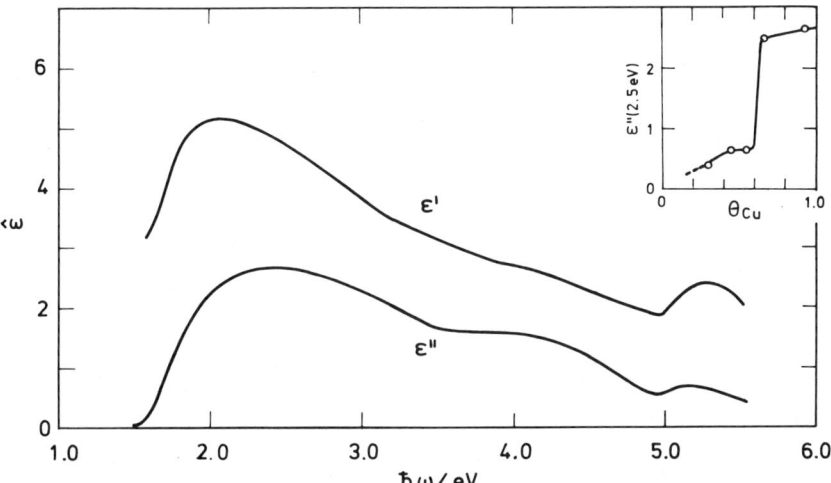

Fig. 32. Frequency dependence of real and imaginary parts of the complex dielectric constant of a Cu monolayer on Pt: $d = 3$ Å; $\varphi_1 = 60°$. Coverage dependence of ε'' at 2.5 eV is shown as insert (158).

The optical properties of lead on gold have been determined by ellipsometry for two different coverages in the visible range (191). The optical constants again differ markedly from bulk Pb values and exhibit a pronounced coverage dependence. However, since the substrate shows a pronounced electroreflectance effect, the measured reflectance change is most likely not only due to the adsorbate, as is discussed in the text that follows.

A first attempt has been made to determine the anisotropy of metal monolayer optical constants. For Pb on Ag(111) single-crystal surfaces, the angle dependence of the ($\Delta R/R$) spectra was determined within the range $5° \leq \varphi_1 \leq 85°$ for photon energies $\hbar\omega < 3.5$ eV to eliminate electroreflectance effects of the substrate (47). Preliminary results indicate a clear anisotropy in the dielectric function for the complete Pb monolayer, showing a much larger absorption tangential to the surface (i.e., within the Pb monolayer) than normal to it ($\varepsilon_t'' > \varepsilon_n''$).

Adsorption of any species onto a metal surface will change to some extent the substrate surface properties by adsorptive interaction. Although the bare substrate surface may already have properties different from those of the bulk due to consequences arising from the lattice termination (192), we will assume that the optical properties of the bare metal surface can well be approximated by those of the bulk. Then the normalized reflectance change, ($\Delta R/R$), which is observed with adsorption or film formation on a substrate as difference in reflectance between the bare and the covered surface, is commonly assigned to the film phase. To derive the film optical constants, $\hat{\varepsilon}_2$, from the experimentally determined ($\Delta R/R$) values the three-phase model as shown in Fig. 33a is usually assumed. However, in many cases the situation will be more complex, since the deposited film may change the substrate surface properties by an amount, $\Delta\hat{\varepsilon}_3$, with respect to the unperturbed bulk substrate values, $\hat{\varepsilon}_3$, and this change may contribute significantly to the measured ($\Delta R/R$). Now the electrode-electrolyte interface is described more adequately by a four-phase model as depicted in Fig. 33b. When the film optical constants are calculated from the experimental ($\Delta R/R$) values using the three-phase model, both sets of dielectric functions, $\hat{\varepsilon}_2$ and $\hat{\varepsilon}_3 + \Delta\hat{\varepsilon}_3$, are incorporated in the apparent film dielectric constant. Therefore, care has to be taken in the interpretation of the ($\Delta R/R$) spectra as well as the apparent film optical properties, since the contributions from the change in the substrate surface properties may be dominant and erroneously assigned to the adsorbate optical properties.

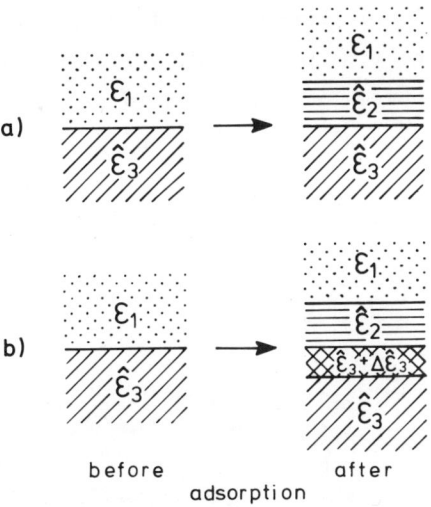

Fig. 33. Model of electrode-electrolyte interface before and after adsorption. ε_1, $\hat{\varepsilon}_2$, $\hat{\varepsilon}_3$, and $\hat{\varepsilon}_3 + \Delta\hat{\varepsilon}_3$ are dielectric constants of electrolyte, film, substrate, and substrate surface layer, respectively; no change (a) and change (b) in the substrate surface properties by adsorption (193).

Kolb, Leutloff, et al. have investigated in some detail the underpotential deposition of Pb, Tl, and Cu on Au electrodes by differential reflectance spectroscopy (193). The ($\Delta R/R$) spectrum for a Tl monolayer on Au, measured at an angle of incidence $\varphi_1 = 45°$

between 1.6 and 5.2 eV, is shown in Fig. 34a. The apparent dielectric functions, however, of all three metal monolayers as evaluated from the three-phase model clearly exhibit characteristic features of a distorted Au surface (193). This strongly supports the assumption that the measured reflectance change ($\Delta R/R$) during monolayer formation results from the change in the Au surface optical properties on adsorption (194,195), and not from the adsorbate itself. This of course makes it very difficult in such cases to determine the true monolayer optical constants (156,180, 181,191). Similar problems are anticipated when determining the optical properties of metal adsorbates on Ag. In Fig. 34b the

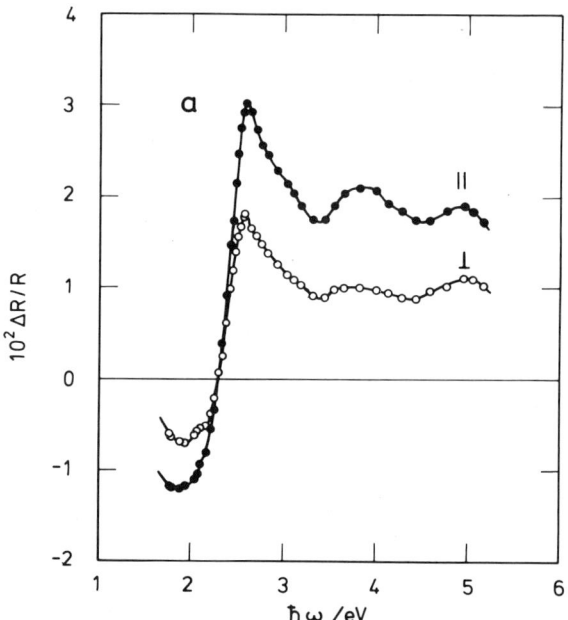

Fig. 34a. Relative reflectance changes ($\Delta R/R$) as function of photon energy $\hbar\omega$, caused by deposition of a Tl monolayer on Au. $[Tl^+] = 2 \times 10^{-4}$M. $\varphi_1 = 45°$ (193).

($\Delta R/R$) spectra for a Tl monolayer on Ag are shown. Here the influences of adsorbate and adsorbate-induced changes in the substrate will have to be separated before the prominent structure in $(\Delta R/R)_\parallel$ around 3.9 eV can be explained.

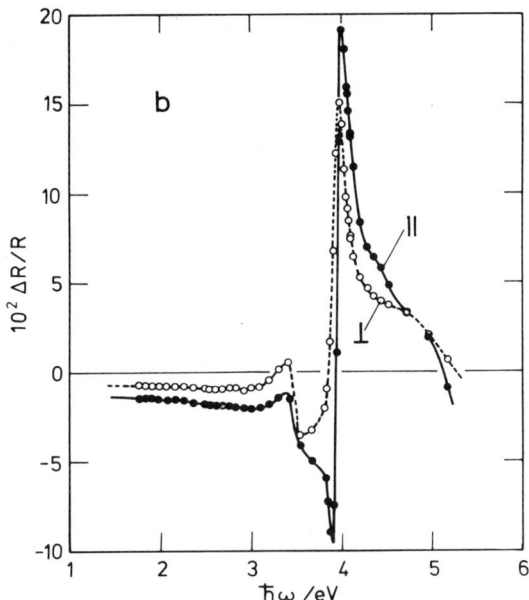

Fig. 34b. ($\Delta R/R$) spectra for a Tl monolayer on Ag. Same conditions as in Fig. 34a.

The very strong influence of the adsorbate on the substrate surface properties was interpreted as <u>enhanced electroreflectance</u>, due to the adatoms' partial charge. This is in analogy to effects observed for anion adsorption at metal electrodes, for example, Cl^- on Au, where a relatively large reflectance change ($\Delta R/R$) is found (163,196-199). Here ($\Delta R/R$) usually cannot be accounted for by the optical properties of the adsorbate itself but is explained by an adsorbate-induced electroreflectance (ER) effect of the substrate (197). However, it has been pointed out that at those wavelengths where the substrate ER effect is negligibly small ($\hbar\omega < 2.0$ eV for Au and $\hbar\omega < 3.0$ eV for Ag), ($\Delta R/R$) should be solely due to the optical constants of the monolayer proper, and hence the "true" monolayer dielectric function may be calculated from the respective ($\Delta R/R$) values. However, at present there is further need for experimental data in order to establish an exact separation of monolayer and monolayer-induced optical effects.

For adsorption of molecules with weak lateral interaction it is expected that the mean adsorbate thickness d in the submonolayer range can be expressed in terms of coverage Θ and monolayer thickness d_o as

$$d = \Theta \cdot d_o \qquad (35)$$

because the adsorbate optical constants are considered Θ-independent. Hence

$$(\Delta R/R)_\Theta = (\Delta R/R)_{\Theta=1} \cdot \Theta \qquad (36)$$

Thus optical methods can be used to establish adsorption isotherms (163,197). When metal atoms are deposited the film optical constants should depend strongly on the coverage due to the increasing interaction between the adatoms themselves, resulting in a clear deviation from the linear dependence ($\Delta R/R$) versus d. A typical example is Ag on Pt (see Fig. 28). Surprisingly, it is found that for metal deposition on Au and Ag in the submonolayer region a linear relationship between ($\Delta R/R$) and Θ is still observed (47,156,180,181). For example, when lead is deposited onto Au in various submonolayer amounts a nearly perfect linear correlation between $(\Delta R/R)_\perp$ and Θ (expressed in terms of charge) is found for $\lambda = 4800$ Å over the whole monolayer range, whereas for $\lambda = 7000$ Å a clear deviation from the initial slope is seen at higher coverages (180) (Fig. 35). It is believed that this be-

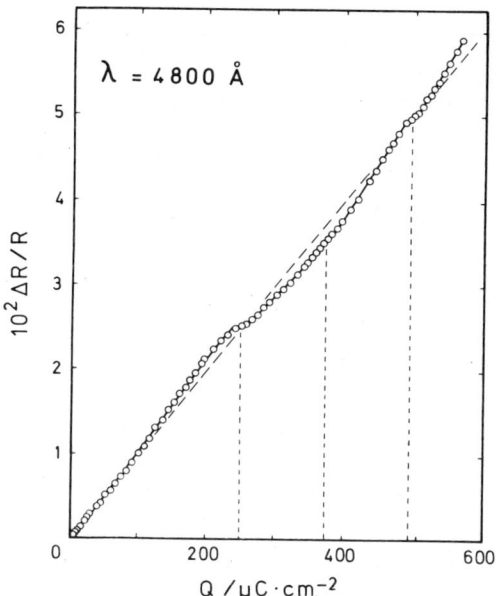

Fig. 35a. Relative reflectance change ($\Delta R/R$) for Au electrode in 1M $HClO_4$ + 1 × 10^{-3}M Pb^{2+} as function of charge equivalent of underpotentially deposited Pb. A complete monolayer corresponds to about 500 – 600 $\mu C/cm^2$. $\varphi_1 = 10°$. s-polarization (180).

havior of ($\Delta R/R$) can be explained by adsorbate induced effects in the substrate, since both metals, Au and Ag, show strong ER effects. At 4800 Å, where the ER of Au is most prominent (145), the ($\Delta R/R$) caused by Pb deposition may be due to changes in the optical properties of the Au surface (therefore the linear dependence of ($\Delta R/R$) on charge Q), while at 7000 Å, where the ER of the substrate is very small, ($\Delta R/R$) should reflect the optical properties of the Pb adatoms, indicating a strong change in the adsorbate optical constants with coverage Θ.

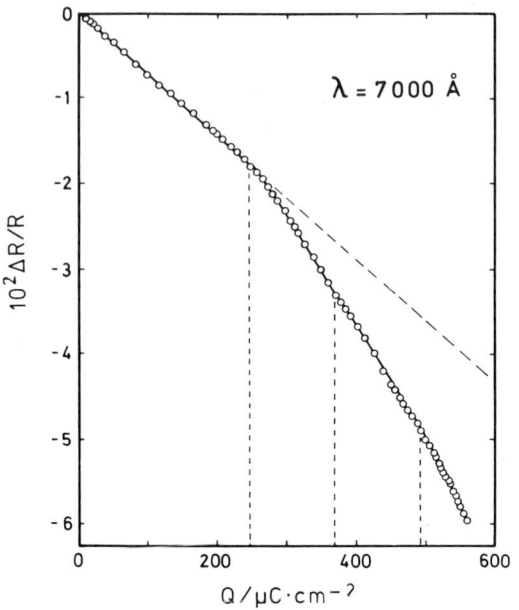

Fig. 35b. Relative reflectance change as in Fig. 35a, but at different wavelength (180).

4.4. Electroreflectance

The electroreflectance (ER) effect of bare metal surfaces has been discussed extensively by McIntyre (145,146) and is not reviewed here. The ER spectra are usually obtained by superimposing a small sinoidale potential modulation to the electrode potential and recording the resulting reflectance change. Electroreflectance studied as a function of $\hbar\omega$ or bias potential E_o certainly will yield valuable information on metal surface optical properties, although at present the ER effect is far from being understood quantitatively. Measurements on single-crystal surfaces (200) may be well suited to refine the existing theoretical models. We report a few experiments on ER from metal surfaces covered with monolayer or submonolayer deposits. Due to the extremely low penetration depth of the applied static electric fields in the order of 1 Å, this method is ideal for surface studies. When a substrate metal electrode is covered with a metal deposit in the underpotential range and a modulating potential is applied, the optical response $(1/R)(\Delta R/\Delta E)$ will in general be caused by superimposition of two effects (156):

$$\frac{1}{R} \cdot \frac{\Delta R}{\Delta E} = \frac{1}{R} \cdot \frac{\partial R}{\partial \Theta} \cdot \frac{\partial \Theta}{\partial E}\bigg|_q + \frac{1}{R} \cdot \frac{\partial R}{\partial q} \cdot \frac{\partial q}{\partial E}\bigg|_\Theta \qquad (37)$$

the first term on the right hand side describing the change in reflectance with coverage $\Theta(E)$, the second term representing the true ER, mainly caused by the surface excess charge (145,146,201, 202). To avoid this complication, at constant Θ has to be measured, which is done simply by introducing a "scavenger" electrode, at which all metal ions in solution are reduced and deposited after having allowed monolayer deposition at a certain potential E onto the working electrode. Then the ER spectrum of the monolayer covered substrate surface can be measured at bias potentials more negative than the deposition potential E (which determines Θ). Preliminary results on ER spectra of Au electrodes covered with various amounts of Tl are shown in Fig. 36. A drastic difference is noted between the ER spectrum of a bare Au surface and that of a Tl-covered surface. With increasing Tl coverage the peak at 2.5 eV, characteristic of the bare Au spectrum, is substantially reduced and shifted toward higher energies while the second peak around 3.5 eV raises strongly. Similar to the $(\Delta R/R)$ spectra obtained from differential reflectance spectroscopy, the

ER spectra exhibit a change in sign for low photon energies (ca. 2.3 - 2.4 eV) (see Fig. 34a). Similar results have been reported for Pb on Au (156). It has been noted that the shapes of the various ER spectra closely resemble those of the energy-loss function for charged Au surfaces (193). Furthermore, it should be mentioned that approximately the same spectral shape is obtained when recording the spectra without a scavenger electrode, that is, with the respective metal ions in solution. Therefore,

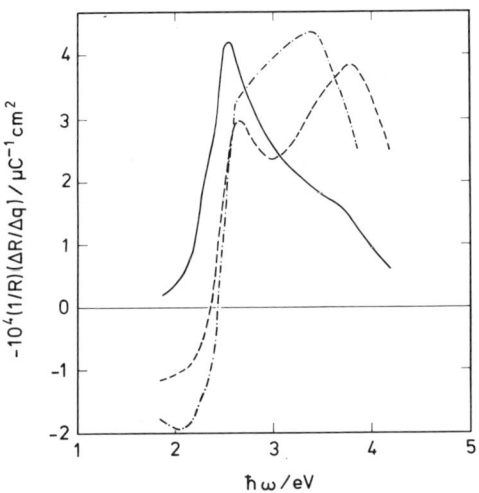

Fig. 36. Electroreflectance spectra of Au electrode in 0.5M $NaClO_4$ (pH 3), covered with various amounts of Tl. p-polarization. $\varphi_1 = 45°$. f = 265 Hz. Modulating charge $\Delta q = 0.6$ $\mu C/cm^2$ (peak to peak). Bias potentials (vs. SCE) $E_o = -0.2$ V (——, bare surface); -0.4 V (---, half a monolayer Tl); -0.7 V (—•—, one monolayer Tl). The Tl^+-ions in solution were deposited onto a scavenger electrode before recording the ER spectra in the Tl^+-free solution.

it may be concluded that at least for Tl and Pb on Au the second term on the right-hand side in Eq. 37 contributes significantly to the total ($\Delta R/R$). However, for a quantitative interpretation of these ER spectra much more experimental work is required.

4.5. Related Studies

4.5.1. SURFACE CONDUCTANCE MEASUREMENTS. The conductance along a metal surface will be changed to some degree when adsorption takes place. Since this change occurs within a very small volume near the surface, it shows up in the total conductivity only when the bulk conductance is kept as small as possible. Therefore, the substrate metal electrode is usually prepared as a thin film with a thickness ranging between approximately 50 and 300 Å. The lower limit in thickness is given by the evaporation technique, since the preparation of ultrathin homogeneous, island-free metal films on nonconductive substrates such as glass or quartz, is not readily feasible. In addition, size effects become significant when the film thickness is comparable with the mean free path of the electrons. In a typical metal such as Au the latter is of the order of 300 to 400 Å at room temperature. Therefore, various scattering mechanisms have to be taken into account when comparing the film conductance with that of the bulk metal.

Since the electrical conductivity σ of a bulk metal is determined by the number N of free electrons per unit volume, their effective mass m^*, and the collision time τ ($\sigma = Ne_o^2\tau/m^*$), a change in conductance $\Delta\sigma$ on adsorption may be related to a change in the free electron concentration N (196, 203):

$$\frac{\Delta\sigma}{\sigma} = \frac{\Delta N}{N} = \alpha \frac{n}{N} \tag{38}$$

where n is the total number of adsorbed atoms or molecules per unit area and α a proportionality constant, which is assumed to correspond directly to the fraction of an electron transferred between the adsorbate and the metal. Thus information on the direction and the amount of charge transfer within the adatom-substrate complex should theoretically be gained if this simple approach were satisfactory. However, a number of other factors will also influence $\Delta\sigma$. For example, the adatom will act as a scattering center on the surface, which may reduce the conductivity without any appreciable charge transfer (202). Since localized surface charges on a metallic film are strong scatterers for conduction electrons, ion adsorption will even further complicate the situation. Anderson and Hansen (202) have shown that adsorption of J^- and Cs^+ on gold decrease the conductance regardless of the sign of the charge. It must also be considered that the adsorbate may change the metal film thickness when a reaction with the substrate occurs or when another layer is plated (e.g., in metal deposition). These effects, which are quite difficult to separate, all determine the measured conductance change $\Delta\sigma$ on adsorption and often make it impossible to obtain detailed information on the adsorption mechanism. Therefore, it seems that in general surface conductance measurements alone will not yield any new information on adsorbate properties and adsorption behavior other than that already obtained by optical methods.

Fujihira and Kuwana (203) have studied the underpotential deposition of copper onto a Pt electrode by monitoring the relative conductance change ($\Delta\sigma/\sigma$) as a function of potential. It was found that the deposition of the Cu monolayer increases the conductance of the Pt film, which is interpreted as due to a partial electron transfer from the monolayer to the Pt electrode in accordance with electrochemical measurements. The proportionality factor α between

($\Delta\sigma/\sigma$) and ($\Delta q/q$) is determined as 0.65 which, however, is by far too large to be interpreted as that fraction of an electron donated from the Cu. Other effects besides the increase in free electrons obviously contribute also to the increase in conductance.

4.5.2. PHOTOEMISSION INTO ELECTROLYTES. In situ photoemission experiments (204,205) on metal electrodes covered with submonolayer amounts of foreign metal atoms seem to become an interesting tool for studying the structures of metal monolayers. Since photoemission from the substrate can be markedly changed by a regular array of adatoms (Surface-Umklapp) (206), information on the existence of ordered structures in the adsorbate may be gained. Experiments with Cu deposited on Au(111) surfaces at underpotentials indeed reveal a pronounced minimum in the photoemission yield at $\Theta_{Cu} = 0.25$ (206), suggesting the presence of a superlattice of Cu adatoms at that coverage.

4.5.3. MÖSSBAUER SPECTROSCOPY. Mössbauer spectroscopy studies of metal monolayers have been reported only once, although this technique can be used in situ and will furnish valuable insight into the nature of the adatom-substrate bond. Bowles and Cranshaw (46) investigated the properties of underpotentially deposited ^{119}Sn on a Pt black electrode and found similar spectra for a fully and a partially covered surface, and for bulk Sn, the latter one being recorded for comparison. This result was one of the first direct proofs for the underpotentially formed species being metallic rather than ionic (complete discharge). A chemical shift has been observed and a decrease in the area under the spectrum in the sequence: fractional monolayer, full monolayer, and bulk metal. The latter observation was explained by the fact that the Sn-Pt bond is stronger than the respective Sn-Sn bond and that the Sn-Pt bond strength increases with decreasing coverage in accordance with electrochemical considerations. From the chemical shift it was concluded that the Sn adatoms at low coverage lie closer to the Pt surface than in a complete monolayer, a result, also obtained for Cu on Pt from electrochemical measurements (see Section 2.5).

5. CATALYTIC EFFECTS OF METAL MONOLAYERS ON ELECTROCHEMICAL REACTIONS

Although a high catalytic activity of metal adatoms, monolayers, or small nuclei has already been observed (207,208), systematic investigations on the catalytic properties of metal monolayers have been undertaken only rather recently (209-221), with increased understanding of underpotential deposition. Some typical systems, for which the influence of underpotentially formed metal adsorbates has been studied, are: (a) oxidation of CO on Pt electrodes covered with As, Sb, Bi (210), or Ru (215), (b) oxidation of HCOOH on Pt electrodes covered with Hg (209), As, Sb, Bi (210), Pb, Tl, Bi, Cd, Cu, or Ag (218,219), (c) oxidation of HCOOH on various group VIII metal electrodes covered with Pb (220), (d) oxidation of methanol on Pt and Pd electrodes covered with Au and Ru (213,214), (e) oxygen reduction on Au and Pt electrodes covered with Pb, Tl, Bi, Cd, Cu, or Ag (217), (f) hydrogen evolution on Pt covered with Ag (211) and on Au covered with Pd (212), and (g) the redox reactions Fe^{2+}/Fe^{3+} and Ti^{3+}/Ti^{4+} at Au electrodes covered with Ag and Cu, respectively (216). Measurements were usually performed with and without the respective metal ions in the electrolytic solution. Consequently, the studies were confined to systems where the catalytic reaction takes place in the underpotential region of that metal. Since the coverage of the metal adatoms varies with potential, the electrocatalytic reaction often could not be studied at constant adatom coverage. Despite this lack of a controlled and constant adatom coverage, interesting results were obtained. In Fig. 37 the influence of As adatoms on Pt is demonstrated for the CO oxidation (210). The overpotential for this reaction on a bare Pt electrode is found to be markedly reduced already by very small amounts of adsorbed As. An investigation of the effects of underpotentially deposited As, Sb, and Bi revealed that the catalytic activity increased in the order Bi < Sb < As, consistent with the tendency of these metals to adsorb OH radicals. Since the oxidation of CO on bare Pt does not start before about + 0.8 V (NHE), where O is beginning to adsorb on Pt, the strong influence of adatoms like As is explained by the fact that As adsorbs O or OH radicals much earlier [in the potential range from 0.2 to 1.0 V (NHE)] (210). The enhancement of the oxidation rate is then explained by the cooperation of substrate sites and adatoms; while CO adsorbs on Pt, a OH radical is adsorbed on an adatom adjacent to the CO. It may be interesting to note that here underpotential deposition essentially allows study of the catalytic properties of a metal in a potential range, where that metal as bulk phase is not stable. Although we know that the chemical behavior of adatoms may differ from that of the bulk metal (e.g., see Section 2.8), first-order information may be gained from the monolayer properties. Another interesting example is the oxidation of formic acid on bare and monolayer covered Pt electrodes. In Fig. 38 cyclic current-potential curves are shown for the oxidation of formic acid on a bare Pt electrode surface and on a Pt surface covered with Pb adatoms. In the latter case, a tremendous increase in the oxidation current by about one order of magnitude is observed. In a systematic investigation it has

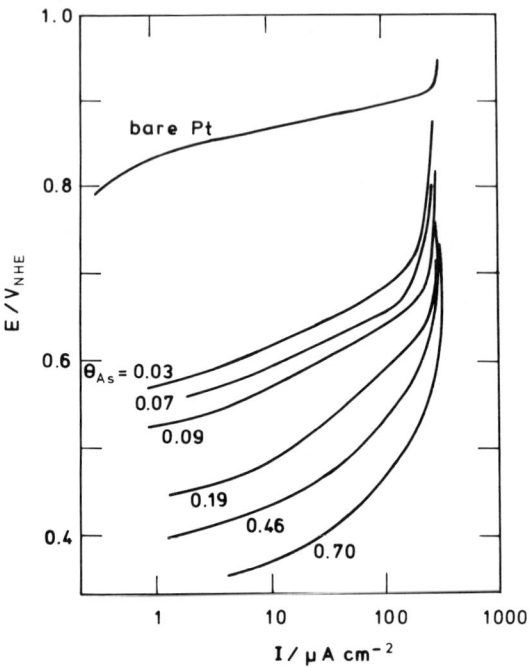

Fig. 37. Potential-current curves for CO oxidation on bare and As-covered Pt electrodes in 1N H_2SO_4. T = 40°C (210).

been shown (218-220) that the catalytic effect of the metal monolayers arises here not so much from the specific chemical nature of the various metal adsorbates, such as Cd, Pb, Tl, or Bi, but merely from their ability to block substrate surface sites. It is assumed that during the oxidation of HCOOH intermediates are formed that adsorb strongly on Pt and thus inhibit the further HCOOH oxidation (209). The foreign metal adsorbates on Pt prevent poisoning of the electrode without changing much of the catalytic properties of the substrate.

The oxygen reduction on Au and Pt may also be discussed. Adžić

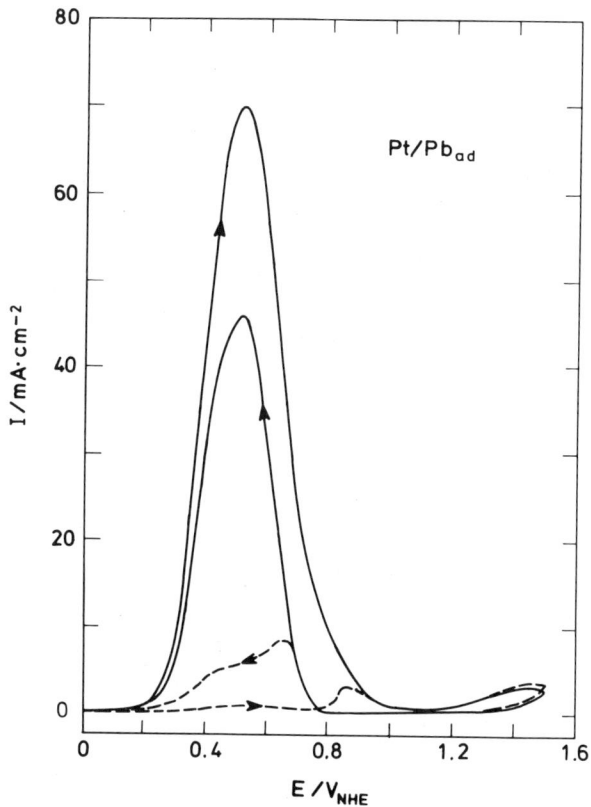

Fig. 38. Oxidation of HCOOH on bare and Pb-covered Pt electrodes. 1M $HClO_4$ + 0.265M HCOOH. $[Pb^{2+}] = 0$ (---) and 1×10^{-3}M (——). Sweep rate, 50 mVs^{-1} (219,220).

and Despić (217) have investigated the influence of various metal adsorbates on the oxygen reduction on Au and Pt electrodes. It was found that adatoms of Tl, Pb, and Bi on Au can increase the oxygen reduction significantly, the enhancement effect being largest for small coverages ($\Theta \approx 0.1$), whereas for Cd no influence was found and for Cu on Au, inhibition was observed (decrease in current). When oxygen reduction is studied on Pt, only inhibition is found for the underpotential deposits Cu, Pb, Bi, and Ag. The strong inhibitory effect of Ag on Pt for the oxygen reduction is worthwhile to note, because it has been shown earlier (27) that oxygen would not adsorb on top of a Ag monolayer on Pt. This clearly contrasts the behavior of bulk Ag.

Furuya and Motoo (212) have studied the influence of Pd adatoms on a Au electrode on the hydrogen evolution reaction. Although no conclusive evidence is presented for the formation of a uniform Pd layer on Au, the experimental results suggest strongly such a monolayer formation. An analysis of the exchange current density i_o and the Tafel factor for the hydrogen evolution reaction as a function of Θ_{Pd} shows that in the range $0 \leqslant \Theta_{Pd} \leqslant 0.2$, i_o increases rapidly from the value of Au to nearly that of Pd, while the Tafel factor remains about 120 mV/decade independent of coverage. Thus one may conclude that 0.2 of a Pd monolayer on Au exhibits the highest catalytic activity for this reaction. Assuming a regular array of adatoms, this coverage value would correspond to that case, where the highest density of single adatoms is reached without appreciable Pd-Pd interaction.

Finally, an interesting observation reported by McIntyre and Peck (221) should be mentioned briefly, although it deals not with catalysis in the usual sense, but in greater detail the influence of trace amounts of metal impurities, such as Pb or Tl, on the deposition behavior of gold. They found what has been phenomenologically known for quite some time, namely, that these trace amounts induce deposition of bright gold films of uniform coverage. This effect may be correlated with the underpotential deposition behavior of the impurity metals on gold (221). Obviously, the underpotentially deposited adatoms can act as nucleation centers for gold deposition, but they will not be burried in the gold during the deposition process. They will remain on the surface, since the surface site is energetically more favorable than a site in the bulk gold. For instance, it has been shown that in the underpotential range Pb will not form an alloy with gold but will stay as adatom on the surface (52). Therefore, by choosing an electrode potential, that yields a high surface concentration of adatoms, a large number of nucleation centers are provided, which will lead to the formation of rather uniform deposits.

6. MODELS EXPLAINING PREDEPOSITION

6.1. The Pauling Model

According to Pauling the bond between two dissimilar atoms in a diatomic molecule A-B consists of two parts, a covalent contribution represented by the arithmetic or geometric means of the respective bond strengths in the molecules A-A and B-B (as given by their dissociation energies), and an ionic contribution arising from the difference in electronegativities. The bond strength is given by Pauling (222):

$$D(A-B) = \frac{1}{2} \left[D(A-A) + D(B-B) \right] + (\chi_A - \chi_B)^2 \qquad (39)$$

with D in eV and χ being Pauling's electronegativity values. Although this equation was largely derived from experimental evidence rather from theoretical considerations, it turned out to be quite successful and is widely used. A molecular type of approach was frequently chosen to describe the chemisorption of single atoms on metal surfaces. The adatom-substrate complex was approximated by a surface molecule consisting of the adatom and only a few next neighbor substrate atoms that can form chemical bonds with the adsorbate. The heats of adsorption for hydrogen atoms on various metal substrates have been calculated in such a way and they show good agreement with the experimental values (223).

As a first-order approach Pauling's concept should also be applicable to metal monolayer adsorption. The difference in heats of adsorption of a metal adatom on a foreign substrate and on a surface of its own kind should then be given by (78,224):

$$ze_o \Delta E_p \approx - \Delta H_{ads} = \frac{1}{2} \left[D(S-S) - D(A-A) \right] + (\chi_S - \chi_A)^2 \qquad (40)$$

where $D(S-S)$ and $D(A-A)$ denote the single bond strength in the respective bulk of substrate and adsorbate metals, for which choosing the proper values is quite a difficult task. The single bond strength in a metal has usually been obtained by dividing the heat of sublimation by half of the coordination number (223, 225). When calculating now the differences in the heats of adsorption - ΔH_{ads} according to Eq. 40 and comparing them with the experimental determined values $ze_o \Delta E_p$, a rather poor correlation is found (78). For $\Delta \chi < 1$ the calculated values are persistently smaller than the experimental ones, but for $\Delta \chi > 1$ the reverse is true (Fig. 39). This systematic deviation suggests a linear correlation between $- \Delta H_{ads}$ and $\Delta \chi$ rather than Pauling's quadratic dependence, because the ionic contribution largely determines the shape of the curve (78).

Since Pauling's equation fails to account properly for the observed experimental data on the metal atom-metallic surface systems, it may be of interest to briefly consider results on metallic diatomic molecules. Diatomic molecules composed of different metals have been studied extensively in the last few years by several groups (226-228). Their dissociation energy $D(A-B)$ has

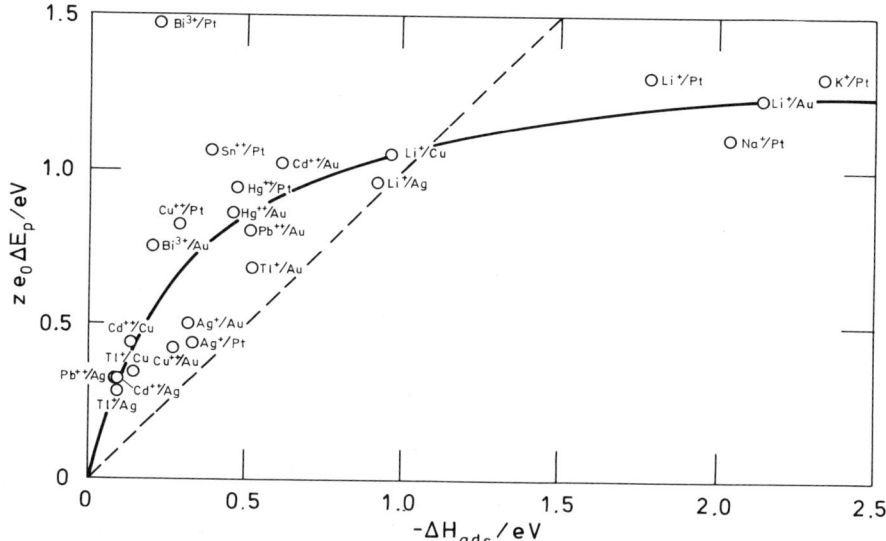

Fig. 39. Difference in heats of adsorption, as approximately given by $ze_o\Delta E_p$, plotted against the difference in heats of adsorption, $-\Delta H_{ads}$, calculated from Eq. 40 (78).

been determined by thermodynamic measurements and compared to those predicted by Eq. 39. Here again it turns out that the validity of Pauling's approach is rather limited. It was shown that for pairs with relatively large electronegativity differences such as AgLi, Eq. 39 obviously overestimates the ionic contribution (228). For pairs containing transition-metal atoms the Brewer-Engel theory, which takes multibond formation by d-electrons into account, has to be used instead of Eq. 39 (229). Therefore, it may not be surprising that Pauling's model does not describe properly the results from underpotential deposition. Schultze and Koppitz (84), however, have shown that the partial charge of adsorbates (see Section 2.5) can be related to the difference in

Pauling's electronegativities, $\Delta\chi$, of substrate and adsorbate, when assumptions on the geometric factor g are made. They were able to establish a correlation between electrosorption valency γ and $\Delta\chi$, in which the total range in charge transfer (from metallic to ionic bonds) is covered.

6.2. Correlation between Underpotential Shifts and Work-Function Differences

It was first pointed out by Kolb, Przasnyski, et al. (19) that any chemisorption model explaining the underpotential deposition should be based on a sound physical relation between the experimentally determined energy gain, $e_o \Delta E_p$, and parameters pertinent to the system. It is generally accepted that the adatom-substrate bond will gain energy with increasing ionicity of the bond due to a change in electron distribution. The partial charge that is transferred from the adatom to the metal is proportional to the difference in electronegativities of both entities. However, it was demonstrated first by Gordy and Thomas (230) and more recently by Trasatti (231,232) that there is a linear correlation between Pauling's electronegativity χ_M of a metal atom and the work function ϕ of the respective bulk metal. The empirically found correlation is (231):

$$\chi_M = 0.5 \phi - \text{const} \qquad (41)$$

with const = 0.29 for sp-metals and 0.55 for transition metals. Thus it had been suggested to replace for the metal Pauling's value χ_M (222) by that obtained from Eq. 41 when calculating the heats of adsorption for hydrogen on metals (233).

Kolb, Przasnyski, et al. (19,96) have shown that $e_o \Delta E_p$, the "excess" binding energy, depends linearly on $\Delta\phi$, the difference in work functions of substrate and adsorbate material, over a wide range in $\Delta\phi$ (Fig. 40):

$$e_o \Delta E_p = 0.5 \Delta\phi \qquad (42)$$

This correlation, which is obeyed so far by about 30 different metal pairs when working with polycrystalline substrates, shows astonishingly little scattering, which might be somewhat fortuitous when considering the ambiguities in the choice of work function and E_p values.

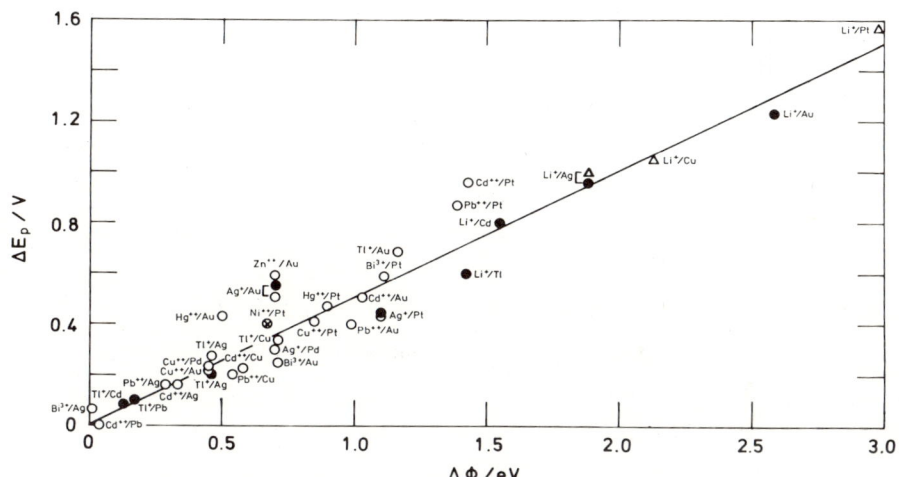

Fig. 40. Underpotential shift ΔE_p between monolayer and bulk stripping peak as a function of $\Delta\phi$, the difference in work functions of substrate and adsorbate material: (o) aqueous solution, (●) acetonitrile, (▲) propylene carbonate, (⊗) salt melt. The ΔE_p values are taken from Table I, the work function values from Trasatti (231), except for the alkali (234) and for Au (ϕ = 5.0).

Since $\Delta\phi \sim \Delta\chi$ represents the ionic contribution to the total adsorbate-substrate bond, the extremely good linear relation between $e_o \Delta E_p$ and $\Delta\phi$ suggests that the covalent part of the adatom-substrate bond obviously does not differ appreciably from the bond strength between the adatom and the surface of its own material and thus cancels. This should not be true in those cases where a strong covalent bond formation due to d-orbital overlap is possible. Indeed, Plummer and Rhodin have shown (128) that the adsorption energies of the period 6 transition metals on tungsten are much higher than those predicted by Eq. 42 and would not follow any trend with work function differences. In this case the relative binding energies can be correlated with the number of unpaired d-electrons in the adatoms. Hence Eq. 42 should be operative essentially only for sp-metals.

Although in many cases the work function of a complete metal monolayer is already that of the bulk material (82,235), it might sound strange to describe adatoms in the submonolayer region by the respective bulk work function, a property usually applicable only to crystals with broad energy bands. Two reasons, among several, may be given to justify this. First, the empirically found correlation between χ_M and ϕ (Eq. 41) shows that the Pauling electronegativity of a metal atom as derived from data of diatomic molecules can be expressed by the work function of that metal. Equation 42 also suggests that the electronegativity of a metal atom interacting with a metal surface is characterized best by its electronegativity in a metal lattice, namely, its work function. Second, a metal atom approaching a metallic surface has been shown to experience level broadening due to strong interaction with the continuum set of electron states in the substrate and level shifting (236-240). The latter can be explained in terms of a classical image potential and of an electron-correlation energy (see Fig. 41). Here, a metal atom far away from the metal surface at r = 0 is characterized by two sharp electronic niveaus at the ionization potential I_A and the electron affinity E_A. When the atom approaches the surface, I_A is shifted upward while E_A is pushed down by about the same amount, both being lifetime-broadened. This is most conveniently described by "effective" ionization potentials and electron affinities

$$I_{A,eff}(r) = I_A - \frac{e_o^2}{4r} \tag{43a}$$

and

$$E_{A,eff}(r) = E_A + \frac{e_o^2}{4r} \tag{43b}$$

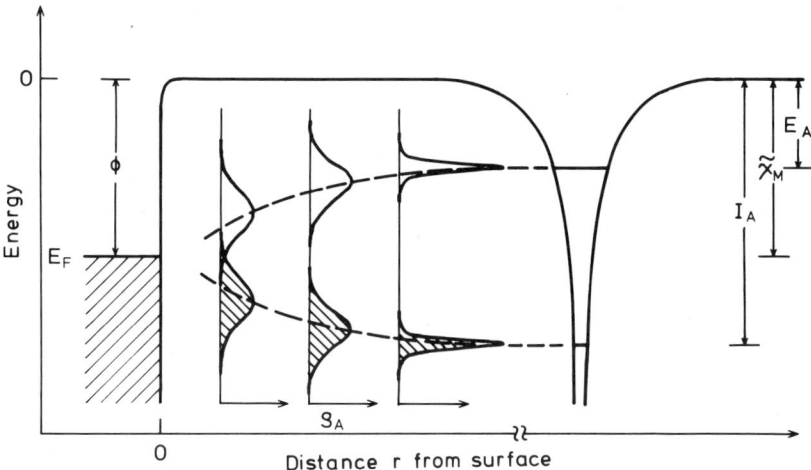

Fig. 41. Schematic representation of energy-level shifts and broadening for a metal atom approaching a surface of that metal. I_A, ionization potential; E_A, electron affinity; $\tilde{\chi}_M$, Mullikan's electronegativity; ρ_A, virtual level density of states.

For instance, removing an electron from an atom located at a distance r in front of a metal surface requires a lower energy, $I_{A,eff}(r)$, because of the energy gain due to the arising attraction between ion and image charge. For the very same reason the effective electron affinity, $E_{A,eff}(r)$, is higher than that of the free atom. To first order the shift for both levels is the same, neglecting differences in the spatial distribution of the electron wave functions that would yield different image potentials. Furthermore, the adsorbate level is shifted due to charge transfer. Figure 42 gives some typical values of level shift $\Delta\varepsilon$ and broadening Γ as a function of distance r for Na on a metal. For r around 2 Å, what might be the usual equilibrium distance for chemisorbed adatoms (other than hydrogen), the separation between $I_{A,eff}$ and $E_{A,eff}$ and the level widths Γ are such that sufficient overlap between both density-of-states functions will occur, creating a Fermi level E_F^{ad} in the adatom about half way

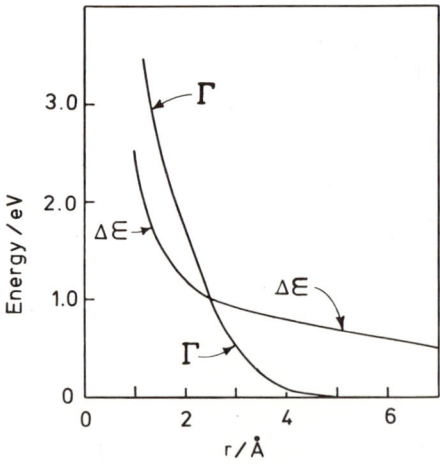

Fig. 42. Typical values of level shift $\Delta\varepsilon$ and broadening Γ for Na atom approaching a metal surface (82).

between I_A and E_A. The chemisorption behavior, such as the direction and the amount of a partial charge transfer, will now be governed mainly by $E_F^{ad} \approx -\frac{1}{2}(I_A + E_A)$, which corresponds to Mullikan's electronegativity $\tilde{\chi}_M = \frac{1}{2}(I_A + E_A)$. The ionization potential, I_A, of the free atom, sometimes used to characterize the adatom, is no longer of any significance. Again, one can show that $\tilde{\chi}_M$ (which is linearly correlated with Pauling's value χ_M) and the work function ϕ are about the same, which justifies also the use of ϕ of a bulk metal to describe a metal adatom.

A physical model, however, which can explain Eq. 42 quantitatively, has not yet been offered. Trasatti has demonstrated (241), using an energy cycle as given by Levine and Gyftopoulos (242) and Carroll and May (243), that for an adsorbed metal ion the difference in bond strengths for two different substrate materials is given by their work-function difference $\Delta\phi^+$. Trasatti can show by replotting the data from Kolb and colleagues for $\Theta \to 0$ that a correlation

$$e_o \Delta E_{\Theta \to 0} = \Delta\phi \tag{44}$$

may be established; however, the assumption of an adion is unrealistic for the pairs under consideration. It has been suggested by several authors that even for metal pairs with extremely large electronegativity differences such as the alkali metals on W and Ni the partial ionic charge at the adatom in only a fraction of unity charge, even at low coverages [e.g., ≈ 0.3 for K on W at $\Theta = 0.1$ (83)]. For most of the pairs shown in Fig. 40 the partial charge at the adatom will not exceed 0.1, which invalidates the above-proposed model (19). The $e_o \Delta E_p - \Delta\phi$ correlation might favor-

---

+ It should be noted that $e_o \Delta E_p$ may also be interpreted as the difference in binding energy between any "test" adatom and two different metal substrates (241).

ably sustain Lang's "functional density" approach derived for simple metals, which correlates the work function of a jellium with the electron density (244-248). Thus far insufficient data are available for heats of adsorption as calculated from this model to see any success or failure. However, it is remarkable that this is the only theory that derives bond energies from quantities such as work functions or Wigner-Seitz radii without taking the heats of sublimation into account, which is in agreement with the information gained from the data in Fig. 40.

When trying to interpret the experimental data for single-crystal surfaces along the lines of Eq. 42, certain difficulties are encountered. The adsorption isotherms have in general a rather complex form, strongly dependent on the substrate crystal face. Here the choice of an appropriate (i.e., characteristic and comparable) ΔE_p value seems more difficult. First attempts to correlate single-crystal ΔE_p values with $\Delta\phi$ gave a rather poor result, when $\Delta\phi$ was taken as difference between the work function of the substrate single-crystal face and that of the polycrystalline adsorbate metal (93,97). This, however, is not necessarily a meaningful approach. It should not be overlooked that the work function ϕ was chosen as a more realistic measure of the metal adatom's electronegativity, an <u>atomistic</u> quantity to describe the properties of a single adatom-substrate complex (19). These metal electronegativities of course cannot and should not reflect symmetry properties and band structures of solids as the ϕ-values for single-crystal surfaces do. Adatom-adatom attraction as observed on (111)-faces already at low coverages is also not accounted for in this simple $\Delta\phi$-$\Delta\chi$ picture. Furthermore, in an atomistic picture of the substrate-adatom complex as a surface molecule the number of bonds between adatom and substrate atoms and thus the total bond energy depends on the crystal plane (see Section 6.3) (249,

250). For example, despite the fact that the work function of a (111)-surface is highest, the bond energy for an adatom on this surface should be considerably lower than on a (110)- or (100)-surface, because interaction with only three substrate atoms is possible on the closest packed (111)-face, whereas four substrate atoms are involved in forming the surface molecule for (110)- and (100)-surfaces (see Fig. 43).

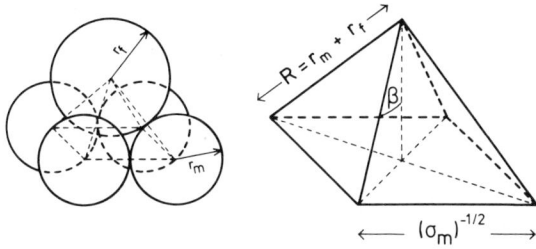

Fig. 43. Schematic representation of an adatom on a (100)- or (110)-surface in position of maximum ligancy; σ_m, number of substrate surface atoms per cm^2; R is in direction of single-link dipole μ_{mf} (249).

Schultze and Dickertmann (97) have reported that Eq. 42 seems to hold fairly well for various metals on a (110)-Au surface, when evaluating $\Delta\phi$ from data for the polycrystalline adsorbate material. Since it is believed that the (110)-face contributes most to a polycrystalline surface, this is understood in the light of the above-mentioned suggestion.

A close inspection of Eq. 42 raises the question as to whether $\Delta\phi$ has to be considered as an absolute value of the difference $\phi_o - \phi_1$. All pairs reported thus far to show underpotential deposition have one common factor: the work function of the substrate, ϕ_o, is always larger than that of the adsorbate metal, ϕ_1. Thus

the dipole of the adatom-substrate bond is positive at the adatom
because of an electron transfer from the adatom to the substrate.
This seems to be a necessary condition for monolayer formation.
Copper on Ag (19), Au on Ag, and Au on Cu (75) are the only pairs
investigated where this dipole is reversed, and these pairs do not
show any underpotential effect[+], despite appreciable $|\Delta\phi|$ values.
This means that Eq. 42 is only valid for $\phi_o > \phi_1$. Consequently,
Ag on Cu is expected to form a monolayer ($\Delta\phi > 0$), although underpotential deposition of Ag on Cu cannot be measured directly for
obvious electrochemical reasons. A possible explanation for this
$\Delta\phi$-dependence is discussed in Section 6.3.

6.3. The Modified Levine and Gyftopoulos Model

Levine and Gyftopoulos have derived quantitative expressions
for work-function changes (249,250) and adsorption energies (242)
for metal adatoms on metallic surfaces on the basis of a chemical
concept for surface molecules. The authors assume that the effective work function $\phi(\Theta)$ of the composited surface consists of two
superimposed barriers, the electronegativity barrier $e(\Theta)$ and the
dipole barrier $d(\Theta)$. The first one changes continuously from ϕ_o
at $\Theta = 0$ to ϕ_1 at $\Theta = 1$, where ϕ_o and ϕ_1 are, respectively, the
substrate metal and the film metal work functions:

$$e(\Theta) = \phi_1 + \Delta\phi \cdot G(\Theta) \tag{45}$$

[+] It has been shown, however, that a monolayer of gold can be
formed on Ag and Cu electrodes (75), when the gold is deposited
onto a surface, which is already covered with a monolayer of Pb.
The lead monolayer will be displaced by Au and readsorbed on top
of it, obviously stabilizing the Au monolayer.

($\Delta\phi = \phi_o-\phi_1$), choosing somewhat arbitrarily $G(\Theta) = 1 - 3\Theta^2 + 2\Theta^3$. The composited surface is believed to be homogenized, consisting of atoms of different electronegativity hence changing the work function with coverage, since electronegativity and work function of a constituent are linearly correlated (see Eq. 41). The dipole barrier $d(\Theta)$ is considered as a direct consequence of the different electronegativities. According to the findings of Pauling (222) and Malone (251), molecules of dissimilar atoms have a dipole moment that is proportional to the electronegativity difference $\Delta\chi$. The adatom-substrate complex is now treated as a surface molecule, the net dipole moment (perpendicular to the surface) of which can be calculated from molecular or atomic data. Assuming an arrangement of adsorbed atom and substrate atoms as shown in Fig. 43 for a (100) or (110) plane allows for evaluation of the surface molecule dipole moment μ_o from the single-link dipoles μ_{mf} by simple geometrical considerations (see Fig. 43):

$$\mu_o = 4 \mu_{mf} \cos \beta = 4 \mu_{mf} \left[1 - \frac{1}{2\sigma_m R^2} \right]^{1/2} \quad (46)$$

with

$$\mu_{mf} = \frac{\tilde{k} \cdot \Delta\chi}{(1 + \alpha/R^3)} = \frac{k \cdot \Delta\phi}{(1 + \alpha/R^3)} \quad (46a)$$

where the proportionality factor $\tilde{k}$ is taken from Pauling (222),[+] and the denominator accounts for the self-depolarization of the dipole, which increases with increasing polarizability α of the adatom-substrate bond (252). The resulting dipole barrier is then

---

[+] The k-value differs somewhat from that due to Gyftopoulos and Levine (249) since we have chosen the proportionality factor between ϕ and χ as 2.0 (231) instead of 2.27 (230).

given by:

$$d(\Theta) = -\frac{4\pi \mu_o \sigma_f \Theta G(\Theta)}{(1 + 9\alpha \sigma_f^{3/2} \Theta^{3/2}/\varepsilon)} \qquad (47)$$

where the denominator accounts for the depolarization due to dipole-dipole interaction assuming a uniform distribution of the dipoles in a square array (253), σ_f is the number of adsorbed atoms per cm^2 in a monolayer, and ε is the dielectric constant effective for the depolarizing field between the dipoles. The final expression is given by (242,249):

$$\phi(\Theta) = \phi_1 + \Delta\phi \cdot G(\Theta)\left[1 - \frac{16\pi k \sigma_f \Theta \cos\beta}{(1 + \alpha/R^3)(1 + 9\alpha \sigma_f^{3/2} \Theta^{3/2}/\varepsilon)}\right] \qquad (48)$$

with $16\pi k = 0.86 \times 10^{-14}$ cm^2. The authors were able to reproduce experimentally measured work-function changes as a function of coverage astonishingly well with this equation (249,250), despite the rather parametrical approach (235). This aspect is discussed in greater detail by Kolb (254).

Levine and Gyftopoulos (LG) extended their concept of a chemical bond in the surface molecule to calculate the adsorption energy, assuming that the energy arises from a partly ionic and partly covalent bond (242):

$$-H_{ads} = H_{ii} + H_{cc} \qquad (49)$$

While the covalent part, H_{cc}, of the adsorption energy is essentially given by the geometrical means of the heats of sublimation of the constituents in analogy to Pauling's expression for diatomic molecules with some correction factors for angular efficiency of the overlapping orbitals and charge efficiency, the ionic part H_{ii} is estimated from a cyclic process as follows. It is assumed that H_{ii} arises from a partial charge, δe_o, which is transferred from the adatom to the substrate. To remove the adatom from the surface to infinity, three energy contributions are involved: (a) removing the partially charged adatom to infinity overcoming the image potential ($\delta^2 e_o^2/2R$), (b) removing the charge δe_o from the substrate metal, $\delta\phi(\Theta)$, and (c) combining it with the atom, thus gaining the energy $\delta^2 I_A$, where I_A is the ionization potential of the adatom. Hence

$$H_{ii} = \delta \cdot \phi(\Theta) + \frac{\delta^2 e_o^2}{2R} - \delta^2 I_A = \delta \cdot \phi(\Theta)\left[1 + \eta\right] \tag{50a}$$

To a good approximation we find

$$H_{ii} \approx \delta \cdot \phi(\Theta), \text{ since } \eta = \frac{\delta}{\phi(\Theta)}\left[\frac{e_o^2}{2R} - I_A\right] \ll 1 \tag{50b}$$

The fractional charge δ is defined by the ratio of the actual dipole $\mu(\Theta)$ and $e_o R \cos\beta$, the dipole due to a complete electron transfer.

$$\delta = \frac{\mu(\Theta)}{e_o R \cos\beta} = \frac{0.48 \, \Delta\phi \, G(\Theta)}{R(1 + \alpha/R^3)(1 + 9\alpha \sigma_f^{3/2} \Theta^{3/2}/\varepsilon)} \tag{51}$$

for R given in Å and $\Delta\phi$ in eV.

When applying the procedure described in the preceding paragraph for calculating the underpotential shift ΔE_p,

$$ze_o \Delta E_p \approx -\Delta H_{ads}(\Theta_p) = H_{ii} + H_{cc} - D_f \tag{52}$$

(D_f: heat of sublimation of the film material), the calculated adsorption energies are usually far too large due to the large contribution of H_{cc}. Here again it appears that the covalent bond obviously does not contribute significantly to the underpotential deposition, which means the covalent bond has to be about the same for the underpotentially adsorbed adatom on the foreign substrate and on a like surface. However, we may consider to a first approximation (see Eq. 42)

$$e_o \Delta E_p \approx H_{ii}(\Theta_p) \quad (52a)$$

Since we have seen that the peak potential corresponds to a rather low coverage (typically between 0.1 and 0.2; see Section 2.6), the H_{ii} values have been calculated according to the LG model for $\Theta_p = 0.1$, roughly corresponding to an ordered structure with p(2 x 2) for the (most anodic) monolayer peak. The result, shown in Fig. 44, appears to be extremely favorable, which might be somewhat for-

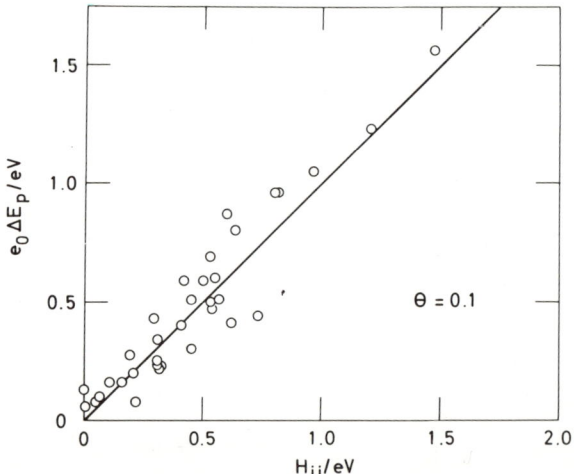

Fig. 44. Underpotential shift, $e_o \Delta E_p$, plotted against H_{ii}, the ionic contribution to the adatom-substrate bond, as calculated from Eq. 50b for $\Theta_p = 0.1$; ΔE_p values from Table I.

tuitous when considering that only one Θ_p value is chosen for all systems. Choosing higher Θ_p values reduces H_{ii}, hence shifting the points generally to the left of the theoretically predicted line of unity slope. This would imply that some small contributions from covalent bonding should then be taken into account.

From the LG model two interesting consequences can be derived. First, when considering Eqs. 50a and 50b, it becomes evident that the main part contributing to the energy gain on adsorption is $\delta\phi(\Theta)$, when bringing the fractional charge into the substrate. Choosing now a deposit with a work function $\phi_1 > \phi_o$ (hence $\chi_1 > \chi_o$), the reversed dipole will result, leaving the adatom with a negative partial charge. This, however, will not lead to a monolayer formation, because H_{ii} is then usually negative (that is no energy gain).

$$H_{ii} = -\delta \cdot \phi(\Theta) + \frac{\delta^2 e_o^2}{2R} + \delta^2 I_A \qquad (53)$$

Only systems with a very low $\phi(\Theta)$ and a rather high I_A value (e.g., Hg on Na at low coverages), which can usually not be rationalized in electrochemical systems, may exhibit underpotential deposition, that is, an energy gain on adsorption. This would explain the finding that Cu on Ag, Au on Ag, and Au on Cu, thus far the only systems studied that fulfill the condition $\phi_1 > \phi_o$, do not show underpotential deposition. Second, the LG model allows to take differences in substrate-surface structures into account, which is important when dealing with single-crystal surfaces. For example, calculating H_{ii} ($\Theta = 0.1$) for Tl on Ag(110), (100), and (111) faces one finds 0.19 eV for (110) and (100) (as in the case of polycrystalline Ag), but 0.15 eV for Tl on Ag(111) [assuming the work function $\phi(\Theta)$ to be the same for all planes]. The latter value arises because only three surface atoms can form bonds with the adatom, whereas at (110) and (100) the adatom has four substrate atoms to interact with. It is worthwhile to emphasize that the bond energy is deduced from molecular or atomic data of the surface molecule, such as electronegativities (expressed in terms of work functions), and atomic radii. The specificy of different substrate planes is again reflected mainly by atomic data (number of surface bonds, geometry), and not so much by the difference in work function of the single-crystal planes themselves (see, e.g., respective data in Table 1 of ref. 97).

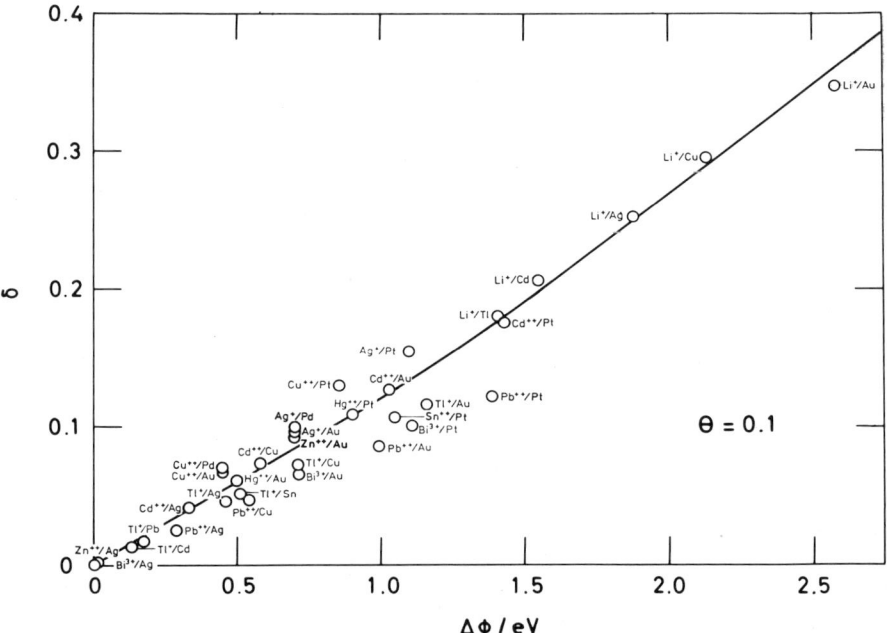

Fig. 45. Fractional charge δ at the adatom for $\Theta = 0.1$ as given by Eq. 51 as a function of $\Delta\phi$, the difference in work functions between substrate and adatom metal.

In Fig. 45 the partial charge δ of the adatoms at $\Theta = 0.1$ as calculated from Eq. 51 is given in fractions of one electronic charge for the studied systems, which are characterized by the respective $\Delta\phi$ values. This is quite instructive since it gives an idea of the ionicity of the adatom-substrate bond at low coverages. The result supports the well-known fact that even for such extreme couples as Li on Pt and at low coverages the adatom is still far from being adsorbed as an ion. This is quite important since for small δ-values ($\delta \ll 1$) any coulombic type of energy, such as image potential or ionisation potential, will not significantly contribute to the bonding due to the dependence on δ^2, while noncoulombic interaction will be predominant. For ionic adsorption ($\delta \approx 1$) the reverse is true.

Finally, we may calculate the difference in adsorption energy at $\Theta = 0$ and 1, since this quantity governs the half width of the monolayer adsorption isotherm. From Eqs. 48, 50, and 51 we find

$$\Delta H_{ii} = H_{ii}(\Theta = 0) - H_{ii}(\Theta = 1) = \frac{0.48 \cdot \Delta\phi \cdot \phi_o}{R(1 + \alpha/R^3)} \tag{54}$$

Introducing the relation $\Delta E_p = 0.5 \Delta\phi$ (Eq. 42) into Eq. 54 yields

$$\Delta H_{ii} \approx \frac{\phi_o}{R(1 + \alpha/R^3)} \cdot \Delta E_p \tag{55}$$

Since the factor $\phi_o/R(1 + \alpha/R^3)$ may not be so different for the metal pairs under consideration, a rough linearity between ΔH_{ii} and ΔE_p is seen, as suggested by the experimental data on half widths of the monolayer isotherms (19).

7. REFERENCES

1. G. V. Hevesy, Physik. Z., 13, 715 (1912).
2. K. F. Herzfeld, Physik. Z., 14, 29 (1913).
3. L. B. Rogers, D. P. Krause, J. C. Griess, Jr., and D. B. Ehrlinger, J. Electrochem. Soc., 95, 33 (1949).
4. R. C. De Geiso and L. B. Rogers, J. Electrochem. Soc., 106, 433 (1959).
5. J. T. Byrne, L. B. Rogers, and J. C. Griess, Jr., J. Electrochem. Soc., 98, 452 (1951).
6. L. B. Rogers and A. F. Stehney, J. Electrochem. Soc., 95, 25 (1949).
7. M. Haissinsky, J. Chim. Phys., 43, 21 (1946).
8. A. Coche and M. Haissinsky, C. R. Acad. Sci., 222, 1284 (1946).
9. M. Haissinsky, Experientia, 8, 125 (1952).
10. M. M. Nicholson, J. Am. Chem. Soc., 79, 7 (1957).
11. M. M. Nicholson, Anal. Chem., 32, 1058 (1960).
12. S. S. Lord, Jr., R. C. O'Neill, and L. B. Rogers, Anal. Chem., 24, 209 (1952).
13. T. Mills and G. M. Willis, J. Electrochem. Soc., 100, 452 (1953).
14. G. I. Finch, H. Wilman, and L. Yang, Disc. Faraday Soc., 1, 144 (1947).
15. W. J. Lorenz, H. D. Hermann, N. Wüthrich, and F. Hilbert, J. Electrochem. Soc., 121, 1167 (1974).
16. B. J. Bowles, Electrochim. Acta, 15, 589 (1970).
17. B. J. Bowles, Electrochim. Acta, 10, 717, 731 (1965).
18. B. J. Bowles, Electrochim. Acta, 15, 737 (1970).
19. D. M. Kolb, M. Przasnyski, and H. Gerischer, J. Electroanal. Chem., 54, 25 (1974).

20. W. J. Lorenz, J. Moumtzis, and E. Schmidt, J. Electroanal. Chem., 33, 121 (1971).
21. W. J. Lorenz, Chem.-Ing.-Techn., 45, 175 (1973).
22. G. W. Tindall and S. Bruckenstein, J. Electroanal. Chem., 22, 367 (1969).
23. L. B. Anderson and C. N. Reilley, J. Electroanal. Chem., 10, 295 (1965).
24. L. B. Anderson and C. N. Reilley, J. Electroanal. Chem., 10, 538 (1965).
25. J. W. Schultze, Ber. Bunsenges. Phys. Chem., 74, 705 (1970).
26. H. D. Hermann, N. Wüthrich, W. J. Lorenz, and E. Schmidt, J. Electroanal. Chem., 68, 273, 289 (1976).
27. G. W. Tindall and S. Bruckenstein, Electrochim. Acta, 16, 245 (1971).
28. S. H. Cadle and S. Bruckenstein, Anal. Chem., 43, 1858 (1971).
29. D. P. Sandoz, R. M. Peekema, H. Freund, and Ch. F. Morrison, Jr., J. Electroanal. Chem., 24, 165 (1970).
30. A. R. Nisbet and A. J. Bard, J. Electroanal. Chem., 6, 332 (1963).
31. S. Stucki, J. Electroanal. Chem., 80, 375 (1977).
32. G. W. Tindall and S. Bruckenstein, Anal. Chem., 40, 1051 (1968).
33. S. H. Cadle and S. Bruckenstein, Anal. Chem., 43, 932 (1971).
34. M. W. Breiter, J. Electrochem. Soc., 114, 1125 (1967).
35. M. W. Breiter, Trans. Faraday Soc., 65, 2197 (1969).
36. G. W. Tindall and S. Bruckenstein, Anal. Chem., 40, 1637 (1968).
37. N. Furuya and S. Motoo, J. Electroanal. Chem., 72, 165 (1976).
38. M. Z. Hassan, D. F. Untereker, and S. Bruckenstein, J. Electroanal. Chem., 42, 161 (1973).

39. S. Stucki, private communication.
40. E. Schmidt and N. Wüthrich, J. Electroanal. Chem., 40, 399 (1972).
41. F. Mikuni and T. Takamura, Denki Kagaku, 38, 113 (1970).
42. S. H. Cadle and S. Bruckenstein, Anal. Chem., 44, 1993 (1972).
43. F. Mikuni and T. Takamura, Denki Kagaku, 39, 579 (1971).
44. S. Motoo, private communication.
45. A. N. Frumkin, G. N. Mansurov, V. E. Kazarinov, and N. A. Balashova, Collect. Czech, Chem. Commun., 31, 806 (1966).
46. B. J. Bowles and T. E. Cranshaw, Phys. Lett., 17, 258 (1965).
47. D. M. Kolb, unpublished results.
48. F. Mikuni and T. Takamura, Denki Kagaku, 37, 852 (1969).
49. F. Mikuni and T. Takamura, Denki Kagaku, 39, 237 (1971).
50. E. Schmidt and S. Stucki, J. Electroanal. Chem., 39, 63 (1972).
51. E. Schmidt, P. Beutler, and W. J. Lorenz, Ber. Bunsenges. Phys. Chem., 75, 71 (1971).
52. E. Schmidt and N. Wüthrich, J. Electroanal. Chem., 34, 377 (1972).
53. E. Schmidt and H. R. Gygax, J. Electroanal. Chem., 14, 126 (1967).
54. E. Schmidt and H. R. Gygax, J. Electroanal. Chem., 13, 378 (1967).
55. V. A. Vicente and S. Bruckenstein, Anal. Chem., 45, 2036 (1973).
56. J. W. Schultze, F. D. Koppitz, and M. M. Lohrengel, Ber. Bunsenges. Phys. Chem., 78, 693 (1974).
57. S. H. Cadle and S. Bruckenstein, J. Electrochem. Soc., 119, 1166 (1972).
58. E. Schmidt, H. R. Gygax, and Y. Cramer, Helv. Chim. Acta, 53, 649 (1970).

59. E. Schmidt and H. R. Gygax, J. Electroanal. Chem., 12, 300 (1966).
60. V. A. Vicente and S. Bruckenstein, Anal. Chem., 44, 297 (1972).
61. E. Schmidt and N. Wüthrich, J. Electroanal. Chem., 28, 349 (1970).
62. E. Schmidt and H. R. Gygax, Helv. Chim. Acta, 48, 1178 (1965).
63. D. J. Astley, J. A. Harrison, and H. R. Thirsk, J. Electroanal. Chem., 19, 325 (1968).
64. E. Schmidt and H. R. Gygax, Helv. Chim. Acta, 48, 1584 (1965).
65. E. Schmidt, M. Christen, and P. Beyeler, J. Electroanal. Chem., 42, 275 (1973).
66. E. Schmidt, J. Electroanal. Chem., 47, 441 (1973).
67. B. J. Bowles, Nature, 212, 1456 (1966).
68. M. Breiter, Electrochim. Acta, 7, 25 (1962).
69. F. G. Will, J. Electrochem. Soc. 112, 451 (1965).
70. S. P. Perone, Anal. Chem., 35, 2091 (1963).
71. K. Z. Brainina, N. F. Zakharchuk, D. P. Synkova, and I.G. Yudelevich, J. Electroanal. Chem., 35, 165 (1972).
72. I. Morcos, J. Electroanal. Chem., 66, 250 (1975).
73. B. H. Vassos and H. B. Mark, Jr., J. Electroanal. Chem., 13, 1 (1967).
74. M. Przasnyski and D. M. Kolb, unpublished results.
75. S. Stucki, J. Electroanal. Chem., 78, 31 (1977).
76. R. D. Gretz and G. M. Pound, Appl. Phys. Lett., 11, 67 (1967).
77. E. Schmidt and H. R. Gygax, Helv. Chim. Acta, 49, 1105 (1966).
78. D. M. Kolb and H. Gerischer, Surface Sci., 51, 323 (1975).
79. J. O. Besenhard and H. P. Fritz, Electrochim. Acta, 20, 513 (1975).
80. I. Fried and H. Barak, J. Electroanal. Chem., 30, 279 (1971).

81. G. J. Hills, D. J. Schiffrin, and J. Thompson, J. Electrochem. Soc., 120, 157 (1973).
82. R. L. Gerlach and T. N. Rhodin, Surface Sci., 19, 403 (1970).
83. L. Schmidt and R. Gomer, J. Chem. Phys., 42, 3573 (1965).
84. J. W. Schultze and F. D. Koppitz, Electrochim. Acta, 21, 327 (1976).
85. F. D. Koppitz and J. W. Schultze, Electrochim. Acta, 21, 337 (1976).
86. J. W. Schultze, in Proc. 2nd Internatl. Summer School on Quantum Mechanical Aspects of Electrochemistry, Ohrid, Yugoslavia, 1972.
87. K. J. Vetter and J. W. Schultze, Ber. Bunsenges. Phys. Chem., 76, 920, 927 (1972).
88. J. W. Schultze and K. J. Vetter, J. Electroanal. Chem., 44, 63 (1973).
89. G. Horanyi, J. Electroanal. Chem., 55, 45 (1974).
90. E. Schmidt, H. R. Gygax, and P. Böhlen, Helv. Chim. Acta, 49, 733 (1966).
91. A. N. Frumkin, Z. Physik. Chem., 116, 466 (1925).
92. B. E. Conway and E. Gileadi, Trans. Faraday Soc., 58, 2493 (1962).
93. A. Bewick and B. Thomas, J. Electroanal. Chem., 65, 911 (1975).
94. A. Bewick and B. Thomas, J. Electroanal. Chem. (in press, 1977).
95. K. Jüttner, G. Staikov, W. J. Lorenz, and E. Schmidt, J. Electroanal. Chem., 80, 67 (1977).
96. H. Gerischer, D. M. Kolb, and M. Przasnyski, Surface Sci., 43, 662 (1974).
97. J. W. Schultze and D. Dickertmann, Surface Sci., 54, 489 (1976).

98. D. Dickertmann, F. D. Koppitz, and J. W. Schultze, Electrochim. Acta, 21, 967 (1976).
99. A. Bewick and B. Thomas, J. Electroanal. Chem., 70, 239 (1976).
100. F. Hilbert, C. Mayer, and W. J. Lorenz, J. Electroanal. Chem., 47, 167 (1973).
101. D. M. Kolb, M. Przasnyski, and H. Gerischer, Elektrokhimiya, 13, 700 (1977).
102. S. Stucki, Ph.D. thesis, U. Bern, Switzerland, 1973.
103. G. Horányi, J. Solt, and G. Vértes, J. Electroanal. Chem., 32, 271 (1971).
104. G. Horányi and G. Vértes, J. Electroanal. Chem., 45, 295 (1973).
105. J. W. Schultze and K. J. Vetter, Electrochim. Acta, 19, 913 (1974).
106. E. Schmidt and N. Wüthrich, Helv. Chim. Acta, 50, 2058 (1967).
107. D. O. Raleigh, Nato Summer School, Corsica, 1975.
108. M. M. Nicholson, J. Electrochem. Soc., 121, 734 (1974).
109. J. O. Besenhard, Ph.D. thesis, Technical U. Munich, 1973.
110. S. D. James, J. Electrochem. Soc., 122, 921 (1975).
111. A. Ölander, J. Am. Chem. Soc., 54, 3819 (1932).
112. M. J. Nicol and H. I. Philip, J. Electroanal. Chem., 70, 233 (1976).
113. K. S. Kim, N. Winograd, and R. E. Davis, J. Am. Chem. Soc., 93, 6296 (1971).
114. J. S. Hammond and N. Winograd, J. Electroanal. Chem., 78, 55 (1977).
115. J. S. Hammond and N. Winograd, J. Electroanal. Chem., 80, 123 (1977).
116. J. S. Hammond and N. Winograd, J. Electrochem. Soc., 124, 826 (1977).

117. E. Bauer and H. Poppa, Thin Solid Films, 12, 167 (1972).
118. E. Bauer, H. Poppa, and G. Todd, Thin Solid Films, 28, 19 (1975).
119. E. Bauer, H. Poppa, G. Todd, and F. Bonczek, J. Appl. Phys., 45, 5164 (1974).
120. J. P. Biberian and G. E. Rhead, J. Phys. F, 3, 675 (1973).
121. J. Perdereau, J. P. Biberian, and G. E. Rhead, J. Phys. F, 4, 798 (1974).
122. J. B. Hudson and Ch. Ming Lo, Surface Sci., 36, 141 (1973).
123. J. P. Jones, Proc. Royal Soc. (London), 284, 469 (1965).
124. J. P. Jones, Nature, 211, 479 (1966).
125. O. Nishikawa and T. Utsumi, J. Appl. Phys., 44, 945, 955 (1973).
126. L. D. Schmidt and R. Gomer, J. Chem. Phys., 45, 1605 (1966).
127. C. J. Todd and T. N. Rhodin, Surface Sci., 42, 109 (1974).
128. E. W. Plummer and T. N. Rhodin, J. Chem. Phys., 49, 3479 (1968).
129. J. M. Bermond, B. Felts, and M. Drechsler, Surface Sci., 49, 207 (1975).
130. P. W. Palmberg and T. N. Rhodin, J. Chem. Phys., 49, 134 (1968).
131. J. Polanski and Z. Sidorski, Surface Sci., 40, 282 (1973).
132. J. Henrion and G. E. Rhead, Surface Sci., 29, 20 (1972).
133. J. Perdereau and I. Szymerska, Surface Sci., 32, 247 (1972).
134. F. Delamare and G. E. Rhead, Surface Sci., 35, 172, 185 (1973).
135. W. F. Egelhoff, Jr., D. L. Perry, and J. W. Linnett, Surface Sci., 54, 670 (1976).
136. H. D. Hagstrum, Surface Sci., 54, 197 (1976).
137. D. R. Lloyd, C. M. Quinn, and N. V. Richardson, J. Chem. Soc. Faraday II, 72, 1036 (1976).

138. J. P. Jones and N. T. Jones, Thin Solid Films, 35, 83 (1976).
139. A. Cetronio and J. P. Jones, Thin Solid Films, 35, 113 (1976).
140. C. M. Lo and J. B. Hudson, Thin Solid Films, 12, 261 (1972).
141. E. Bauer, private communication.
142. G. Ertl and J. Küppers, in Low Energy Electrons and Surface Chemistry, Verl. Chemie, D 694 Weinheim, 1974, p. 7.
143. D. E. Eastman and W. D. Grobman, Phys. Rev. Lett., 30, 177 (1973).
144. H. Neddermeyer, Habilitationsschrift, U. Munich, 1976.
145. J. D. E. McIntyre, in Advances in Electrochemistry and Electrochemical Engineering, Vol. 9, R. H. Müller, ed., Wiley-Interscience, New York, 1973, p. 61.
146. J. D. E. McIntyre, Surface Sci., 37, 658 (1973).
147. J. D. E. McIntyre and D. M. Kolb, Symp. Faraday Soc., 4, 99 (1970).
148. A. Bewick and A. M. Tuxford, Symp. Faraday Soc., 4, 114 (1970).
149. W. J. Plieth, Symp. Faraday Soc., 4, 137 (1970).
150. T. Takamura, K. Takamura, W. Nippe, and E. Yeager, J. Electrochem. Soc., 117, 626 (1970).
151. B. D. Cahan, J. Horkans, and E. Yeager, J. Electrochem. Soc., 118, 1322 (1971).
152. M. J. Dignam and M. Moskovits, J. C. S., Faraday II, 69, 65 (1973).
153. J. D. E. McIntyre and W. F. Peck, Jr., Disc. Faraday Soc., 56, 122 (1973).
154. D. M. Kolb and H. Gerischer, Electrochim. Acta, 18, 987 (1973).
155. S. Gottesfeld and B. E. Conway, J. C. S., Faraday I, 70, 1793 (1974).

156. R. Adžić, E. Yeager, and B. D. Cahan, J. Electrochem. Soc., 121, 474 (1974).
157. J. D. E. McIntyre, in Optical Properties of Solids--New Developments, B. O. Seraphin, ed., North-Holland, Amsterdam, 1976, p. 555.
158. D. M. Kolb and R. Kötz, Surface Sci., 64, 698 (1977).
159. J. Anderson, G. W. Rubloff and P. J. Stiles, Solid State Commun., 12, 825 (1973).
160. J. Anderson, G. W. Rubloff, and P. J. Stiles, Surface Sci., 37, 75 (1973).
161. G. W. Rubloff, J. Anderson, M. A. Passler, and P. J. Stiles, Phys. Rev. Lett., 32, 667 (1974).
162. J. D. E. McIntyre and D. E. Aspnes, Surface Sci., 24, 417 (1971).
163. T. Takamura, K. Takamura, and F. Watanabe, Surface Sci., 44, 93 (1974).
164. D. M. Kolb and J. D. E. McIntyre, Surface Sci., 28, 321 (1971).
165. W. J. Plieth and K. Naegele, Surface Sci., 50, 53 (1975).
166. K. Naegele and W. J. Plieth, Surface Sci., 50, 64 (1975).
167. W.-K. Paik and J. O'M. Bockris, Surface Sci., 28, 61 (1971).
168. J. Horkans, B. D. Cahan, and E. Yeager, Surface Sci., 46, 1 (1974).
169. B. D. Cahan, J. Horkans, and E. Yeager, Surface Sci., 37, 559 (1973).
170. M. J. Dignam, M. Moskovits and R. W. Stobie, Trans. Faraday Soc., 67, 3306 (1971).
171. M. J. Dignam and M. Moskovits, J. C. S., Faraday II, 69, 56 (1973).
172. C. S. Strachan, Proc. Cambridge Phil. Soc., 29, 116 (1933).
173. D. V. Sivukhin, Soviet Phys.-JETP, 3, 269 (1956).

174. J. R. MacDonald and C. A. Barlow, Jr., J. Chem. Phys., 44, 202 (1966).
175. J. R. MacDonald and C. A. Barlow, Jr., Surface Sci., 4, 381 (1966).
176. B. Caroli, Phys. Kondens. Mater., 1, 346 (1963).
177. B. Kjollerstrom, Phil. Mag., 19, 1207 (1969).
178. A. J. Bennett and D. Penn, Phys. Rev. B, 11, 3644 (1975).
179. G. A. Bootsma and F. Meyer, Surface Sci., 14, 52 (1969).
180. T. Takamura, F. Watanabe, and K. Takamura, Electrochim. Acta, 19, 933 (1974).
181. T. Takamura and Y. Sato, J. Electroanal. Chem., 47, 245 (1973).
182. T. Takamura, Y. Sato, and K. Takamura, J. Electroanal. Chem., 41, 31 (1973).
183. T. Takamura and U. Moriyama, Denki Kagaku, 40, 300 (1972).
184. T. Takamura, Y. Sato, and U. Moriyama, in 5th Internatl. Congr. Metallic Corrosion (Japan), 1974, p. 168.
185. R. S. Sennett and G. D. Scott, J. Opt. Soc. Am. 40, 203 (1950).
186. P. L. Clegg, Proc. Phys. Soc. (London), 65B, 774 (1952).
187. S. Yoshida, T. Yamaguchi, and A. Kinbara, J. Opt. Soc. Am., 61, 62 (1971).
188. S. Yoshida, T. Yamaguchi, and A. Kinbara, J. Opt. Soc. Am., 61, 463 (1971).
189. O. S. Heavens, Optical Properties of Thin Solid Films, Dover, New York, 1965, p. 177.
190. F. Watanabe, K. Takamura, and T. Takamura, Denki Kagaku, 43, 469 (1975).
191. J. Horkans, B. D. Cahan, and E. Yeager, J. Electrochem. Soc., 122, 1585 (1975).
192. F. Meyer and G. A. Bootsma, Surface Sci., 16, 221 (1969).

193. D. M. Kolb, D. Leutloff, and M. Przasnyski, Surface Sci., $\underline{47}$, 622 (1975).
194. D. M. Kolb, Disc. Faraday Soc., $\underline{56}$, 138 (1973).
195. D. M. Kolb, D. Leutloff, and M. Przasnyski, paper presented at the Dechema Annual Meeting, Frankfurt/M., 1974.
196. W. J. Anderson and W. N. Hansen, J. Electroanal. Chem., $\underline{43}$, 329 (1973).
197. T. Takamura, K. Takamura, and E. Yeager, J. Electroanal. Chem., $\underline{29}$, 279 (1971).
198. W.-K. Paik and J. O'M. Bockris, Surface Sci., $\underline{27}$, 191 (1971).
199. T. Takamura and K. Takamura, J. Electroanal. Chem., $\underline{39}$, 478 (1972).
200. D. M. Kolb and R. Kötz, Surface Sci., $\underline{64}$, 96 (1977).
201. J. D. E. McIntyre, Symp. Faraday Soc., $\underline{4}$, 50, 55, 61 (1970).
202. W. J. Anderson and W. N. Hansen, J. Electroanal. Chem., $\underline{47}$, 229 (1973).
203. M. Fujihira and T. Kuwana, Electrochim. Acta, $\underline{20}$, 565 (1975).
204. A. M. Brodsky and Y. V. Pleskov, in Progress in Surface Science, Vol. 2, S. G. Davison, ed., Pergamon, New York, 1972, p. 1.
205. J. K. Sass and H. Gerischer, in Photoemission from Surfaces, B. Fitton et al., eds., Wiley, New York, 1977, Chapter 16.
206. J. K. Sass, H. Laucht, and S. Stucki, in Proc. Internatl. Symp. Photoemission, R. F. Willis et al., eds., ESA, Paris, 1976, p. 83.
207. M. Fleischmann, J. Koryta, and H. R. Thirsk, Trans. Faraday Soc., $\underline{63}$, 1261 (1967).
208. R. D. Giles, J. A. Harrison, and H. R. Thirsk, J. Electroanal. Chem., $\underline{20}$, 47 (1969).
209. H. Angerstein-Kozlowska, B. MacDougall, and B. E. Conway, J. Electrochem. Soc., $\underline{120}$, 756 (1973).

210. M. Watanabe and S. Motoo, in Proc. Japan-USSR Seminar on Electrochemistry, Tokyo, 1974, p. 239.
211. N. Furuya and S. Motoo, Denki Kagaku, $\underline{41}$, 307 (1973).
212. N. Furuya and S. Motoo, Denki Kagaku, $\underline{41}$, 364 (1973).
213. M. Watanabe and S. Motoo, J. Electroanal. Chem., $\underline{60}$, 259 (1975).
214. M. Watanabe and S. Motoo, J. Electroanal. Chem., $\underline{60}$, 267 (1975).
215. M. Watanabe and S. Motoo, J. Electroanal. Chem., $\underline{60}$, 275 (1975).
216. R. R. Adžić and A. R. Despić, J. Chem. Phys., $\underline{61}$, 3482 (1974).
217. R. R. Adžić and A. R. Despić, Z. Phys. Chem., N. F., $\underline{98}$, 95 (1975).
218. R. R. Adžić, D. N. Simić, D. M. Dražić, and A. R. Despić, J. Electroanal. Chem., $\underline{61}$, 117 (1975).
219. R. R. Adžić, D. N. Simić, A. R. Despić, and D. M. Dražić, J. Electroanal. Chem., $\underline{65}$, 587 (1975).
220. R. R. Adžić, D. N. Simić, A. R. Despić, and D. M. Dražić, J. Electroanal. Chem., $\underline{80}$, 81 (1977).
221. J. D. E. McIntyre and W. F. Peck, Jr., J. Electrochem. Soc., $\underline{123}$, 1800 (1976).
222. L. Pauling, The Nature of the Chemical Bond, 3rd ed., Cornell U. P., Ithaca, N.Y., 1960.
223. D. D. Eley, Disc. Faraday Soc., $\underline{8}$, 34 (1950).
224. A. K. Vijh, Surface Sci., $\underline{46}$, 282 (1974).
225. A. K. Vijh, Electrochemistry of Metals and Semiconductors, Dekker, New York, 1973, Chapter 1.
226. D. L. Cocke and K. A. Gingerich, J. Phys. Chem., $\underline{75}$, 3264 (1971).
227. V. Piacente and K. A. Gingerich, High Temp. Sci., $\underline{4}$, 312 (1972).

228. A. Neubert and K. F. Zmbov, J. C. S., Faraday I, $\underline{70}$, 2219 (1974).
229. L. Brewer, Science, $\underline{161}$, 115 (1968).
230. W. Gordy and W. J. O. Thomas, J. Chem. Phys., $\underline{24}$, 439 (1956).
231. S. Trasatti, J. Electroanal. Chem., $\underline{33}$, 351 (1971).
232. S. Trasatti, J. C. S., Faraday I, $\underline{68}$, 229 (1972).
233. D. P. Stevenson, J. Chem. Phys., $\underline{23}$, 203 (1955).
234. D. E. Gray, ed., American Institute of Physics Handbook, 3rd ed., McGraw-Hill, New York, 1972, pp. 9-173.
235. Th. A. Flaim and P. D. Ownby, Surface Sci., $\underline{32}$, 519 (1972).
236. A. J. Bennett and L. M. Falicov, Phys. Rev., $\underline{151}$, 512 (1966).
237. J. W. Gadzuk, Surface Sci., $\underline{6}$, 133 (1967).
238. J. W. Gadzuk, Surface Sci., $\underline{43}$, 44 (1974).
239. J. W. Gadzuk, J. K. Hartman, and T. N. Rhodin, Phys. Rev. B, $\underline{4}$, 241 (1971).
240. B. J. Thorpe, Surface Sci., $\underline{33}$, 306 (1972).
241. S. Trasatti, Z. Phys. Chem., N. F., $\underline{98}$, 75 (1975).
242. J. D. Levine and E. P. Gyftopoulos, Surface Sci., $\underline{1}$, 171 (1964).
243. C. E. Carroll and J. W. May, Surface Sci., $\underline{29}$, 60 (1972).
244. N. D. Lang and W. Kohn, Phys. Rev. B, $\underline{3}$, 1215 (1971).
245. N. D. Lang, Sol. State Commun., $\underline{9}$, 1015 (1971).
246. S. C. Ying, J. R. Smith, and W. Kohn, Phys. Rev. B, $\underline{11}$, 1483 (1975).
247. N. D. Lang and A. R. Williams, Phys. Rev. Lett., $\underline{34}$, 531 (1975).
248. N. D. Lang, in Solid State Physics, Vol. 28, F. Seitz and D. Turnbull, eds., Academic, New York, 1973, p. 225.
249. E. P. Gyftopoulos and J. D. Levine, J. Appl. Phys. $\underline{33}$, 67 (1962).

250. E. P. Gyftopoulos and D. Steiner, in Report 27th Annual Conference on Physical Electronics, Cambridge, Mass., 1967, p. 169.
251. J. G. Malone, J. Chem. Phys., $\underline{1}$, 197 (1933).
252. E. S. Rittner, J. Chem. Phys., $\underline{19}$, 1030 (1951).
253. J. Topping, Proc. Roy. Soc. (London), $\underline{A114}$, 67 (1927).
254. D. M. Kolb, Ber. Bunsenges. Phys. Chem. (in press, 1978).
255. W. N. Hansen and W. A. Abdou, J. Opt. Soc. Amer. (in press, 1977); J. de Physique (in press, 1977); and references therein.

The Zinc Electrode

JAMES McBREEN
Brookhaven National Laboratory
Upton, New York

ELTON J. CAIRNS
General Motors Research Laboratories
Warren, Michigan

1.	Introduction	275
2.	Brief History of Some Zinc Cells	277
3.	Solution Chemistry of Zinc in Battery Electrolytes	283
4.	Anodic Processes at the Zinc Electrode	298
5.	Cathodic Processes at the Zinc Electrode	305
	5.1. Zinc Deposition	306
	5.2. Hydrogen Evolution	315
	5.3. Oxygen Reduction	317
6.	Cycling Behavior of the Zinc Electrode	317
7.	Zinc Electrode Structure and Preparation	326
8.	Zinc Cells: A Status Report	329
	8.1. Primary Cells	329
	8.1.1. The Zinc/Manganese Dioxide Cell	329
	8.1.2. The Zinc/Mercuric Oxide Cell	331
	8.1.3. The Zinc/KOH/Air Cell	333
	8.2. Secondary Cells	335
	8.2.1. The Zinc/KOH/Silver Oxide Cell	335
	8.2.2. The Zinc/KOH/Nickel Oxide Cell	335

8.2.3.	The Zinc/$ZnCl_2$/Chlorine Cell	338
8.2.4.	The Zinc/$ZnBr_2$/Bromine Cell	339
9. Concluding Remarks		341
10. Acknowledgement		341
11. References		342

1. INTRODUCTION

Zinc is one of the metals longest known to man. A zinc idol (87.5% Zn) was found in the prehistoric Dacian settlement at Dordosch, Transylvania. Two zinc-filled bracelets were found in the ruins of Cameros (500 B.C.), and part of a fountain in the ruins of Pompeii (79 A.D.) was covered with zinc. In Asia, Kazvini (who died in 630 A.D.) indicated that the Chinese used zinc for coins and mirrors (1).

The use of zinc grew steadily; for example, in the 14th century it was mass-produced in India, where zinc ore was reduced by charcoal in clay retorts heated over a fire. By 1730 the technology of zinc production was brought to England from China. Until about 1758, zinc was produced mainly from oxide ores; after that time, sulfide ores came into prominence. In the United States, zinc was first produced in 1835. The basic process remained the same for a long period, with various modifications of furnace and retort designs. In the case of sulfide ores, they were roasted in air to produce oxides, which were then treated by the conventional process of heating in the presence of charcoal in retorts. During the period 1880 to 1930 some producers used electrically heated retorts.

During World War I the electrolytic production of zinc was used to meet the increased demand. The electrolytic process used the basic steps of: (1) dissolving the zinc in sulfuric acid ($ZnO + H_2SO_4 \rightarrow ZnSO_4 + H_2O$), (2) purification of the $ZnSO_4$ solution, and (3) electrolytic deposition of zinc:

$$ZnSO_4 + H_2O \rightarrow Zn + \tfrac{1}{2} O_2 + H_2SO_4. \qquad (1)$$

The product sulfuric acid is recycled to dissolve more zinc.

Zinc has been an important item of commerce for a long time and has a number of high-volume uses, as can be seen from Table I. The forms of zinc that are used in batteries are the oxide and rolled sheet. Since batteries represent a small segment of the zinc market, this use could grow significantly without a noticeable impact. This is particularly true in view of the declining demand for zinc die castings in the automotive industry (2).

TABLE I

Consumption and Uses of Zinc
in the United States, 1975 (2)

End use	Percent
Die casting	40
Galvanizing	38
Brass	13
Zinc oxide	4
Rolled zinc	3
Other uses	2

Total U.S. consumption of slab
zinc in 1975 = 925,293 tons.

In view of the long history of zinc and its many applications, it is not feasible in a single chapter to consider all of the important chemistry and electrochemistry of zinc. This chapter discusses information that relates to the use of zinc in electrochemical energy storage and production, including a brief history

of the use of zinc in primary and secondary cells, the chemistry
and electrochemistry of zinc in the electrolytes used in these
cells, information on the preparation of zinc electrodes, and a
status report on zinc cells of current interest. Rather than
attempting to give complete treatment to any of the topics mentioned, only selected portions of the material available are discussed. An attempt has been made to favor the more recent results, since various aspects of the zinc electrode have been treated in monographs published as recently as 1971 (3-5).

2. BRIEF HISTORY OF SOME ZINC CELLS

Count Alessandro Volta's work reported in 1800 (6) represents the
solid forerunner of zinc cells and batteries. Volta's piles were
constructed of disks of (a) zinc or tin as negative electrodes
and (b) copper, brass, or silver as positive electrodes, arranged
as a bipolar stack with disks of cardboard, leather, or other
absorbents to hold electrolyte (water or other). The preferred
electrodes were zinc and silver. In addition, a "crown of cups"
arrangement was used, with the individual cells connected in
series. The main reactions were undoubtedly the oxidation of zinc
to form zinc oxide and the reduction of water to yield hydrogen.
At the beginning of discharge, any oxide on the silver was reduced,
giving a much higher voltage than the main reaction. The fact
that access of air to the silver could help to maintain the higher
voltage was already known to the early users of the Volta pile.

In 1801 Gautherot (7) worked with two electrodes of the same
metal in sulfuric acid and reported them to hold a difference in
potential after charge was passed between them. This formed the
starting point for subsequent work leading to secondary (rechargeable) cells.

During the first three decades following Volta's work, almost all of the primary cells utilized the main reaction of the Volta cell:

$$Zn + H_2O \rightarrow ZnO + H_2 \qquad E \simeq 0.4 \text{ V} \qquad (2)$$

There was significant concern about reducing the hydrogen overvoltage at the positive electrode. Many unusual means for mechanically lowering the hydrogen overvoltage were practiced. During the 1830s a chemical approach was taken to this problem, leading to cells in which a reactant (other than water) was provided at the positive electrode.

One of the most successful of the early zinc cells provided with a cathode reactant was the Daniell cell (8) introduced in 1836. The initial version consisted of an amalgamated zinc rod* in dilute H_2SO_4 inside a tube of animal membrane, immersed in a copper cup of $CuSO_4$ solution. The overall reaction was:

$$Zn + CuSO_4 \rightarrow ZnSO_4 + Cu \qquad E = 1.08 \text{ V} \qquad (3)$$

Excess solid $CuSO_4$ was usually in contact with the catholyte, providing additional capacity. This cell provided much greater capability for sustaining significant currents and became very popular. Many versions of the cell were used, including some with the electrolytes immobilized in blotter paper, those in which the H_2SO_4-$ZnSO_4$ anolyte floated on the more densely saturated solution of $CuSO_4$ (the "gravity cell") and those with porous ceramic dia-

---

*It is interesting to note that before this time the amalgamation of zinc to prevent corrosion was already well established.

phragms. Simple shapes of electrodes were used, usually prepared from rod or sheet. The gravity cell proved very popular in telegraph and railway use.

At about the same time that the Daniell cell became popular, the Grove cell (9) was also used. This cell, introduced in 1838, had the general form:

$$Zn(Hg) \mid \text{acid or alkali} \vdots HNO_3 \mid Pt \qquad E \simeq 1.9 \text{ V} \qquad (4)$$

A porous clay vessel served to prevent mixing of the HNO_3 with the anolyte, which ranged from alkali to acid. This work marks the beginning of the use of alkaline electrolytes. The relatively high voltage of this cell was responsible for its popularity, especially in the Grove-Bunsen form of 1841, which used carbon to replace the platinum. This cell was capable of high currents, but the fumes from the HNO_3 and its reduction were a disadvantage.

The HNO_3 of the Grove-Bunsen cell was replaced with chromic acid by Poggendorff in 1842 (10), giving a cell of the type:

$$Zn(Hg) \mid H_2SO_4 \vdots H_2SO_4 + Na_2Cr_2O_7 \mid C \qquad E = 2.0\text{-}2.1 \text{ V} \qquad (5)$$

The high voltage and good current delivery capability made it suitable for heavy-duty application, including some military uses. Other cells with two fluids include the Darimont cell:

$$Zn \mid NaCl \vdots FeCl_3 \mid C \qquad (6)$$

About this same time the important use of solid cathode reactants in conjunction with zinc anodes was introduced as a result of de la Rive's work (11). His cell scheme was as follows:

$$\text{Zn(Hg)} \mid \text{H}_2\text{SO}_4 \mid \text{PbO}_2 \qquad E \simeq 2.5 \text{ V} \qquad (7)$$

Platinum was used as a current collector in the PbO_2 electrode. Various forms of this cell reached commercial application, and in the 1880s it was used as a power source for street cars. Some efforts were devoted to the development of a rechargeable cell of this type, but the difficulty of recharging the zinc electrode prevented success.

In the 1860s Georges Leclanché developed the highly important $\text{Zn(Hg)} \mid \text{NH}_4\text{Cl} \mid \text{MnO}_2$ cell (12,13). This cell used an amalgamated zinc rod anode and MnO_2 mixed with carbon surrounding a carbon current collector, all in a porous container as a cathode. The electrolyte was a saturated solution of NH_4Cl. The cell voltage was about 1.5 V. These cells, still widely used, had a relatively low self-discharge rate. Among the various improvements made in the electrode structure and cell design over the years, the introduction of bonded and compacted MnO_2 electrodes deserves special mention. It came to be recognized that oxygen from air was also reduced at the cathode, and under high overvoltage conditions, hydrogen was evolved. The MnO_2 was even considered unnecessary in low-current applications. These cells were the forerunners of today's zinc/air cells.

Lalande and Chaperon (14) in the early 1880s patented the cell:

$$\text{Zn(Hg)} \mid \text{NaOH} \mid \text{CuO} \qquad E = 1.06 \text{ V} \qquad (8)$$

which became commercially important, replacing to a very large extent the Daniell cell. It was the first important cell with alkaline electrolyte, one that could be recharged. Also, the discharged CuO electrode could be heated in air to reform the CuO, allowing the electrode to be discharged again. This cell took many forms, using loose powder electrodes, pressed porous electrodes, and so on. Potassium hydroxide was used as the electrolyte in some versions. In the 1880s and 1890s use of this cell as a primary cell was widespread. Recharging was difficult because of the relatively high solubilities of zinc and copper species in the electrolyte. The Zn/CuO cell was manufactured until the 1960s for primary low-drain applications, such as railroad signals (5).

The Maiche cell

$$Zn(Hg) \left| \begin{matrix} NH_4Cl \\ or \\ NaCl \end{matrix} \right| O_2 \text{ (Pt on C)} \tag{9}$$

was used in telegraph service, as was the Walker-Wilkins cell:

$$Zn(Hg) \;|KOH|\; O_2 \text{ (C)} \tag{10}$$

Both cells obtained oxygen directly from air.

The use of metal halides and halogens as cathode reactants in zinc-anode primary cells with alkaline electrolyte was studied in the 1880s and 1890s. Carbon was the most common electrode for reduction of the halogens, although in some cases a platinum catalyst was added to the carbon.

At the start of the 20th century the almost exclusive use of zinc as the negative electrode in all important cells declined with the introduction of rechargeable alkaline cells using cadmium, iron, copper, and other negative electrodes. A vigorous court contest over the patent rights to these cells took place between Edison and Jungner (4). The patent literature reveals that Michalowski (15) was the first to discuss the Zn |KOH| NiOOH cell (in 1899) and that both Jungner (16) and Edison (17) patented certain designs of this cell. Apparently the zinc/nickel oxide cell was not found sufficiently attractive for commercialization, even though it had a voltage of 1.75 V. Drumm (18) worked on this system later (in the 1920s and 1930s), and the zinc/nickel oxide cell found some application in Ireland during this period. Popularity of this cell was not widespread, even though the required materials were plentiful and relatively inexpensive. The problems of short cycle life due to difficulties in zinc redeposition, and the deleterious effects of zincate on the pocket-type NiOOH electrode that was used were probably responsible for the loss of interest in this system. The predominance of the zinc electrode in primary cells continued, however, with the widespread use of the Leclanché and Lalande cells. In the 1920s many radios were battery operated, providing a good market for Zn |KOH| air cells.

In the late 1920s Henri André started his extensive investigations on the Zn |KOH| AgO cell. Not until 1941 (19) did he report that the short cycle life of this cell could be improved by the use of appropriate separators, cellulosic in nature. A great deal of work was performed by André and others on the subject of separators for Zn/AgO cells, with the objective of preventing the migration of active materials and the formation of zinc dendrites, which usually shorted the cells. The cellulosic separator materials, coupled with suitable means of applying them to the electrodes, provided a significant improvement in cycle life over previous results. André had a great interest in electric automobiles and built a zinc/silver oxide battery that was installed in an automobile converted to electric propulsion. Additional work on the zinc/silver oxide system was performed at the Yardney International Corporation and was enhanced by André's joining the company in 1948.

The last zinc cell mentioned in this section is one that has come to prominence only relatively recently:

$$\text{Zn(Hg)} \mid \text{KOH} \mid \text{HgO} \qquad E = 1.35 \text{ V} \qquad (11)$$

Early reports of this system (20,21) did not result in the development of a commercial cell; it was developed for practical use in World War II by Samuel Ruben (22,23). The Ruben cell has a very stable voltage under load at significant current densities, and certain versions can be recharged a number of times.

The early period of battery development was marked by inventiveness and empiricism. Advances in battery technology were made at an impressive rate, without a corresponding advance in the scientific knowledge of the chemistry and electrochemistry of the cells. It was not until the era of Nernst, Oswald, Van't Hoff, and other great scientists around the turn of the century that the basis for scientifically guided advancements was established. It seems appropriate to discontinue the historical discussion of the early development of zinc cells at this point and to introduce a discussion of the chemistry and electrochemistry of the zinc electrode, followed by a synopsis of the current status of a number of zinc cells.

3. SOLUTION CHEMISTRY OF ZINC IN BATTERY ELECTROLYTES

The zinc electrode-discharge products are very soluble in battery electrolytes. The high solubility inhibits zinc passivation and contributes to the high-rate capabilities of the electrode; however, it causes difficulties in controlling zinc morphology on charge and zinc redistribution on cycling. The three electrolytes considered here are KOH, $ZnCl_2$, and $ZnBr_2$.

On anodic oxidation, zinc loses its $4S^2$ electrons and the ion forms Sp^3-hybridized orbitals that yield tetrahedrally coordinated complex ions in concentrated battery electrolytes.

Potassium hydroxide battery electrolytes have concentrations in the range of 4 to 12 M. Figure 1 is a phase diagram for aqueous KOH solutions (24). Figure 2 is a distribution diagram for zinc hydroxide complexes (25). The zincate ion, $Zn(OH)_4^=$, is the only one of importance in KOH battery electrolytes. The tetrahedral structure of this complex ion has been confirmed by spectroscopic measurements (26).

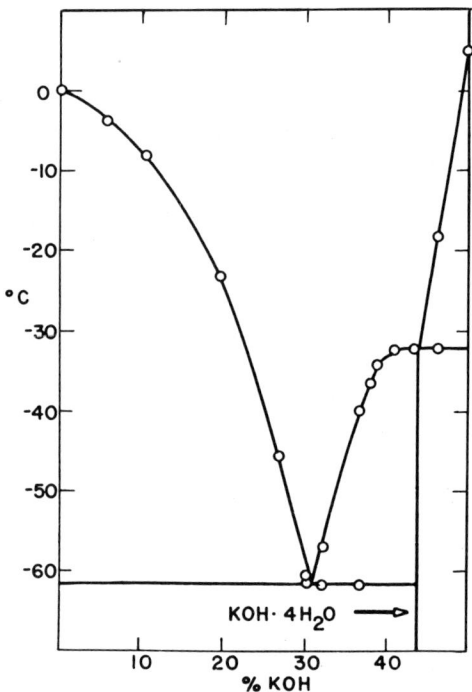

Fig. 1. Phase diagram of aqueous KOH solutions (24).

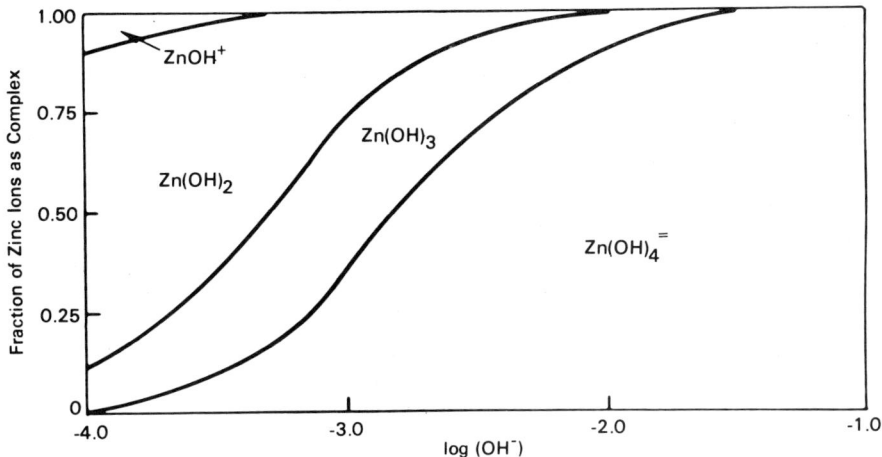

Fig. 2. Distribution diagram for zinc-hydroxide system (25).

The solubility of ZnO in KOH solutions does not vary noticeably with temperature. A composition diagram for saturated KOH-ZnO solutions is given in Fig. 3 (24). More recently it has been found that this solubility invariance with temperature extends from -66°C to 145°C (27,28). The presence of zincate in the electrolyte results in a slight decrease in electrolyte conductivity. Figure 4 shows how the conductivity of various concentrations of KOH decrease with addition of ZnO (27).

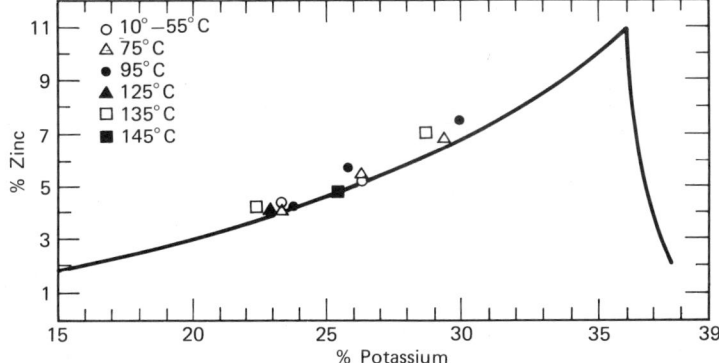

Fig. 3. Composition of saturated KOH-ZnO solutions (27).

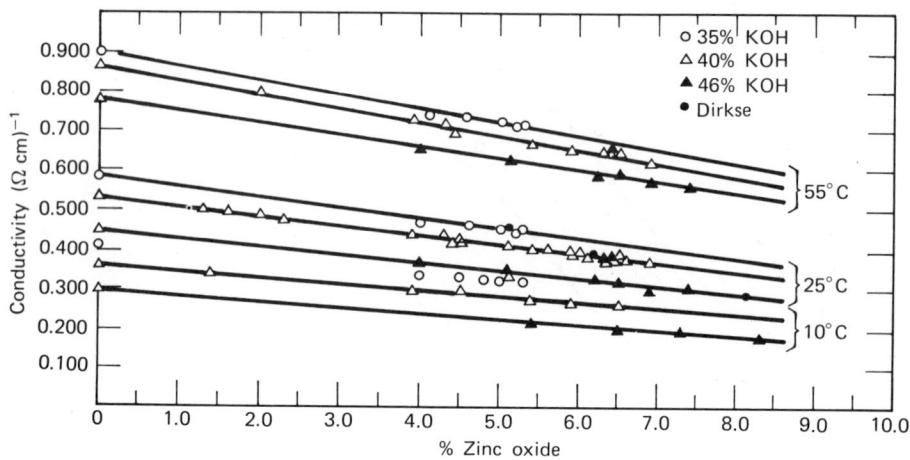

Fig. 4. Conductivity of KOH-ZnO solutions (27).

Supersaturated zincate solutions can be formed by electrodissolution of zinc into alkaline electrolytes (29). Figure 5 gives the experimental range of solubilities of electrochemically generated zincate (29). These solutions, however, are not supersaturated in the ordinary sense of the term. Neither seeding nor shock causes precipitation of ZnO (30,31). Instead, the ZnO is precipitated slowly in a manner resembling a decomposition reaction. The presence of Li^+ and $SiO_3^=$ retard the process (32). Silicate has been added to the electrolyte of zinc/air batteries with slurry-type zinc electrodes for this purpose (33). The decomposition process for these supersaturated solutions is very slow, and it takes about a year for the solution to reach the equilibrium solubility value for ZnO. Nuclear magnetic resonance studies of supersaturated zincate solutions indicate that these solutions contain the same solute zincate species as solutions unsaturated and saturated with respect to solid ZnO (34,35). No evidence was found from these measurements to indicate the presence of a colloid or solid phase.

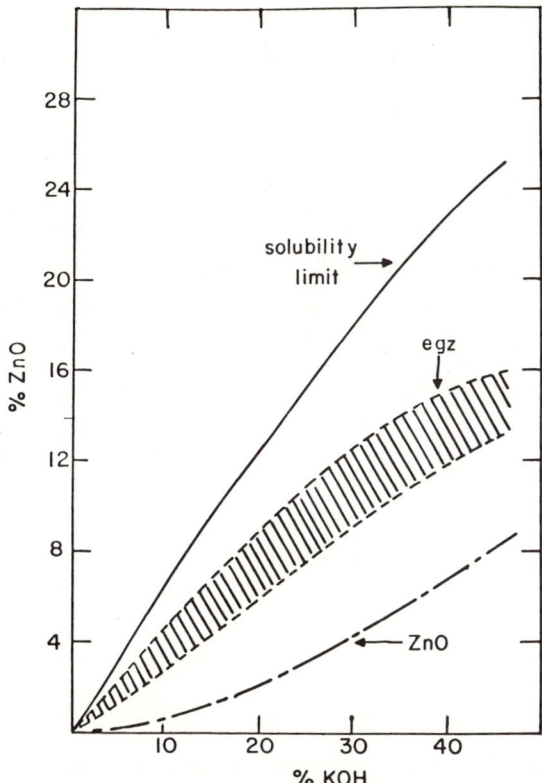

Fig. 5. Zincate solubility in KOH solutions at room temperature. Solid line, solubility limit of electrolytically generated zincate; lowest line, solubility of ZnO; cross-hatched band, experimental range of solubilities of electrolytically generated zincate (29).

The degree of supersaturation and the stability of the solutions are such that this behavior must have an effect on cell performance and subsequent zinc deposition on charge. Unfortunately, no investigations on these matters have been reported.

The diffusivity of zincate in KOH at various concentrations and temperatures has been determined using polarographic techniques (36,37). With this technique, however, problems are encountered because of polarographic maxima of the second kind. There is general agreement that the measured diffusivity remains relatively constant with KOH concentrations up to ∿7.0 M. Above this concentration the diffusivity rapidly decreases with increasing KOH concentration. These data are presented in Tables II (37) and III (36). Unfortunately, there is very little data for diffusivity determinations using other techniques. One investigation used a capillary technique (38); measurements were made using zincate solutions (0.02 to 0.4 M) in 9.9 M KOH. At 20°C the diffusivity was 3.03×10^{-6} cm^2/s, and at 25°C it was 3.24×10^{-6} cm^2/s. The diffusivity was independent of zincate concentration in this concentration range.

TABLE II

Diffusion Coefficients, (cm^2/sec) $\times 10^6$, of Zincate Ions in KOH Solutions (37); concentration of $Zn(OH)_4^=$: 1.08×10^{-3}M

t,°C	Molarity of KOH						
	1.00	2.24	2.94	4.40	4.91	7.19	10.15
45	18.72	19.82	15.80	17.22	12.90	9.58	5.74
25	11.94	9.7	--	8.79	--	5.85	3.43
0	6.47	5.16	--	5.27	--	2.69	1.45
-30	--	--	--	--	--	0.009	0.002

TABLE III

Diffusion Coefficients of Zincate Ions
in KOH Solutions (36), (cm^2/sec) × 10^6;
Concentration of $Zn(OH)_4^=$: 3×10^{-3}M

	Molarity of KOH					
t,°C	0.928	5.4	6.85	8.30	9.89	11.55
0	2.38	1.95	1.55	1.34	0.648	0.269
25	6.86	6.99	7.23	4.82	2.53	1.52
35	8.67	9.21	9.06	8.50	5.23	2.36
50	8.11	9.73	9.55	10.50	7.61	4.93

Boden, Wylie, et al. (39) have determined activity coefficients for zincate in 7.03 M KOH at 25°C. Their results are given in Table IV (39).

Figure 6 is a distribution diagram for zinc chloride complexes (25). For most of the charge-discharge cycle in a zinc/chlorine battery, only the tetrahedral $ZnCl_4^=$ complex would be present. However, toward the end of charge other complexes could be present in varying amounts, depending on the design and operation of the battery.

A phase diagram for the zinc chloride-water system is shown in Fig. 7 (40). Zinc chloride is soluble in water at ambient temperatures to over 70% by weight. However, concentrations greater than 35% are rarely encountered in batteries. Resistivity data for various concentrations of zinc chloride and temperatures are given in Fig. 8 (41). Additions of HCl reduce the electrolyte resistivity (Fig. 9) (42). However, the presence of HCl in $ZnCl_2$ battery electrolytes causes pH fluctuations during cell operation. These are illustrated in Fig. 9 (42). For example, a 25 w/o $ZnCl_2$

TABLE IV

Activities of Zincate in 7.03 M KOH (25°C) (39)

Zincate concentration mol/ℓ	$\log a_{Zn(OH)_4^=}$
0.0126	−0.6606
0.0250	−0.3535
0.0372	−0.1710
0.0609	0.0617
0.0837	0.2182
0.1136	0.3868
0.1571	0.5501
0.1952	0.6821
0.2477	0.8206
0.2953	0.9350
0.3380	1.0280
0.3785	1.1095
0.4260	1.1991
0.4800	1.2945
0.5680	1.4383
0.6400	1.5483

Solution Chemistry of Zinc in Battery Electrolytes

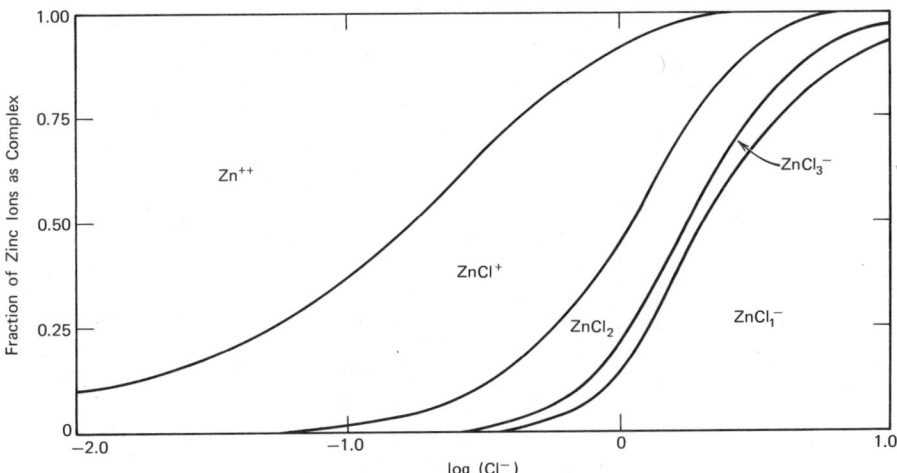

Fig. 6. Distribution diagram for zinc-chloride systems (25).

solution containing 2.5 g HCl/l will have a pH of 3.5 at 25°C. Reduction of the $ZnCl_2$ concentration to 15 w/o by electrodeposition reduces the pH to 1.3. This variation of pH with zinc-chloride concentration in the presence of hydrochloric acid can possibly be explained by a reduction in hydrogen-ion activity by association with the complex species $ZnCl_4(H_2O)_2^=$ at concentrations above 10 w/o. This effect is smaller at higher temperatures (42).

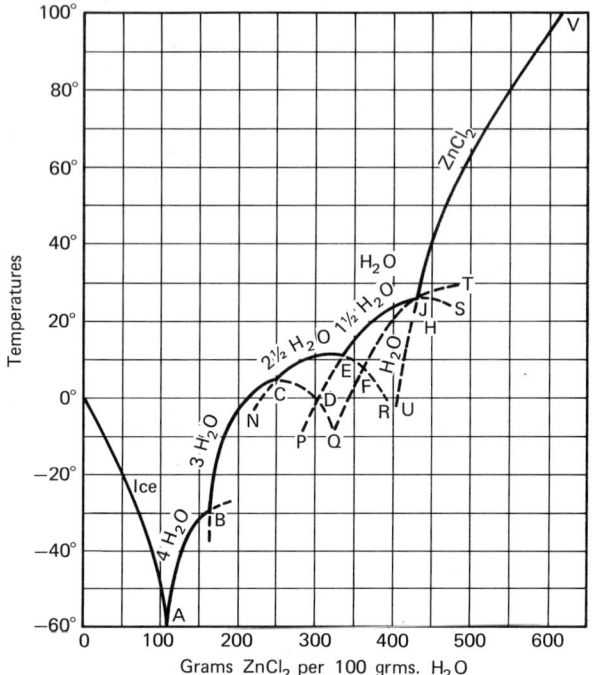

Fig. 7. Phase diagram of $ZnCl_2$ + H_2O (40).

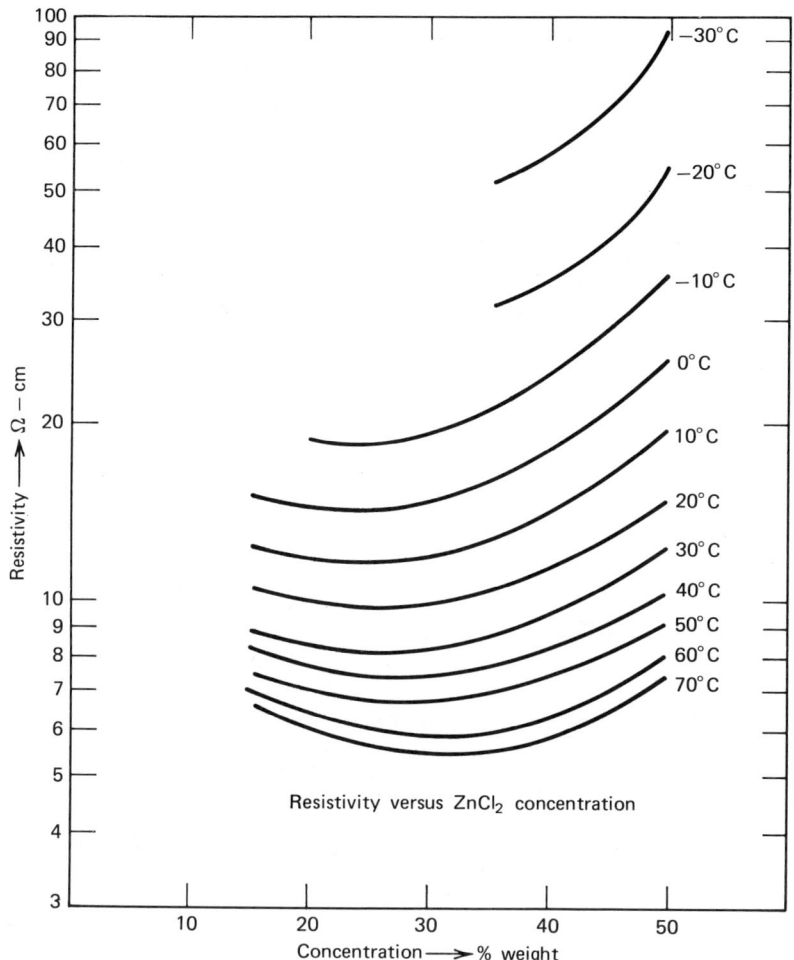

Fig. 8. Resistivity versus $ZnCl_2$ concentration (41).

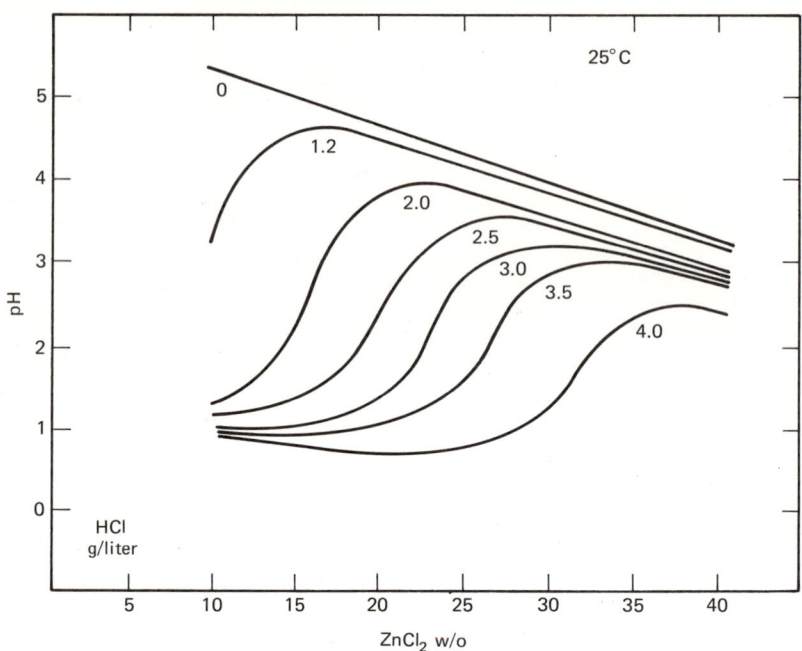

Fig. 9. Variations in pH of $ZnCl_2$-H_2O solutions containing small quantities of HCl (42).

Figure 10 is a distribution diagram for zinc bromide complexes. A phase diagram for the zinc bromide-water system is given in Fig. 11 (43). Like $ZnCl_2$, it is very soluble in water. Resistivity data for various concentrations, at two temperatures have recently been published (44) and are shown in Fig. 12 (44).

The zinc electrode is unique in that it has been successfully used as a battery electrode in electrolytes as varied as chromic acid (pH < 1), ammonium chloride (pH 4-9), and 12 M KOH (pH > 15). The reversible behavior of zinc is summarized in the potential-pH diagram shown in Fig. 13 (45). This diagram was constructed by assuming ε-$Zn(OH)_2$ to be the stable allotrope for the hydroxide. Several allotropes of $Zn(OH)_2$ have been identified (46). The only two of importance in batteries are ε-$Zn(OH)_2$ and γ-$Zn(OH)_2$. The

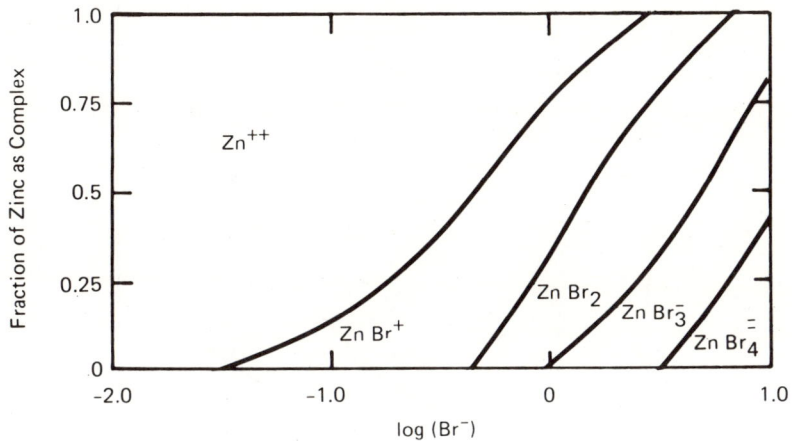

Fig. 10. Distribution diagram for the zinc-bromide system.

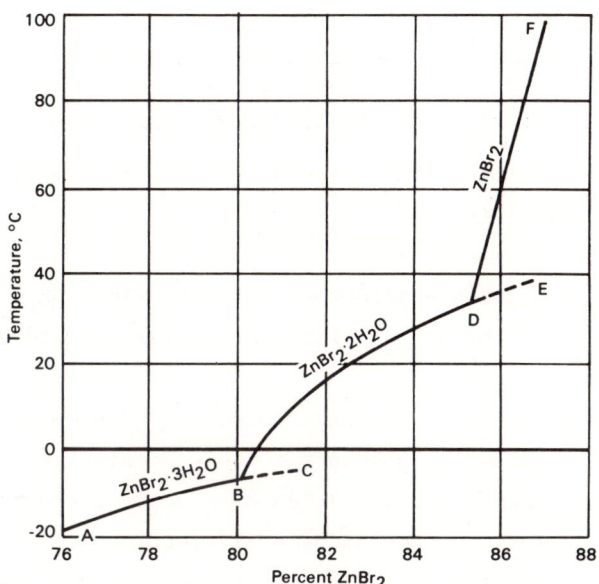

Fig. 11. Solubility curves of zinc bromide (43).

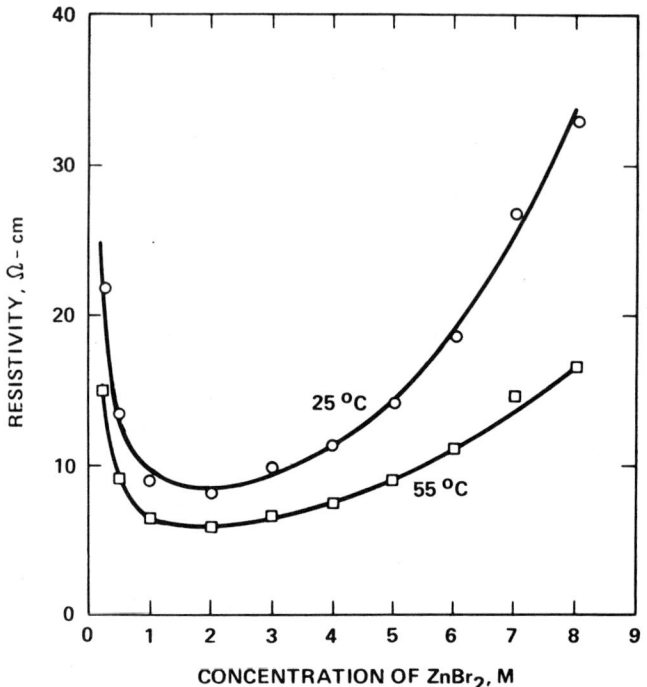

Fig. 12. Resistivity of zinc bromide solutions at various temperatures (44).

former is the most stable allotrope at 28°C. In high concentrations of alkali, γ-Zn(OH)$_2$ can form particularly at temperatures below 20°C (47).

The Leclanché cell electrolyte represents a special case in that other solid reaction products such as diamines (e.g., ZnCl$_2$·2HN$_3$) can form. Brouillet and Jolas have constructed a potential-pH diagram for zinc in a Leclanché cell electrolyte (Fig. 14) (48). In actual cell operation the electrolyte pH can vary between 4 and 9 (49). Below a pH of 5.8 the discharge product is zinc ion. In the pH interval 5.8 to 7.85 crystalline diamine (ZnCl$_2$·2NH$_3$) can form. Above pH 7.85 the diamine dissolves to form Zn(NH$_3$)$_4^{++}$ ions.

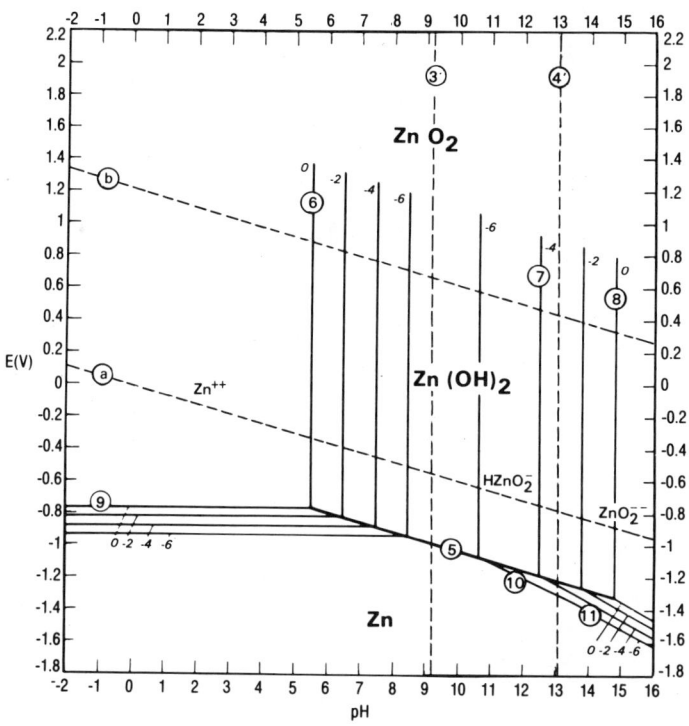

Fig. 13. Potential-pH equilibrium diagram for the system zinc-water at 25°C [established by considering ε-Zn(OH)$_2$] (45).

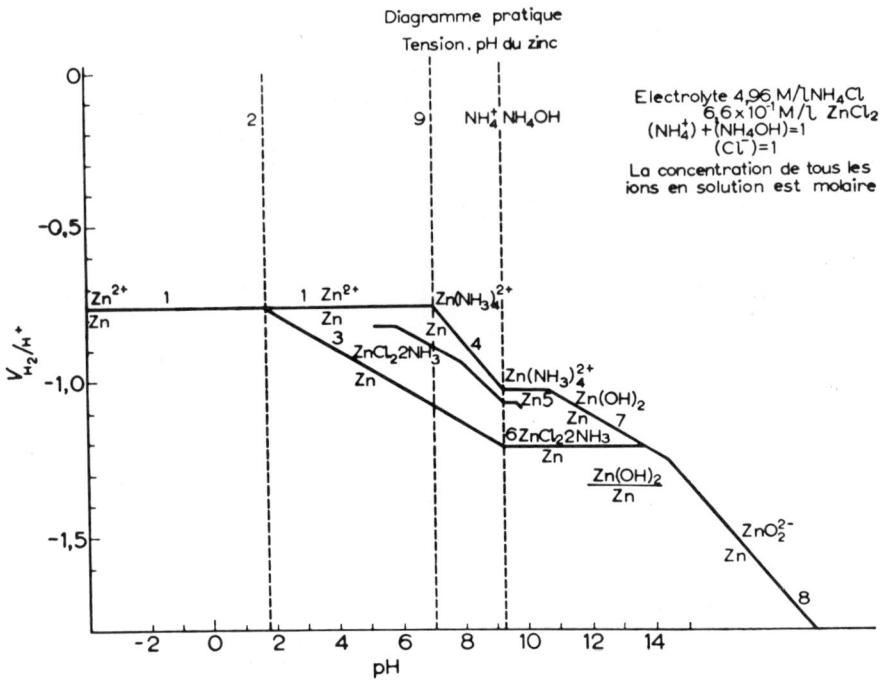

Fig. 14. Potential versus pH for zinc in a $ZnCl_2/NH_4Cl$ electrolyte as used in Leclanché cell manufacture (48).

4. ANODIC PROCESSES AT THE ZINC ELECTRODE

Most of the literature on anodic processes at the zinc electrode concern behavior in alkaline electrolyte. Historically there have been very good reasons for this. In cells with acidic electrolytes rate limitations of the zinc electrode on discharge present no problem, whereas sustained high discharge rates have generally been unnecessary in the Leclanché cell. The incentive for the work in alkaline electrolytes has been the need for developing high-rate primary cells, as zinc passivation can present a problem.

The literature concerning the anodic processes on the zinc electrode is even more confusing and controversial than that for the electrodeposition processes. Early work in this area involved galvanostatic measurements on zinc sheets (50-56). In KOH electrolytes it has been found that when a zinc sheet is anodized at a particular current the overpotential will remain relatively constant or increase slightly for some time. With the onset of zinc passivation a rapid, large increase in the overpotential occurs. In general, the results conform to the relationship

$$(i - i_1) t^{1/2} = k \qquad (12)$$

where i is the applied current density and t is the time to passivation. Both i_1 and k are dependent on KOH concentration and on the nature of the movement of the electrolyte.

Even though this interrelationship yields practically no information on the passivation mechanism, there has been considerable controversy over the values of i_1 and k. All this interrelationship indicates is that diffusion and migration are the primary modes of mass transfer. In cases where there is very significant convection there are deviations from this equation (54). However, Eq. 12 is instructive for indicating the amount of zinc utilization that can be expected at various KOH concentrations and temperatures. Plots of k for various KOH concentrations are given in Fig. 15 (53). The capacity discharged per unit area prior to passivation (Q) is:

$$Q = kt^{1/2} + i_1 t \qquad (13)$$

Values of k give an indication of the degree of utilization to be expected at various KOH concentrations. This curve appears to have a maximum at approximately 7 M KOH. More recently, it has been found that the limiting current for zinc dissolution goes through a maximum at this concentration (57).

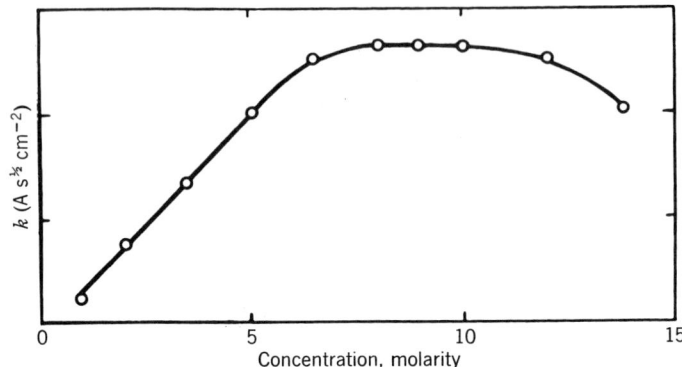

Fig. 15. Plot of k from $(i - i_1)t^{1/2} = k$ versus concentration (53).

Investigations using voltage sweep techniques (47,58,60), visual and microscopic observations (60,62,63,65), X-ray diffraction (67), and identification of product species by the ring-disk technique (61,62), lead to the tentative conclusion that three types of processes may inhibit zinc dissolution in alkaline electrolytes:

1. At overvoltages of about 30 mV passive films with thicknesses of about a monolayer can form and inhibit dissolution (61,62, 68,69).
2. On continued dissolution these films apparently can grow to greater thicknesses (60). As they become thicker they tend to crack and peel from the zinc substrate (60,63). Powers and Brieter refer to these thicker films as Type II films (63).
3. In quiescent electrolytes a nonadherent deposit of zinc oxide can form on the electrolyte side of the Type II film. These are referred to as Type I films (63).

Two mechanisms have been proposed for the formation of passivating monolayers. One is an adsorption mechanism (68) and the other two-dimensional nucleation (69). A critical experiment to distinguish between these mechanisms has not been performed. There now appears to be general agreement that Type II films are not formed by a dissolution-precipitation mechanism, whereas Type I films do appear to be.

Current-voltage curves for voltage sweeps on zinc are complex and correspond to the occurrence of many processes. An example is given in Fig. 16 (63). At low overvoltages the metal is brightly etched; at this stage there is no film and dissolution occurs along screw dislocations (63,70). At higher overpotentials (point B in the figure) a nonadherent (Type I) film forms. As the potential goes in the positive direction the color of the surface beneath the Type I film changes progressively from silvery to black to silvery again.

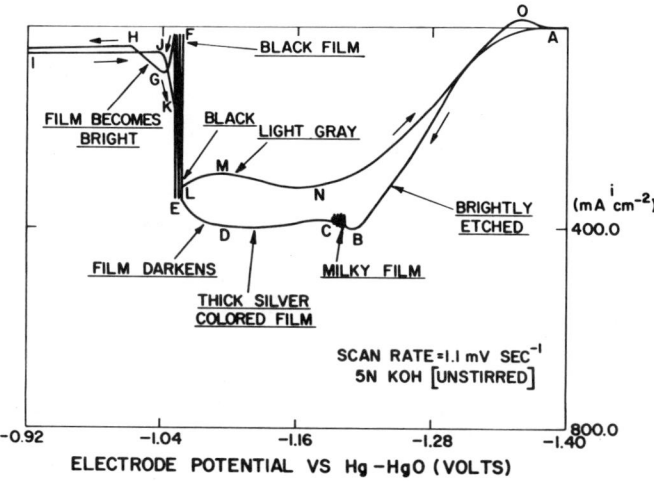

Fig. 16. Current-voltage curve of a zinc wire electrode in unstirred 5 N KOH recorded at a scan rate of 1.1 mV s^{-1} (62).

The current-potential curves have been discussed in detail by Bobker (71). The major feature of the curves is that they have two current peaks that become more distinct at lower sweep rates (65). Prior to passivation, current oscillations are usually observed (61,62,72). Sharpe has pointed out that care must be taken in interpreting current-potential curves in potential regions where both anodic and cathodic processes can occur (73). It is well known that anodic films on zinc catalyze hydrogen evolution (58,65). Hence, the double peaks are probably due to the combined anodic and cathodic processes. This is consistent with the fact that only one current peak is observed for the anodic sweep on amalgamated electrodes (61) where hydrogen evolution would be inhibited.

Several empirical correlations have been proposed between zinc utilization and electrode structure in porous electrodes (74-76). Plots of the discharge current versus the square root of the time to passivation yield straight lines (74). The slopes of these lines, which vary with the surface area of the zinc, are given in Table V (74).

Electrode porosity has a major influence on zinc utilization (75,76). Figure 17 (75) is a plot of zinc utilization as a function of porosity. Pore plugging does not appear to be a major problem in zinc electrodes. The discharge product is mainly ZnO (67,77); this results in only a 30% volume expansion of the active material during discharge. This is why utilizations as high as 80% can be obtained with electrodes 5.3 mm thick (78).

In porous electrodes the discharge processes are very similar to those observed on sheet zinc in quiescent electrolyte. Most of the discharge product forms a porous "carpet-like" structure (79) similar to the Type I films. The current distribution in the electrode is consistent with the formation of a thin resistive film beneath the porous oxide layer (79).

TABLE V

Slopes of i versus $t^{1/2}$ Plots for Porous Zinc Electrodes (Porosity = 70%) (74)

B.S.* mesh size	Methylene blue surface area ($m^2 g^{-1}$)	BET(N_2) Surface area ($m^2 g^{-1}$)	Gradient of I vs. $t^{1/2}$ (A $s^{-1/2}$)
+22	2.27	0.77	14.8
-22 +44	2.78	0.85	12.2
-44 +60	2.94	0.93	11.9
-60 +85	3.40	0.99	9.3
-85	4.46	1.80	9.1

*Brown and Sharp

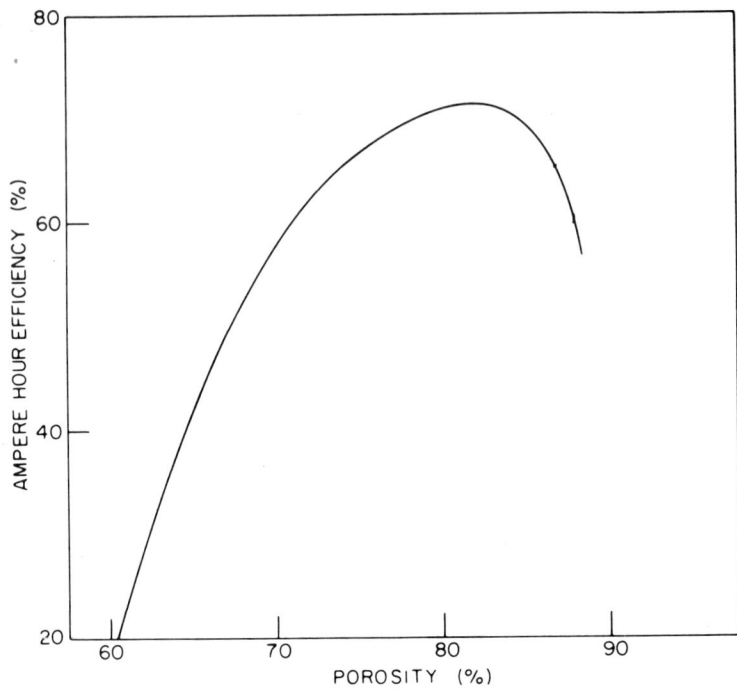

Fig. 17. Zinc utilization as a function of porosity for electroplated zinc electrodes. Discharge current = 0.73 A/cm^2, 0.292 g Zn/cm^2, and 2.5 A/g Zn (75).

Amalgamation of zinc improves the utilization of both sheet (80) and porous electrodes (81). On the basis of recent work on the effect of silicate additions (82-85), it can be concluded that silicate promotes zinc passivation at high KOH concentrations (82). However, there apparently exists at about 3 M KOH (85) a critical composition in the KOH-K$_2$SiO$_3$-ZnO ternary system where passivation is inhibited.

Zinc electrodes in alkaline cells passivate more rapidly at lower temperatures. Figure 18 (81) is a plot of the limiting current density versus temperature (81). Nikitina has investigated the reaction products as a function of discharge temperature using X-ray diffraction (47). She found ε-$Zn(OH)_2$ at 25°C and γ-$Zn(OH)_2$ at temperatures below 20°C. Passivation at low temperatures may be due to the formation of γ-$Zn(OH)_2$ or to a decrease in the rate of the dehydration reaction to form ZnO (86).

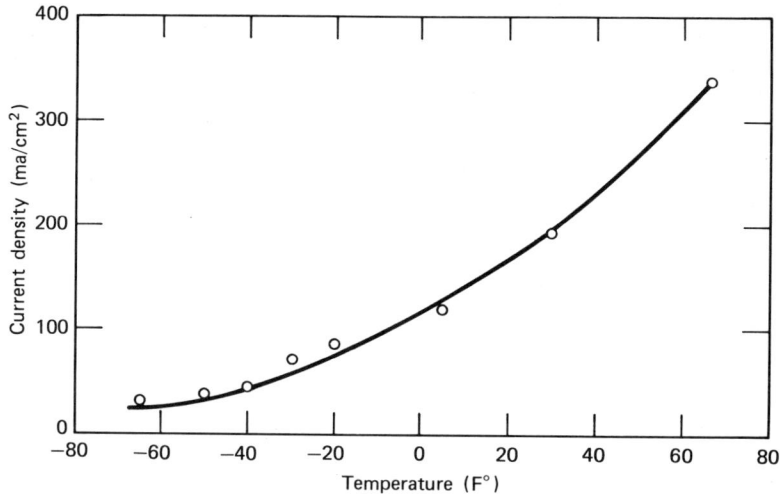

Fig. 18. Limiting current density as a function of temperature, zinc electrode, 2% Hg, acetate method (81).

5. CATHODIC PROCESSES AT THE ZINC ELECTRODE

For the zinc electrode the important cathodic processes are zinc deposition, hydrogen evolution, and oxygen reduction. Controlled zinc deposition during charge-discharge cycling is the key to long cell life. In sealed cells hydrogen evolution should be avoided and oxygen reduction enhanced (87).

5.1. Zinc Deposition

Practically all the investigations related to zinc deposition in batteries have been carried out in alkaline electrolyte. When zinc is deposited from alkaline solutions, three types of deposits are encountered (88):

1. Smooth deposits are formed at low overpotentials in vigorously stirred electrolytes.
2. In quiescent concentrated hydroxide-zincate electrolytes dark gray porous deposits [also referred to as mossy (88) or bulbous (89) deposits] are formed at low overpotentials (<75 mV). These deposits are also found in cells with paste-type ZnO electrodes.
3. In zincate electrolyte at higher overpotentials (>75 to 85 mV), the deposit is invariably dendritic.

Mossy zinc deposits resemble the powdery deposits typically formed by other metals at very high overpotentials (90). Formation of mossy zinc deposits apparently is not controlled by mass transfer but rather is related to the steps of the charge-transfer reaction (91-93). Since the early 1960s several attempts have been made to elucidate the mechanism of the zinc-deposition reaction in alkaline electrolyte (91-97). Most of this work has revolved around the experimental determination electrodic parameters (70, 94-97) (see Table VI). Because of the reported independence of exchange current density on zincate concentration in the late 1960s there was considerable confusion as to what the rate-determining step was. Bockris, Nagy, et al. suggested that the measured independence of exchange current density on zincate concentration was due to the lack of correction for ohmic resistance

TABLE VI

Comparison of Experimental Data for Zinc Deposition

Parameter	Refs. 95 and 96	Ref. 69	Ref. 94	Ref. 97
i_0 (mA/cm^2)	40 to 250	--	--	8.370
$\left[\dfrac{\partial \log i_0}{\partial \text{pH}}\right]_{C_{Zn(OH)_4^=}}$	0.2	--	--	0.14
$\left[\dfrac{\partial \log i_{an}}{\partial \text{pH}}\right]_E$	--	ca 3.5	3	2.6
$\left[\dfrac{\partial \log i_0}{\partial \log C_{Zn(OH)_4^=}}\right]_{\text{pH}}$	0	--	--	0.67
C_{dl} (μF/cm^2)	30 to 600	--	--	40 to 120
Anodic Tafel slope (mV/decade)	65 ± 10 (at low η)[a] 320 (at high η)[a]	42 ± 5	30	49 ± 13
Cathodic Tafel slope (mV/decade)	55 ± 8 (at low η)[a] 280 ± 40 (at high η)[a]	--	--	113 ± 30

[a]Calculated from the reported transfer coefficients.

in the solution (97). This is important when considering highly reversible reactions. After these corrections are made it is found that the exchange current density is indeed dependent on zincate concentration. Analysis of the electrodic parameters indicates that the rate-determining step is:

$$Zn(OH)_3^- + e^- \rightarrow Zn(OH)_2^- + OH^- \qquad (14)$$

Because of the absence of any pseudocapacitance and on the basis of estimations of the activation energies involved, Bockris and colleagues proposed the following overall mechanism (97):

$$\begin{aligned} Zn(OH)_4^= &\rightleftarrows Zn(OH)_3^- + OH^- \\ Zn(OH)_3^- + e^- &\xrightarrow{rds} Zn(OH)_2^- + OH^- \\ Zn(OH)_2^- &\rightleftarrows Zn(OH) + OH^- \\ Zn(OH) + e^- &\rightarrow Zn + OH^- \end{aligned} \qquad (15)$$

More recently Epelboin, Ksouri, et al. have attempted to elucidate the steps involved in the electrocrystallization process (92). They have determined current-potential curves in the absence of transport limitations on rotating zinc cylinders. In addition, current efficiency determinations and impedance measurements were carried out as a function of frequency down to very low frequencies (10^{-3} Hz to 50 kHz). Their overvoltage and impedance results are given in Fig. 19 (92). The impedance diagrams have four faradiac time constants. A model has been constructed to explain the results. In acid solutions this is based on the following reaction scheme:

$$
\begin{align}
H^+ + e^- &\xrightarrow{K_1} H_{ads} \tag{a}\\
H^+ + H_{ads} + e^- &\xrightarrow{K_2} H_2 \tag{b}\\
Zn^{++} + Zn^+_{ads} + e^- &\underset{K'_3}{\overset{K_3}{\rightleftarrows}} 2Zn^+_{ads} \tag{c}\\
Zn^+_{ads} + H_{ads} &\xrightarrow{K_4} H^+ + Zn \tag{d}\\
Zn^+_{ads} + e^- &\xrightarrow{K_5} Zn \tag{e}\\
Zn^{++} + H_{ads} + e^- &\xrightarrow{K_6} H_{ads} + Zn^+_{ads} \tag{f}
\end{align}
\tag{16}
$$

The first two reactions correspond to hydrogen adsorption and the electrochemical desorption reaction, respectively. These reactions are probably important only in acidic media. Reaction 16 is the autocatalytic adsorption of the reactants. Reaction 16 is an alternate scheme for catalyzing this reaction. The other reactions are the various routes for final reduction to the metal. Autocatalytic reactions can give rise to multiple steady states and S-shaped current potential curves. The model is dependent on the presence of adsorbed H, Zn^+, and adsorbed anions. Experimentally S-shaped current-potential curves have been observed, and the impedance diagrams have time constants corresponding to the relaxation of the partial coverages of the electrode by the three adsorbed species. However, all that can be concluded from these results is that adsorbed species are involved and that an autocatalytic step is an important part of the overall process. The nature of the intermediates cannot be determined from these measurements. The situation is similar to cases where the zinc ion is complexed.

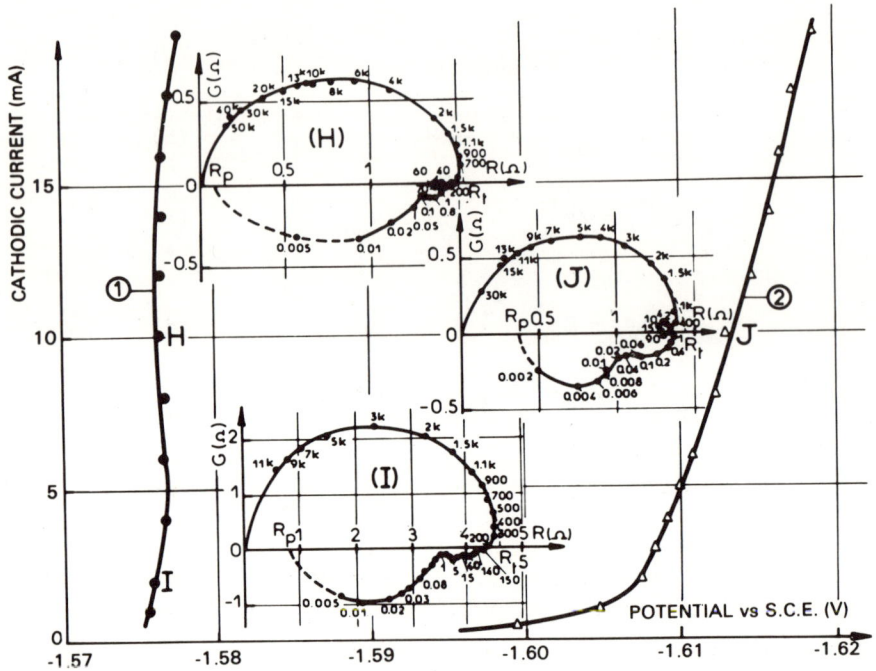

Fig. 19. Influence of the concentration x of ZnO in 7 M KOH on the steady-state current-potential curves relative to the electrocrystallization of zinc obtained with a rotating-disk electrode (area, 0.28 cm^2) at 26° ± 0.5°C. Curve 1, x = 1.24 M, rotating speed Ω = 3000 r.p.m.; curve 2, x = 0.25 M; Ω, 5000 r.p.m. H (10 mA, - 1.576 V/SCE); I (2 mA, - 1.5755 V/SCE); J(10 mA, - 1.613 V/SCE). Impedance diagrams (H), (I), and (J) obtained under the same conditions as for points I, J, and H (frequency in Hz) (92).

Epelboin, Kasouri, et al. have also postulated that coupling between the discharge reactions and surface diffusion of the adsorbed Zn^+ ions can lead to variations in current density over the surface of the electrode and can lead to the formation of mossy zinc (98,99). Experimentally they found that mossy zinc was deposited in nodules on the surface at low current densities (3.5 mA/cm^2). These deposits were formed on rotating discs at 3000 r.p.m. where there was no mass transport limitation. At higher current densities (350 mA/cm^2) dendritic zinc was formed. The dendrites were formed at lower current density when lower rotation speeds were used. There was a narrow current region between the region of moss formation and dendrite formation where the deposit was smooth. The current density versus voltage curve was S-shaped. As the current density increased, the slope of the i versus E curve went from positive to negative to positive again. Epelboin and colleagues were able to show by taking into account electron and materials balances and surface diffusion that the current distribution on the surface with increasing current density was varying in such a manner that these different deposits would be formed. The above argument appears quite reasonable when one considers the more recent results with lead additives (93). It is well known that small additions of lead inhibit moss formation (71,73). Lead additions to the electrolyte in the range of 5×10^{-5} to 3×10^{-4} M eliminate the S-shaped current density versus voltage curves and drastically change the double-layer capacity and the impedance diagram. The results are consistent with a decrease in the rate of one of the elementary reactions. Since small additions of lead do not affect mass transport in the electrolyte it must be concluded that moss formation is related to the mechanism of the discharge reaction; thus the arguments presented by Epelboin and his co-workers appear quite plausible.

Deposits from zincate solutions at overpotentials greater than 75 to 80 mV are usually dendritic. Dendritic deposits have a fern-like structure and differ from mossy deposits, which usually form as nodules on the substrate. The transition from moss to dendrites corresponds to the onset of mass-transport control and is characterized by a critical current density that is temperature dependent (88) and is very dependent on the hydrodynamic conditions at the electrode surface (89).

Bockris and his co-workers (100,101) have carried out an extensive investigation of the factors controlling dendrite initiation and dendrite growth. These studies were carried out in 0.01 to 0.2 M zincate in 10 w/o KOH. Even though the KOH concentration is lower than that used in batteries, the results are probably applicable. Potentiostatic techniques were used, and current variation and dendritic growth were observed. A telescope-microscope system was used to observe the dendrites and measure their rate of growth. At low overpotentials the deposit was mossy zinc. At higher overvoltages (85 to 140 mV) dendrites were observed. When the electrode was set at a potential above 85 mV the current initially remained constant for a transition time t_i; thereafter the current increased with the square of time. The time required for the dendrite tip to approach the outer boundary of the diffusion layer had an inverse relationship to overpotential, zincate concentration, and temperature. Dendritic growth started after a transition time t_d that was always less than t_i.

The model that Bockris and his co-workers have proposed is that dendrites originate from the tips of pyramids arising as the result of the rotation of a screw dislocation. With this type of growth, stepped pyramids are formed and the step width decreases with increasing height. Thus as a pyramid grows its radius of curvature decreases, and eventually the tip becomes a point for spherical diffusion. This marks the onset of dendritic growth, and thereafter the dendrite will continue to grow. The growth is linear with time and it occurs under activation control. Although this theory satisfactorily explains dendrite growth, such a phenomenon hasn't been observed to occur from the tips of pyramids. For instance, Mansfeld and Gilman observed pyramids but found that dendrites were more frequently initiated at the bases of the pyramids (102). Impurity centers (102) and two-dimensional nucleation sites have been suggested (103) as the sites at which dendrites form.

In later work Bockris and his co-workers investigated zinc deposition on single grains of known orientation in polycrystalline electrodes (97). At an overpotential of 100 mV they observed nodules of zinc. On further deposition a small fraction of these (0.1) develop into dendrites. Similar results have been observed by Naybour under galvanostatic conditions (104). It has been suggested that this is due to the limitations imposed by the total current; the number of dendrites is equivalent to the amount that can be supported by the current.

In cells where the discharge products are completely soluble in the electrolyte, dendrite growth will more likely occur at the end of charge. By then the substrate will have deposits of mossy zinc and nodules that can act as dendrite precursors and initiation sites. It is probably unnecessary to invoke concepts such as pyramids, impurity centers, or two-dimensional nucleation. In contrast to popular belief more of the difficulties in these types of batteries stems from the formation of nonadherent mossy deposits (105) than from the formation of dendrites. Moss formation can only be prevented by modifying the electrode processes. Hence the use of hydrodynamic methods alone to solve the cycling problems of the zinc electrode has been unsuccessful (106). For this reason it is common to use an additive in conjunction with methods for improving mass transfer (107). Even with good mass transport and with additives it is difficult to maintain good morphologies at high-capacity densities (>200 mA $\cdot$ h/cm^2) (108).

Several additives have a beneficial effect on zinc morphology. Addition of lead to the electrolyte yields smooth deposits (71,73, 109,110). Tin also has a similar effect (111). Tetraalkylammonium ions also improve the quality of the deposit (111,112). However, the effect of tetraalkylammonium ions is smaller than that of lead or tin. Low concentrations of lead or tin are not completely satisfactory in battery applications. Loose particles of zinc can displace the additive from the electrolyte and thereby reduce its concentration. Several proprietary organic additives have been successfully used in zinc/chlorine batteries (107). Unfortunately, organic additives are rapidly oxidized in the battery environment. These oxidation reactions can alter the electrolyte pH and give rise to operational problems. Hence it is often necessary to replace or purify the electrolyte every few cycles when organic additives are used (113). What is needed is an additive that is neither oxidized nor reduced in the battery environment. Impedance measurements provide a rapid method for screening additives (93).

Several claims have been made for the beneficial effects of various charging methods. These include pulse currents, periodic reversals (114), half-wave rectified a.c. (115). Naybour (89) has investigated all such methods and concluded that they all place limitations on the charge rate without achieving an increase in the rate at which an acceptable zinc deposit can be formed. Even though some modifications in deposit structure can be achieved by various charging methods (114-116), it does not appear to be a very promising route for solving the problems of the zinc electrode in free electrolyte systems.

Two mechanisms for zinc deposition in paste-type ZnO electrodes have been proposed. One mechanism assumes that the zinc oxide dissolves and deposition occurs via the zincate ion (31,117); another mechanism considers that direct solid state reduction from

ZnO occurs (118). Drazić and Nagy have investigated zinc deposition on single crystals of ZnO in KOH, K_2SO_4, and zincate electrolytes (119) and have concluded that direct ZnO reduction makes a very small contribution to the overall process. This is consistent with what is observed in batteries where direct reduction from the zincate in abuttal to the separator appears to occur preferentially to reduction of ZnO in contact with the current collector (120).

Zinc penetration of separators occurs by a metal-deposition process (121). Separators prevent zinc penetration to the positive electrode because the overvoltage for zinc deposition in the separator is higher than that in the zinc electrode itself. This could be due to the fact that: (1) there is a lower zinc activity in the separator or (2) the pore size in the separator is such that a high overvoltage is required to nucleate zinc deposition. Another possibility is that the overvoltage is higher because of the lower diffusivity of zincate in the separator (122). Normally zinc will not penetrate a separator unless the zinc oxide in the zinc electrode becomes exhausted during charge (87); in which case the overvoltage increases and zinc deposition occurs in the separator.

5.2. Hydrogen Evolution

There are several investigations of hydrogen evolution on zinc in concentrated alkaline electrolyte (123-125). Table VII (125) is a summary of the electrodic parameters for hydrogen evolution on pure zinc in 6 N and 9 N KOH. No data are available for other battery electrolytes. Apparently the hydrogen overvoltage in both zinc chloride and zinc bromide electrolyte is considerably reduced at pH values below 2.

TABLE VII

Electrodic Parameters for Hydrogen Evolution on Zinc (125)

	6 N KOH	9 N KOH
Tafel slope (mV/decade)	124	124
Transfer coefficient (α)	0.48	0.48
Exchange current density, i_0 (A/cm^2)	8.5×10^{-10}	1.5×10^{-9}
Activation energy, $\Delta H°$ (kcal/mol)	13.4	--

In alkaline cells mercury is usually added to the electrodes to decrease the rate of hydrogen evolution (126,127). The efficacy of mercury in this regard is obvious when one considers the electrodic parameters for Zn-Hg alloys given in Table VIII (127). A similar inverse relationship between hydrogen evolution and mercury content is found in actual cells (126,128,129). Lead additives also decrease the rate of hydrogen evolution in cells. Surprisingly, the rate is independent of lead content in the range of 1.8 to 8.4 w/o (127). Hydrogen evolution does not present a problem in sealed cells if adequate precautions are taken to remove impurities such as iron, and if the mercury content is 2 to 4% (87,126).

Based on the result of gas-collection experiments it has been reported that the presence of nitrate ions increases the rate of hydrogen evolution on zinc in alkaline electrolyte (130). Recently these experiments have been repeated, and it was concluded that the gas collected was mostly ammonia. Hence nitrate ions do not accelerate hydrogen evolution on zinc in caustic electrolyte (87).

TABLE VIII

Electrode Parameters for Hydrogen Evolution on Zn-Hg Alloys (127)

Alloy	Tafel slope (mV/decade)	Transfer coefficient (α)	Exchange current density (i_0 (A/cm^2))
Zn	124	0.48	1.5×10^{-9}
Zn + 2% Hg	116	0.51	2.7×10^{-10}
Zn + 4% Hg	98	0.60	8×10^{-11}
Zn + 8% Hg	86	0.69	6×10^{-12}

5.3. Oxygen Reduction

It has been demonstrated that oxygen reduction occurs rapidly on zinc electrodes in alkaline electrolyte (87). Little is known of the mechanism of reduction except that the reduction yields large amounts of peroxide that can oxidize organic separators such as cellophane in zinc/silver oxide cells.

6. CYCLING BEHAVIOR OF THE ZINC ELECTRODE

Several processes can occur during cycling of a zinc electrode that can lead to operational problems in cells, resulting in short cycle life.

In cells where all the discharge product is dissolved in the electrolyte, smooth adherent zinc deposits can be initially obtained by a judicious choice of additives and hydrodynamic conditions. However, after a few cycles the zinc often changes to a mossy nonadherent deposit. Changes in zinc morphology with cycling have been observed by several authors in zincate electrolyte (105, 106,131). Flowing electrolytes do not eliminate these morphology changes (105). Similar phenomena have been observed on cycling zinc electrodes in $ZnCl_2$ (42) and $ZnBr_2$ electrolytes (108). In $ZnCl_2$ electrolyte, morphology changes have been ascribed to impurities such as iron (42), and it has been claimed that purification of the electrolyte and use of additives improves the zinc morphology (113).

Zinc morphology changes occur less rapidly if the zinc is completely discharged (dissolved) on each cycle. Powers found that a cobweb-like structure remained when anodic films formed on zinc were dissolved (132). These "cobwebs," which are conductive and contain zinc, were found to promote the formation of nonadherent mossy zinc deposits. More recently it has been found that the onset of nonadherent deposits in $ZnBr_2$ electrolyte can be prevented by stripping the zinc from the current collector potentiostatically every few cycles (108). In this case also, it appears that residual material from the previous discharge catalyzes the formation of nonadherent deposits. From the limited information available on this topic, and with our current understanding of zinc deposition, it appears that the morphology changes are related to changes in the rate of some step in the electrocrystallization process. Even though in some instances impurities and pH changes do affect this change in morphology (42,106), there are indications that these changes can be induced by merely cycling the zinc electrode. In alkaline electrolyte the formation of supersaturated zincate solutions may also be a factor.

The formation of nonadherent deposits gives rise to operational problems in cells with circulating electrolytes or moving substrates. These nonadherent deposits take the form of globular particles on the electrode surface. An example of this is given in Fig. 20 (105). Dendrites can grow from these protuberances and short the cell. The loosely adherent material can slough off the surface and block electrolyte ports. In cells with moving substrates the loosely adherent material can be rolled into nodules that short the cell. If battery systems with circulating electrolyte are to operate reliably and have long cycle life, no morphology changes should occur on the zinc electrode on either shallow or deep discharge regimens. To date the most promising solution to this problem appears to be through the use of additives.

In compact cells with separators, mossy zinc deposits are actually desired since they maintain the porosity and high-rate discharge capability of the electrode. Undesirable changes that can occur with cycling are zinc-electrode shape change, zinc densification, passivation, and dendrite formation.

Shape change involves the reduction of the zinc-electrode geometric area on cycling (133), where the zinc electrode active material is removed from the electrode edges and agglomerates toward the plate center. In this process the edge areas of the current collector are completely denuded of zinc. Once initiated, this movement of zinc progresses as cycling continues and results in a reduction of the capacity and useful life of the cell.

The cycle number at which shape change is first noted and the rapidity of the redistribution of the active material are related to several cell-construction and operating parameters. Excess electrolyte and electrolyte level do not have a significant effect on shape change. No significant differences are noted in the behavior of cells in which the electrolyte is completely contained in the porous electrodes and separators, and cells with free electrolytes over the top of the electrodes (120). Also, no significant effects are noticed when provisions are made to

Fig. 20. Electrodeposited zinc on a sponge nickel substrate. Counterelectrode type II air cathode: (a) first charge; (b) fifth charge; (c) ninth charge; (d) eighteenth charge; (e) twenty-sixth charge (105).

maintain the electrolyte at various levels throughout cycling. Cell orientation during cycling, charge, or discharge has no significant effect on shape change (120,134); thus gravitational effects have been ruled out as being major contributors to shape change in tightly packed cells (133,134). Shape change occurs more slowly in a tightly packed cell than in a loosely fitting cell pack (133-137). The rate at which shape change occurs with cycling is reduced by decreasing the depth of discharge, lowering the KOH concentration (138,139) and increasing the stoichiometric ratio of the negative:positive active material in the cell. The rate at which shape change occurs is increased by amalgamation of the zinc and is directly proportional to the mercury content (134). Some recent work has been performed with other heavy metal additives for preventing hydrogen evolution, with the objective of finding an additive that does not accelerate shape change (140). Thallium oxide, lead oxide, cadmium oxide, and combinations of these suppressed hydrogen evolution, and certain combinations reduced the rate of shape change in accelerated tests. The best of the combinations of additives were:

$$1.0 \text{ to } 5.0\% \text{ PbO}$$
$$1.0\% \text{ Tl}_2\text{O}_3 + 2.0 \text{ to } 5.0\% \text{ PbO}$$
$$0.5\% \text{ CdO}$$
$$1.0\% \text{ Tl}_2\text{O}_3 + 0.5\% \text{ CdO}$$

Sample sketches of the zinc electrodes after cycling are shown in Fig. 21 (140). Other factors that promote shape change are high-resistivity membranes, wrinkles in separators, bare spots or cracks in the electrodes (133,134), and nonalignment of positive and negative electrodes (134). The latter effect is rather dramatic (see Fig. 22) (134). Zinc is removed rapidly with cycling from

the edge of the negative electrode if the positive electrode is
larger and more slowly if the positive electrode is smaller.
There does not appear to be any significant effect of charge and
discharge rate on shape change. The parameters having the greatest
effect on shape change are the negative:positive stoichiometric
ratio, the type of separator used, the depth of discharge, the
degree of amalgamation of the zinc, and the alignment of positive
and negative electrodes in the cell pack. In one instance no shape
change was observed on cycling a Zn/NiOOH cell 70,000 times over
a regimen simulating hybrid electric vehicle service. The depth
of discharge varied from 0.5 to 3% (141). The cells had a capacity
of 35 A · h, and the total throughput during cycling was 16,500
A · h. Since this throughput is equivalent to 471 deep cycles,
this test illustrates the importance of the depth of discharge on
shape change.

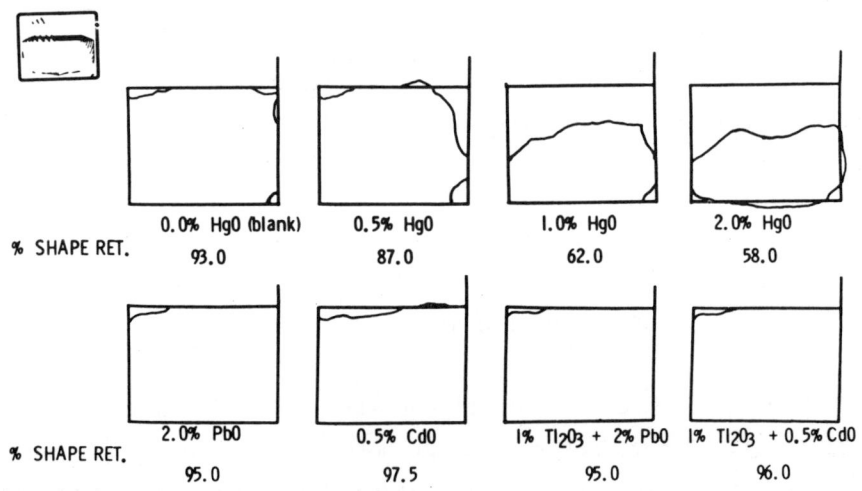

Fig. 21. Percent retention of electrode shape (140).

Fig. 22. Effect of electrode alignment on shape change (134).

Even though shape change is the life-limiting factor in many cells, very few investigations of shape change per se have been made (120,133,142,143,144). Reasons for this are that this type of study is very time consuming, interpretation of the data is difficult, great care must be taken in the experimental design, and considerable skill in cell construction is necessary to yield reproducible results. Lander and Cooper (142) carried out a preliminary study on the effects of current distribution on shape change. They found a reduction in the rate of shape change when the electrode-to-terminal leads were connected at opposite ends of the cell for negative and positive plate groups.

McBreen (133) carried out an extensive experimental study on current distribution on a cycling zinc electrode. He found that the current distribution on discharge was more uniform than on charge and tended to become more uniform the longer the cell was discharged. He postulated that these differences in current distribution could lead to development of concentration cells between the edges and center of the electrode and that these concentration cells could lead to shape change. Potential measurements confirmed the presence of concentration cells. In general, McBreen found that the active zinc material tended to move away from high current density areas and accumulate in low current density areas. He used this mechanism to explain the effects of various cell parameters on shape change, such as high-resistivity separators, wrinkling separators, misalignment of positive and negative electrodes, stoichiometric ratios, and amalgamation.

Recently another mechanism based on convection induced by electroosmosis through the separator has been proposed (144,145). Many separator membranes exhibit electroosmosis effects in alkaline electrolytes. During charge the electroosmosis could induce a flow of electrolyte from the center to the edge of the electrode. On discharge the reverse would occur. It has been proposed that the outward flow would consist of a flow of dilute zincate electrolytes and the inward flow would be a flow of concentrated zincate electrolytes. Hence there would be a net transfer of material from the edge to the center.

A mathematical analysis has been made of the effect of electroosmotic pumping on zincate transport in cells (144). This analysis also took into account porosity variations in the electrodes and variations in the current density over the surface of the electrode and predicted that zinc species would be transferred across the electrode in a manner similar to that observed in operating cells.

An experimental study was also made on Zn/AgO cells with radiation-grafted polyethylene separators. The presence of electroosmosis was confirmed. Experiments were carried out in cells where provisions were made to prevent convection. Shape change was considerably less than in cells where convection was permitted to occur. However, this cannot be regarded as a critical test of the theory since several differences in cycling regimen and cell construction existed between the cells. The cell with the reduced shape change was cycled at about 50% of the depth of discharge used in the control cell and the construction was such that the zinc electrode essentially operated in a "half box." In the control cell current paths existed beyond one edge of the electrode. The theory predicts Zn and ZnO distributions that do not occur in cells. Shape change also occurs in cells with microporous separators. The separator pore sizes are such (0.05 μm) that electroosmosis is unlikely. The critical experiment has yet to be performed to determine the validity of either the concentration cell or electroosmosis mechanism.

Densification and loss of electrode porosity are often observed in electrodes that have undergone shape change. The zinc often displays a metallic luster when scraped with a knife. It is not clear whether this is a separate phenomenon or is merely due to the fact that the shape change transports the material into limited volume. Passivation may accompany densification, resulting from increased current density caused by loss of zinc surface area.

Several methods have been used to mitigate shape change. These are: (1) the addition of binders such as polytetrafluoroethylene (146,147) or polyvinyl alcohol (132) to the zinc electrode, (2) the extension of the negative plate edges beyond the edges of the positive electrode (148), (3) the use of contoured negative electrodes with extra ZnO at the electrode edges (147,149), or

(4) zinc electrodes that are thicker at the plate top than bottom (150). Another promising approach is the use of improved separators in cells as opposed to the cellulosic separators in current use (151,152).

Dendrites alone do not cause any problems in cells. However, their presence indicates that zinc deposition is occurring at high overvoltages -- a condition favorable to zinc deposition in the separator. Dendrites normally will not be formed if there is an excess of reducible zinc species in the zinc electrode. Exhaustion of reducible species can be brought about during cell operation if there are disparities between the charging efficiencies of the positive and negative electrodes (87), such as occurs in vented Zn/NiOOH cells.

Sealed operation compensates for this disparity in charging efficiency, maintains the concentration of reducible species, and prevents dendrite growth (87).

7. ZINC ELECTRODE STRUCTURE AND PREPARATION

Since this chapter is devoted primarily to the zinc electrode as it applies to various battery systems, it should be instructive to briefly consider the zinc electrode structure in various batteries. Primary zinc batteries, developed prior to the turn of the century, had zinc electrodes that consisted of solid zinc rod or sheet. The reason for this was that in most of these primary batteries the electrolyte pH was so low that no zinc passivation occurred (e.g., the Poggendorff cell) or that the rate of the positive electrode reaction was limiting, so that there was no need for a zinc electrode of high surface area (e.g., the Leclanché cell).

Zinc Electrode Structure and Preparation

In 1914 Achenbach (153) proposed a corrugated zinc wire fabric electrode for this alkaline Zn/MnO_2 cell. He was probably the first to recognize the advantages of a porous zinc electrode in primary batteries. In the 1940s several porous zinc electrodes were developed for high-rate alkaline Zn/MnO_2 and Zn/AgO primary cells.

Porous zinc primary-cell electrodes can be prepared by electroplating dendritic deposits of zinc from a zincate solution (154). Such electrodes consist of compacted dendrites with diameters of 10 to 50 μm (155) and have porosities of about 70% and surface areas of 0.2 to 0.15 m^2/g (75).

Electrodes have also been prepared by electrochemical reduction of confined pastes of slightly soluble zinc compounds such as ZnO and $ZnCO_3$ (76,77). These electrodes have higher specific surface areas than electrodes prepared by electroplating. Zinc electrodes prepared by the electroreduction of ZnO pastes consist of a mossy deposit of zinc whiskers with diameters in the range of 0.6 to 1.0 μm (155).

Zinc electrodes for primary Zn/MnO_2, Zn/Ago, and Zn/air cells have been prepared from pastes of zinc powder and binders, such as polyvinylalcohol or carboxymethylcellulose (78,156). Zinc powders between 60 and 200 mesh size are used (78). Surface areas greater than 0.1 m^2/g yield the best results (75). These types of electrodes operate well in alkaline cells. Utilizations as high as 80% have been obtained for 5.3-mm-thick electrodes at current densities in the range of 0.028 to 0.123 A/cm^2 (78).

One interesting feature of porous zinc electrodes is that there is little electronic resistance at porosities below 85%. Above this porosity the metal structure is consumed to a significant extent on discharge, causing the formation of high-resistance paths and crumbling of the metal (76).

The ability of zinc pastes and slurries to conduct current has been used to advantage in cell design. Some zinc/air battery concepts are based on flowing slurries of zinc particles (33,157).

Two types of zinc electrodes are used in rechargeable batteries: (1) zinc electrode (solid electrode; dissolved reactants or products), and (2) solid electrode (solid electrode; solid reactants or products). In type (1) the quantity of electrolytes is such that the zinc discharge products are completely dissolved. With this type of electrode various schemes are often used to enhance mass transport to and from the electrode. These are additives, flowing electrolytes, moving substrates, or substrates consisting of a fluidized bed of finely divided particles. Attempts have been made to convert the zinc electrode to an electrode of type (2). One method is to limit the quantity of electrolyte to such an extent that precipitation of the product occurs. In some cases, additives are incorporated to promote even a greater degree of discharge product precipitation. This type of electrode is used in conventional Zn/AgO and Zn/NiOOH batteries. In electrodes of this type, deposition and dissolution occur in the presence of a number of soluble species such as $Zn(OH)_2$, $Zn(OH)_3^-$, and $Zn(OH)_4^=$. Cells of this type usually have a compact structure. The porosities of zinc electrodes of this type are typically 70%. As in the case of many rechargeable batteries, the zinc-electrode structure is not as dependent on the initial design as it is on the cycling regimen. In free-electrolyte systems the electrode structure is very dependent on the cycling regimen, hydrodynamic conditions, and impurities in the electrolyte. However, the goal always is to plate a compact adherent zinc deposit during the charging process.

8. ZINC CELLS: A STATUS REPORT

Zinc continues to be an extremely important electrode for use in electrochemical energy conversion. It serves as the negative electrode in both primary and rechargeable cells, with both acid and alkaline electrolytes. Despite all of the efforts applied to the development of other negative electrodes, it still offers a very attractive combination of features: (1) reasonably low equivalent weight, (2) low electronegativity (high cell voltage), (3) high exchange current density, (4) rechargeability, (5) low cost, (6) abundance, and (7) ease of fabrication into electrodes.

Improvements continue to be made in the cells that contain zinc electrodes, and zinc cells are used in a wide variety of sizes in many different applications. The status of the performance, lifetime, recent work, and problems for a number of zinc cells is reviewed here.

8.1. Primary Cells

8.1.1. THE ZINC/MANGANESE DIOXIDE CELL. This cell continues to be the most ubiquitous of all zinc cells and is used predominantly in small sizes, from a fraction of 1 to about 10 A · h. As indicated in Table IX, the cell voltage at the start of discharge is 1.54 V. The theoretical specific energy, as calculated from Eq. 17, is $NF\bar{E}/\Sigma MW$. Where $\bar{E}$ is the average reversible potential for the cell reaction; ΣMW is the summation of the molecular weights of the reactants; N is the number of equivalents involved in the reaction; and F is Faraday's constant.

Zinc/manganese dioxide cells with ammonium chloride electrolyte are still used for relatively light-duty applications such as flashlights, and radios, and offer 20 to 50 W · h/kg at low

TABLE IX

The Zinc/Manganese Dioxide Cells
(Zn + 2MnO$_2$ → ZnO·Mn$_2$O$_3$:
E = 1.54 V; 318 W·h/kg Theoretical)

Status	Reference
Specific energy[a]	
NH$_4$Cl + ZnCl$_2$ Electrolyte, 20 to 50 W·h/kg @ 1 W/kg	156,158,159
KOH Electrolyte, 30 W·h/kg @ 10 W/kg	156,159
Specific power[a]	
NH$_4$Cl + ZnCl$_2$ Electrolyte, 2 W/kg @ 2 to 4 W·h/kg	156,158,159
KOH Electrolyte, 6 W/kg @ 8 to 10 W·h/kg	156,159
Cycle life	
NH$_4$Cl + ZnCl$_2$ Electrolyte, one discharge only	
KOH Electrolyte, 50 @ 7 W/kg, 30% depth of discharge	156
Recent work	156,158
Rechargeable cells (KOH electrolyte)	
High-rate, thin-electrode cells	
Improved binders for MnO$_2$ electrodes	
Improved current collection in MnO$_2$	
Additives to electrolyte to p.p.t. Zn	
Problems	
Seals	
High internal resistance	
Poor utilization at high specific power	
Poor rechargeability of MnO$_2$ electrodes	

[a] The values can be converted to energy or power per unit volume by use of the density, 2 g/cm^3.

specific power (~1 W/kg). The use of a potassium hydroxide electrolyte provides a higher specific power (by a factor of ca. 10) without any sacrifice in specific energy, for heavy-duty applications. These cells are almost always used as primary cells, however, there has been a significant effort in the last several years to develop a rechargeable version.

The studies on rechargeability of the $Zn/KOH/MnO_2$ cell have identified MnO_2 as the more irreversible electrode. If this electrode is discharged below $MnO_{1.5}$, it suffers a significant permanent capacity loss due to the accumulation of Mn_3O_4, which cannot be recharged. Cycling to only 30% of the nominal capacity can be performed up to about 50 cycles.

Improvements in the MnO_2 electrode have been made in the areas of current collection and greater strength through improved binders such as polystyrene and latex. The capacity of the zinc electrode has been increased by the addition of materials to precipitate some of the zinc from the electrolyte, thus avoiding saturation. One such additive is calcium, in the form of a soluble salt.

Seals remain a problem, and new designs continue to be developed. The internal resistance of Zn/MnO_2 cells continues to be higher than that for most other cells, contributing to the problem of poor utilization of active materials at high specific power.

8.1.2. THE ZINC/MERCURIC OXIDE CELL. This cell has the unique property of a very stable voltage on discharge (1.35 V) and is capable of 80 to 90% utilization of the active materials. It makes use of a concentrated KOH electrolyte (~40 w/o), and high specific power versions contain porous zinc electrodes in a multiple-electrode arrangement (160). The specific energy achieved in practical cells is almost 40% of the theoretical value at low rates (see Table X). This high specific energy, coupled with a density

TABLE X

The Zinc/KOH/Mercuric Oxide Cell
(Zn + HgO → ZnO + Hg:
E = 1.35 V; 225 W · h/kg Theoretical)

Status	Reference
Specific energy[a]	
45 to 70 W · h/kg @ W/kg	156,160
100 W · h/kg @ 1 W/kg	156,159
Specific power[a]	
10 W/kg @ 15 W · h/kg	156
Cycle life	
Usually used as primary cell; special construction permits up to 300 shallow cycles	161
Recent work	
Compact cells for pacemakers	162
High-rate cells-multiple electrodes	160
Larger cells (55 A · h)	160
Porous Zn electrodes	160
Problems	
Limited rechargeability	
Low specific power	
Mercury hazard	

[a]These values can be converted to energy or power per unit volume by use of the density, 3.5 g/cm^3.

of 3.5 g/cm^3, provides for an unusually high energy per unit volume, making the Zn/HgO cell valuable in volume-limited uses such as power for portable electronic devices, cameras, radios, and watches. Recent work has been devoted to the construction of very high reliability, sealed cells for pacemakers, and special large cells (~55 A · h) for military applications (160).

In secondary operation the current problems with this cell are limited cycle life, depth of discharge, and low specific power. The disposal of the cell can present problems because of the significant health hazard associated with mercury.

8.1.3. THE ZINC/KOH/AIR CELL. The idea of the zinc/air cell has been very attractive because of its high theoretical specific energy (1310 W · h/kg, cf Table XI). Because of the high efficiency of the discharge process for the zinc electrode, specific energies in the range of 300 W · h/kg have been achieved (23% of the theoretical value) (163). These very high specific-energy cells are being evaluated in some applications now being served by the Zn/KOH/HgO cell since they offer longer intervals between replacements. The recharge process is rather difficult and inefficient owing to the problems of zinc deposition and the high overvoltage of the oxygen electrode. These problems result in a low overall efficiency in the vicinity of 40% (164). The provision for recharge involves the use of an interesting zinc slurry electrode. As the cell discharges, the zinc slurry (in KOH) is converted to a zinc oxide slurry. This zinc oxide slurry is reconverted to zinc outside of the cell (164). The demonstrated cycle life is still rather short (165).

Multikilowatt zinc/air batteries have been tested in an electric automobile (157). Work is still needed for the improvement of the overall efficiency and the lifetime, among other variables.

TABLE XI

The Zinc/KOH/Air Cell
($Zn + \frac{1}{2} O_2 \rightarrow ZnO$:
E = 1.6 V; 1310 W · h/kg Theoretical)

Status	Reference
Specific energy	
Primary, 200 to 300 W · h/kg @ 10 W/kg	163
Rechargeable, 70 to 90 W · h/kg @ 20 W/kg	164
Specific power	
Primary, 18 W/kg	163
Rechargeable, 80 W/kg peak	164
Cycle life, <100	165
Recent work	
Improved circulating zinc electrodes	164
Improved air electrodes	164
Multicell batteries built (~3 kW)	157
Zinc recharge unit	164
Problems	
Rechargeable air electrodes	157,166
Control of zinc in slurry	166
Wetting	
Carbonation	
Short life	
Low efficiency (~35%)	

8.2. Secondary Cells

8.2.1. THE ZINC/KOH/SILVER OXIDE CELL. A highly successful primary cell, the zinc/silver oxide system continues to be of interest for secondary applications requiring both high specific energy and high specific power. Table XII shows that the primary version of this cell can deliver 200 W · h/kg. This is important in a number of military applications. Good utilization of the zinc electrode is obtained by providing a porosity of about 70%. These electrodes can be prepared by either electrodeposition and pressing or bonding fine zinc powder to a current collector. For rechargeable cells, bonded structures are also used, based on a rolling and pressing process (167) or a vacuum-table process (151), starting with zinc oxide.

The specific energy of the rechargeable version of this cell is lower than that for the primary version partly because heavier zinc electrodes are necessary to achieve a significant cycle life. The cycle life is limited by the behavior of the zinc electrode discussed earlier in this chapter, namely, shape change, densification, and passivation. This behavior is improved by using additives to the zinc electrode, extending the zinc electrode edges, providing for uniform current density, and using improved separators.

In nonmilitary applications the high cost of silver limits the range of usefulness of this cell. Only special purposes are served where high specific energy and high specific power are essential, and cost is secondary.

8.2.2. THE ZINC/KOH/NICKEL OXIDE CELL. This cell offers a reasonably high specific energy in the range of 75 W · h/kg, with a lower material cost than that of the zinc/silver oxide cell. It

TABLE XII

The Zinc/KOH/AgO Cell
($Zn + AgO + H_2O \rightarrow Zn(OH)_2 + Ag$:
$E = 1.86$; 434 W·h/kg Theoretical)

Status	Reference
Specific energy	
Primary, ~200 W·h/kg	3,4
Rechargeable, 75 to 120 W·h/kg @ 60 W/kg	3,4,167,168
Specific power, 400 W/kg @ 60 W·h/kg	3,4
Cycle life, 100 to 150 @ 100% DOD	4,167
Recent work	
Improved Zn electrodes	
Nonsintered electrode	151,167
Separators--inorganic/organic	169
Problems	
Short cycle life	
High cost	
Zn shape change	
Separators	

is not widely available and is still undergoing development. The specific power is rather high, as shown in Table XIII, and the cycle life is among the best for zinc cells, though not as great as available from the Cd/NiOOH cell.

The zinc electrodes used in this cell are very similar to those of the Zn/AgO cell, and the same design practices are followed. Polymer-bonded electrodes, prepared by the rolling and pressing and the vacuum-table processes are common (151,170,171). One unique

TABLE XIII

The Zinc/KOH/Nickel Oxide Cell
($Zn + 2NiOOH + H_2O \rightarrow ZnO + 2Ni(OH)_2$:
E = 1.74 V; 373 W · h/kg Theoretical)

Status	Reference
Specific energy, 60 to 80 W · h/kg @ 30 W/kg	151,170,173
Specific power, 200 to 300 W/kg @ 35 W · h/kg	174
Cycle life, 100 to 250 @ 25 to 50 W/kg--100% DOD	151,170,173, 171
Recent work	
Inorganic separators	
(e.g., K_2TiO_3, ZrO_2, others)	151,170
Sealed cells	151,87
Polymer-bonded	151,170,171
Vibrating Zn electrodes	172
Problems	
Sealing of cells--O_2 evolution and recombination; H_2 evolution	
Shape change and densification of zinc electrode	
Separators	

approach recently reported involves the use of a densely deposited zinc electrode, prepared by vibrating the zinc electrode in the cell (±~2 mm, 50 Hz) during recharge (172). This allows a flat deposit to be formed, avoiding dendrites and shape change.

The nickel oxide electrodes used in this cell may be either of the type based on a sintered porous nickel plaque (172) (characteristic of Cd/NiOOH cells), or a polymer-bonded electrode (151,170, 171). Recent work on improvement of cycle life has centered around the zinc electrode (cf. comments on the Zn/AgO cell in preceding paragraph) and the separator.

Efforts are being devoted to sealed cells to obtain maintenance-free system (87,173). This means that any oxygen evolved during recharge must be recombined with the zinc, maintaining the state-of-charge balance between the electrodes. This cell is being developed for possible use in electric vehicle propulsion and various military applications.

8.2.3. THE ZINC/$ZnCl_2$/CHLORINE CELL. The zinc/chlorine cell is one of the systems that has attracted interest rather recently (175-178). The zinc electrode in this system is dissolved completely into the circulating zinc chloride electrolyte during discharge and is replated as a dense deposit from the flowing electrolyte during recharge. Additives in the electrolyte maintain a dense deposit; impurities such as iron cause problems, including hydrogen evolution (178), so high electrolyte purity must be maintained. The chlorine formed on recharge is stored as $Cl_2 \cdot 6H_2O$ in a refrigerated compartment. The chlorine moves throughout the system dissolved in the circulating electrolyte (175).

Table XIV shows that a specific energy of 66 W · h/kg has been achieved and a peak specific power of 60 W/kg is available. The best cycle life demonstrated so far in a 1 kW · h Zn/Cl_2 battery is 100 cycles, with electrolyte repurification required every few cycles (1978).

Various materials have been evaluated for the electrode substrates, including carbon and ruthenium oxide-coated titanium. A significant amount of effort has been devoted to the development of system components such as pumps, heat exchangers, the refrigerator needed for the formation of $Cl_2 \cdot 6H_2O$, the solid-liquid phase separator, and other apparatuses. A 30 kW · h system is under

TABLE XIV

The Zinc/ZnCl$_2$/Chlorine Cell
(Zn + Cl$_2$ → ZnCl$_2$:
E = 2.12 V; 825 W · h/kg Theoretical)[a]

Status	Reference
Specific energy, 66+ W · h/kg @ 3 to 4 W/kg	175
Specific power, 60 W/kg for seconds	175
Cycle life, 100+[b]	178
Recent work	
Precious metal on Ti for Cl$_2$ electrode	177
Additives for Zn deposition	175,176
System components	177
Problems	
Impurities in electrolyte	178
Precious metals	
Bulky	
Complex	
Low specific power	
Gaskets	

[a] 460.6 W · h/kg for Zn/Cl$_2$·6H$_2$O

[b] With electrolyte purification every few cycles.

construction for testing late in 1977. One objective of the development efforts is a system for off-peak energy storage. The problems remaining to be solved are indicated in Table XIV.

8.2.4. THE ZINC/ZnBr$_2$/BROMINE CELL. The zinc/bromine cell has been under investigation for a number of years (179) but only recently has reached the stage of engineering development (180). Table XV summarizes the properties of the system.

TABLE XV

The Zinc/ZnBr$_2$/Br$_2$ Cell
(Zn + Br$_2$ → ZnBr$_2$:
E = 1.82 V; 430 W · h/kg Theoretical)

Status	Reference
Specific energy, 61 W h/kg @ 12 W/kg	180
Specific power, 73 W/kg peak	180
Cycle life, 200 @ 90 to 100% DOD	180
Recent work	
Prototype battery of 2.5 kW · h tested	180
Complexing of bromide	
Problems	
Complete discharge on each cycle required	
Flowing electrolyte--complex system	
Low efficiency: 65% W · h	
High self-discharge rate: 50% in 50 h	
Poor Zn adherence; Zn dendrites	
Recharge of bromine electrode	

A prototype battery of 30 cells, producing 2.5 kW · h and 3 kW, was built and tested, yielding 61 W · h/kg, with a peak power of 73 W/kg. This battery demonstrated a cycle life of 200 deep cycles. To avoid zinc dendrites it is important to discharge all of the zinc on every cycle. Some problems were experienced with zinc adherence and recharge of the bromine electrode. Other topics needing attention are indicated in Table XV.

9. CONCLUDING REMARKS

This chapter has attempted to give some perspective on the history and importance of the zinc electrode in electrochemical energy conversion. Zinc is the oldest and most commonly used negative electrode in cells and batteries, and it has continued to hold a very important position in both acid and alkaline cells, primary and rechargeable.

The chemistry and electrochemistry of zinc have been shown to be very complex. Many questions regarding the mechanism and kinetics of the electrode reactions of zinc remain open. Numerous practical issues relevant to the practical design and operation of zinc electrodes (e.g., shape change, densification, the action of additives) are not well understood.

Despite the absence of satisfactory answers to many important questions, progress continues in the development of zinc batteries, some of which may play an important role in our quest to make more effective and efficient use of our dwindling energy resources.

10. ACKNOWLEDGEMENT

Figures 1, 3, 4, 9, 12, 15, 16, 17, 19 and Table VI are published by permission of the publisher, The Electrochemical Society, Inc. Figure 9 was originally presented at the 148th Fall Meeting of The Electrochemical Society, Inc. in Dallas, Texas. Figures 2 and 6 reprinted from Ref.(25), p. 5, by courtesy of Marcel Dekker, Inc.

11. REFERENCES

1. C. H. Mathewson, ed., Zinc, Reinhold, New York, 1959.
2. S. D. Agade and F. B. Leitz, Report of the Electrolytic Industries for the Year 1975, The Electrochemical Society, Princeton, N. J., May 1976.
3. A. Fleischer and J. J. Lander, eds., Zinc-Silver Oxide Batteries, Wiley, New York, 1971.
4. S. U. Falk and A. J. Salkind, Alkaline Storage Batteries, Wiley, New York, 1969.
5. G. W. Heise and N. C. Cahoon, The Primary Battery, Vol. 1, Wiley, New York 1971.
6. A. Volta, Phil. Trans., Roy. Soc. (Lond.), $\underline{90}$, 403 (1800).
7. N. Gautherot, Phil. Mag., $\underline{24}$, 183 (1806).
8. J. F. Daniell, Phil. Mag. III, $\underline{8}$, 421 (1836); Phil. Trans., $\underline{126}$, 106, 125 (1836); $\underline{127}$, 141 (1837).
9. W. R. Grove, Phil. Mag. III, $\underline{13}$, 430 (1838); $\underline{14}$, 388 (1839); $\underline{15}$, 287 (1839).
10. J. Poggendorff, Pogg. Ann. Phys., $\underline{57}$, 101 (1842).
11. A. A. de la Rive, Arch. Electrochem., $\underline{3}$, 112 (1843).
12. G. Leclanché, French Patent No. 71,865 (1866).
13. G. Leclanché, Les Mondes, $\underline{16}$, 532 (1868).
14. F. Lalande and G. Chaperon, U.S. Patent No. 274,110, March 20, 1883.
15. T. de Michalowski, British Patent No. 15,370 (1899).
16. W. Jungner, Swedish Patent No. 15,567 (1901).
17. T. A. Edison, British Patent No. 20,072 (1901).
18. J. J. Drumm, U.S. Patent Nos. 1,955,115 (1934); 2,277,636 (1942).
19. H. André, Bull. Soc. Franc. Electriciens, $\underline{1}$, 132 (1941).
20. C. L. Clarke, U.S. Patent No. 298,175 (1884).

21. H. Aron, German Patent No. 38,220 (1886).
22. S. Ruben, U.S. Patent No. 2,422,045 (1947).
23. S. Ruben, Trans. Electrochem. Soc., $\underline{92}$, 183 (1947).
24. T. P. Dirkse, J. Electrochem. Soc., $\underline{106}$, 154 (1959).
25. R. J. Brodd and V. E. Leger, in Encyclopedia of Electrochemistry of the Elements, Vol. 5, A. J. Bard, ed., Dekker, New York, 1976.
26. J. S. Fordyce and R. L. Baum, J. Chem. Phys., $\underline{43}$, 843 (1965).
27. W. H. Dyson, L. A. Schreier, W. P. Scholette, and A. J. Salkind, J. Electrochem. Soc., $\underline{115}$, 566 (1968).
28. C. T. Baker and I. Trachtenberg, J. Electrochem. Soc., $\underline{114}$, 1045 (1967).
29. T. P. Dirkse, Technical Report No. AFAPL-TR-69-90, Contract No. AF33[615]-3292, Calvin College, Grand Rapids, Mich., December 1969.
30. V. N. Flerov, Zh. Fiz. Khim., $\underline{31}$, 310 (1957).
31. T. P. Dirkse, J. Electrochem. Soc., $\underline{102}$, 497 (1955).
32. V. N. Flerov, Zh. Fiz. Khim., $\underline{31}$, 49 (1957).
33. A. J. Appleby, J. Jacquelin, and J. P. Pompon, SAE Trans., Paper No. 770381 (1977).
34. W. Van Doorne, J. Electrochem. Soc., $\underline{122}$, 1 (1975).
35. A. G. Briggs, N. A. Hampson, and A. Marshall, Trans. Faraday Soc., $\underline{70}$, 1978 (1974).
36. J. McBreen, Report No. N68-15716, Contract No. NAS 5-10231, Yardney Electric Corp., New York, June 1967.
37. T. P. Dirkse, Technical Report No. AFAPL-TR-72-87, Contract No. S33615-70-C-1022, Project 3145, Calvin College, Grand Rapids, Mich., December 1972.
38. L. Nanis, JPL Contract No. 952,543, Report No. N70-23265, University of Pennsylvania, February 1970.
39. D. P. Boden, R. Wylie, and V. J. Spera, J. Electrochem. Soc., $\underline{118}$, 1298 (1971).

40. J. W. Mellor, Comprehensive Treatise on Inorganic and Theoretical Chemistry, Vol. 4, Longmans Green, London, 1969, p. 539.
41. P. C. Symons, Paper presented at Society for Electrochemistry, International Conference on Electrolytes for Power Sources, Brighton, England, December 13 and 14, 1973.
42. A. F. Sammells, in Energy Storage, J. B. Berkowitz and H. P. Silverman, The Electrochemical Society, Princeton, N.J., 1976, p. 121.
43. R. Dietz, Berichte, 32, 95 (1899).
44. H. S. Lim, A. M. Lackner, and R. C. Knechtli, J. Electrochem. Soc., 124, 1154 (1977).
45. N. de Zoubov and M. Pourbaix, Atlas of Electrochemical Equilibria in Aqueous Solutions, National Association of Corrosion Engineers, Houston, Texas, 1974, p. 406.
46. W. Feitnecht, Metaux et Corrosion, 23, 192 (1947).
47. Z. A. Nikitina, Zh. Priklad. Khim., 31, 218 (1958).
48. P. Brouillet and F. Jolas, Electrochim. Acta, 1, 246 (1959).
49. N. C. Cahoon and G. W. Heise, Trans. Electrochem. Soc., 94, 214 (1948).
50. R. Landsberg, Z. Phys. Chem., 206, 291 (1957).
51. R. Landsberg and H. Bartel, Z. Electrochem., 61, 162 (1957).
52. M. Eisenberg, H. F. Bauman, and D. M. Brettner, J. Electrochem. Soc., 108, 909 (1961).
53. N. A. Hampson and M. J. Tarbox, J. Electrochem. Soc., 110, 95 (1963).
54. N. A. Hampson, M. J. Tarbox, J. T. Lilley, and J. P. G. Farr, Electrochem. Technol., 2, 309 (1964).
55. J. P. Elder, J. Electrochem. Soc., 116, 757 (1969).
56. T. P. Dirkse and N. A. Hampson, Electrochem. Acta, 16, 2049 (1971).

References

57. T. P. Dirkse, D. DeWitt, and R. Shoemaker, J. Electrochem., **115**, 442 (1968).
58. I. Sanghi and M. Fleischman, Electrochim. Acta, **1**, 161 (1959).
59. G. S. Vozdvizhenskii and E. D. Kochman, Zh. Fiz. Khim., **39**, 657 (1965).
60. T. I. Popova, N. A. Simonova, and B. N. Kabanov, Elektrokhimiya, **3**, 1419 (1967).
61. M. N. Hull and J. E. Toni, Trans. Faraday Soc., **67**, 1128 (1971).
62. M. N. Hull and J. E. Ellison, and J. E. Toni, J. Electrochem. Soc., **117**, 192 (1970).
63. R. W. Powers and M. W. Brieter, J. Electrochem. Soc., **116**, 719 (1969).
64. M. W. Brieter, Electrochim. Acta, **16**, 1169 (1971).
65. R. W. Powers, J. Electrochem. Soc., **118**, 685 (1971).
66. T. P. Dirkse and N. A. Hampson, Electrochim. Acta, **17**, 387 (1972).
67. R. W. Powers, J. Electrochem. Soc., **116**, 1652 (1969).
68. B. N. Kabanov, Electrochim. Acta, **6**, 253 (1962).
69. R. D. Armstrong and G. M. Bulman, J. Electroanal. Chem., **25**, 121 (1970).
70. F. Mansfeld and S. Gilman, J. Electrochem. Soc., **117**, 588 (1970).
71. R. V. Bobker, Zinc-in-Alkali Batteries, The Society for Electrochemistry, 1973, p. 13, University of Southampton, Great Britain.
72. K. L. Hampartzumian and R. V. Moshtev, in Power Sources, Vol. 3, D. H. Collins, ed., Oriel Press, Newcastle upon Tyne, 1971, p. 495.
73. T. F. Sharpe, J. Electrochem. Soc., **122**, 845 (1975).
74. G. Coates, N. A. Hampson, A. Marshall, and D. F. Porter, J. Appl. Electrochem., **4**, 75 (1974).

75. C. M. Shepherd and H. C. Langelan, J. Electrochem. Soc., 109, 657 (1962).
76. C. M. Shepherd and H. C. Langelan, J. Electrochem. Soc., 114, 8 (1967).
77. P. Ruetschi, in Power Sources, Vol. 4, D. H. Collins, ed., Oriel Press, Newcastle upon Tyne, 1973, p. 381.
78. R. R. Witherspoon, E. J. Zeitner, and H. A. Schulte, in Proceedings 1971 IECEC, Society of Automotive Engineers, pp. 96-102.
79. Z. Nagy and J. O'M. Bockris, J. Electrochem. Soc., 119, 1129 (1972).
80. T. P. Dirkse, in Power Sources, Vol. 2, D. H. Collins, ed., Pergamon, New York, 1970, p. 411.
81. J. Goodkin and F. Solomon, in Batteries, Vol. 2, D. H. Collins, ed., Pergamon, New York, 1965, p. 475.
82. R. W. Lewis and J. Turner, J. Appl. Electrochem., 5, 343 (1975).
83. A. Marshall, N. A. Hampson, and J. S. Drury, J. Electroanal. Chem., 59, 19 (1975).
84. A. Marshall, N. A. Hampson, and J. S. Drury, J. Electroanal. Chem., 59, 33 (1975).
85. A. Marshall, N. A. Hampson, J. S. Drury, and J. P. G. Farr, Surface Technol., 5, 149 (1977).
86. E. C. Jerabek, R. F. Thornton, and J. B. Weinstock, Proc. 26th Power Sources Symposium, PSC Publications Committee, Red Bank, N. J., 1974, p. 123.
87. J. McBreen, Extended Abstracts, The Electrochemical Society, Princeton, N. J., Vol. 77-1, Abstr. No. 356, p. 909.
88. S. Arouete, K. F. Blurton, and H. G. Oswin, J. Electrochem. Soc., 116, 166 (1969).
89. R. D. Naybour, J. Electrochem. Soc., 116, 520 (1969).

90. N. Ibl, in Advances in Electrochemistry and Electrochemical Engineering, Vol. 2, P. Delahay and C. W. Tobias, eds., Interscience, New York, 1962.
91. I. Epelboin, M. Ksouri, E. Lejay, and R. Wiart, Electrochim. Acta, 20, 603 (1975).
92. I. Epelboin, M. Ksouri, and R. Wiart, J. Electrochem. Soc., 122, 1206 (1975).
93. J. Bressan, P. Gaullochet, and R. Wiart, Extended Abstracts, 27th ISE Meeting, Zurich, Switzerland, September 6-11, 1976, Abstr. No. 78.
94. B. N. Kabanov, Izvest. Akad. Nauk SSSR, 980 (1962).
95. J. P. G. Farr and N. S. Hampson, J. Electroanal. Chem., 13, 483 (1967).
96. N. A. Hampson, G. A. Herdman, and R. Taylor, J. Electroanal. Chem., 25, 9 (1970).
97. J. O'M. Bockris, Z. Nagy, and A. Damjanovic, J. Electrochem. Soc., 119, 285 (1972).
98. I. Epelboin, M. Kasouri, and R. Wiart, J. Electroanal. Chem., 58, 488 (1975).
99. I. Epelboin, M. Kasouri, and R. Wiart, J. Electroanal. Chem., 65, 373, (1975).
100. J. W. Diggle, A. R. Despic, and J. O'M. Bockris, J. Electrochem. Soc., 116, 1503 (1969).
101. J. O'M. Bockris, Z. Nagy, and D. Drazic, J. Electrochem. Soc., 120, 30 (1973).
102. F. Mansfeld and S. Gilman, J. Electrochem. Soc., 117, 1521 (1970).
103. R. W. Powers, Electrochem. Technol., 5, 429 (1967).
104. R. D. Naybour, Electrochim. Acta, 13, 763 (1968).
105. R. Thacker, Energy Convers., 12, 17 (1972).

106. D. S. Adams, in Power Sources, Vol. 4, D. H. Collins, ed., Oriel Press, Newcastle upon Tyne, 1973, p. 347.
107. P. C. Symons, SAE Trans., Paper No. 730253 (1973).
108. F. G. Will, Proceedings Twelfth IECEC, Paper No. 779043, American Nuclear Society, La Grange Park, Ill., 1977.
109. R. Yu. Bek and N. T. Kudryavtsev, Zh. Priklad. Khim., $\underline{34}$, 2613 (1961).
110. F. Mansfeld and S. Gilman, J. Electrochem. Soc., $\underline{117}$, 1328 (1970).
111. F. Mansfeld and S. Gilman, J. Electrochem. Soc., $\underline{117}$, 1154 (1970).
112. J. W. Diggle and A. Damjanovic, J. Electrochem. Soc., $\underline{117}$, 65 (1970).
113. P. C. Symons and M. J. Hammond, EPRI Report No. EM-249, Energy Development Associates, Madison Heights, Mich., September 1976.
114. J. E. Oxley, NASA Report No. CR-377, Leesona Moos Laboratories, Great Neck, N. Y., February 1966.
115. V. V. Romanov, Zh. Priklad. Khim., $\underline{34}$, 2692 (1961).
116. J. N. Jovicavic, D. M. Drazic, and A. R. Despic, Electrochim. Acta, $\underline{22}$, 589 (1977).
117. R. W. Powers and M. W. Brieter, J. Electrochem. Soc., $\underline{116}$, 719 (1969).
118. O. Hladic and K. Schwabe, Electrochim. Acta, $\underline{15}$, 635 (1970).
119. D. Drazic and Z. Nagy, J. Electrochem. Soc., $\underline{118}$, 255 (1971).
120. J. McBreen and G. A. Dalin, Extended Abstracts, The Electrochemical Society, New York, Vol. 11, Abstr. No. 45, p. 123 (1966).
121. Z. Stachurski, NASA Report No. N67-26278, Yardney Electric Corp., New York, December 1965.

122. J. McBreen and G. A. Dalin, Extended Abstracts, The Electrochemical Society, New York, Vol. 12, Abstr. No. 36, p. 91 (1967).
123. M. B. Zhoulder and V. V. Steneler, Zh. Priklad. Khim, $\underline{31}$, 711 (1958).
124. Z. A. Jofa, L. V. Komlev, and V. S. Bagotskii, Zh. Fiz. Khim., $\underline{35}$, 1571 (1961).
125. T. S. Lee, J. Electrochem. Soc., $\underline{118}$, 1278 (1971).
126. R. N. Snyder and J. J. Lander, Electrochem. Technol., $\underline{3}$, 161 (1965).
127. T. S. Lee, J. Electrochem. Soc., $\underline{122}$, 171 (1975).
128. P. Ruetschi, J. Electrochem. Soc., $\underline{114}$, 301 (1967).
129. D. P. Gregory, P. C. Jones, and D. P. Redfearn, J. Electrochem. Soc., $\underline{119}$, 1288 (1972).
130. G. Schneider, Electrochim. Acta, $\underline{13}$, 2223 (1968).
131. M. W. Brieter, Electrochim. Acta, $\underline{15}$, 1297 (1970).
132. R. W. Powers, J. Electrochem. Soc., $\underline{116}$, 1652 (1969).
133. J. McBreen, J. Electrochem. Soc., $\underline{119}$, 1620 (1972).
134. G. A. Dalin, in Zinc-Silver Oxide Batteries, A. Fleischer and J. J. Lander, eds., Wiley, New York, 1971, p. 87.
135. T. Z. Palagyi, J. Electrochem. Soc., $\underline{108}$, 904 (1961).
136. F. Solomon and G. Work, in Power Sources, Vol. 2, D. H. Collins, eds., Pergamon, Oxford, 1965, p. 463.
137. H. K. Farmery and W. A. Smith, in Batteries, D. H. Collins, ed., Macmillan, New York, 1963, p. 179.
138. J. J. Lander, in Zinc-Silver Oxide Batteries, A. Fleischer and J. J. Lander, eds., Wiley, New York, 1971, p. 457.
139. G. A. Dalin, M. Sulkes, and Z. Stachurski, Proc. 18th Annu. Power Sources Conf., PSC Publications Committee, Red Bank, N. J., 1964, p. 54.

140. O. Wagner and A. Himy, in Proc. 27th Power Sources Symp. PSC Publications Committee, Red Bank, N. J., 1976, p. 135.
141. M. J. Sulkes, Report on EPA Task 690403, U.S. Army Electronics Command, Fort Monmouth, N. J., March 1972.
142. J. J. Lander and J. E. Cooper, Technical Report No. AFAPL-TR-71-32, Airforce Aero Propulsion Laboratory, Wright Patterson Air Force Base, Ohio, 1971.
143. K. W. Choi, D. N. Bennion, and J. Newman, J. Electrochem. Soc., $\underline{123}$, 1616 (1976).
144. T. W. Choi, D. Hamby, D. N. Bennion, and J. Newman, Report No. AFAPL-TR-76-9, Contract No. S-33615-74-C-2004, School of Engineering and Applied Science, University of California, Los Angeles, March 1976.
145. K. W. Choi, D. Hamby, D. N. Bennion, and J. Newman, J. Electrochem. Soc., $\underline{123}$, 1628 (1976).
146. G. A. Dalin and M. Sulkes, Proc. 19th Annu. Power Sources Conf., PSC Publications Committee, Red Bank, N. J., 1965, p. 69.
147. J. Goodkin, Proc. 22nd Annu. Power Sources Conf., PSC Publications Committee, Red Bank, N. J., 1968, p. 79.
148. J. McBreen, U.S. Patent No. 3,505,115 (1970).
149. J. Goodkin, U.S. Patent No. 3,493,434 (1969).
150. French Patent No. 1,601,216 (1970).
151. E. J. Cairns and J. McBreen, Ind. Res., $\underline{17}$, 56 (1975).
152. Sheibley, NASA Technical Memorandum No. NASA TMX-3465, National Aeronautics and Space Administration, Washington, D.C., December 1976.
153. E. Achenbach, U.S. Patent No. 1,090,372 (1914).
154. J. A. Keralla, in Zinc-Silver Oxide Batteries, A. Fleisher and J. J. Lander, eds., Wiley, New York, 1971, p. 183.

155. Z. Stachurski, Investigation and Improvement of Zinc Electrodes for Electrochemical Cells, NASA Report No. N67-2627.8, CR-84025, Yardney Electric Corp., New York, December 1965.
156. K. V. Kordesch, in Batteries, Vol. 1, K. V. Kordesch, ed., Dekker, New York, 1974, p. 241.
157. H. Baba, SAE Trans., Paper No. 710237, 1971.
158. R. Huber, in Batteries, Vol. 1, Manganese Dioxide, K. V. Kordesch, ed., Dekker, New York, 1974.
159. Eveready Battery Applications Engineering Data, Union Carbide Corp., New York, 1971.
160. F. Fagan, Y. L. Ko, and R. E. Ralston, Proc. 25th Power Sources Conf., PSC Publications Committee, Red Bank, N. J., 1972, p. 21.
161. S. Ruben, in The Primary Battery, Vol. 1, G. W. Heise and N. C. Cahoon, eds., Wiley, New York, 1971, p. 207.
162. H. A. Cataldi, Proc. 27th Power Sources Conf., PSC Publications Committee, Red Bank, N. J., 1976, p. 138.
163. H. R. Espig and D. F. Porter, in Power Sources, Vol. 4, D. H. Collins, ed., Oriel Press, Newcastle upon Tyne, England, 1973, p. 327.
164. A. J. Appleby, J. Jacquelin, and J. P. Pompon, SAE, Paper No. 770381, February 1977.
165. N. P. Yao, in Proc. Symp. and Workshop on Advanced Battery Research and Design, Report No. ANL-76-8, Argonne National Laboratory, Argonne, Ill., March 1976.
166. A. J. Appleby, J. P. Pompon, and M. Jacquier, in Proc. 10th IECEC, IEEE, New York, 1975, p. 811.
167. A. Charkey, Proc. 26th Power Sources Symp., PSC Publications Committee, Red Bank, N. J., 1974, p. 87.

168. H. J. Schwartz and D. G. Soltis, in Power Sources, Vol. 4, D. H. Collins, ed., Oriel Press, Newcastle upon Tyne, 1973, p. 185.
169. F. C. Arrance, R. Greve, and A. Rosa, Douglas Aircraft Report to NASA, No. CR-965, February 1968.
170. A. Charkey, in Proc. 4th Internat. Electric Vehicle Symp., Dusseldorf, 1976, Paper No. 32.3.
171. R. G. Gunther, Extended Abstracts, The Electrochemical Society, Princeton, N. J., Vol. 76-1, Abstr. No. 2, p. 10.
172. O. Von Krusenstierna and M. Reger, SAE, Paper No. 770384, February 1977.
173. E. J. Cairns, Extended Abstracts, The Electrochemical Society, Princeton, N. J., Vol. 77-2, Abstr. No. 35, p. 95.
174. F. P. Kober and A. Charkey, in Power Sources, Vol. 3, D. H. Collins, ed., Oriel Press, Newcastle upon Tyne, 1971.
175. P. C. Symons, SAE, Preprint No. 730253, February 1973.
176. P. C. Symons and H. K. Bjorkman, Jr., Presented at AIChE Meeting, Detroit, Michigan, June 1973.
177. P. C. Symons, in Proc. 3rd Internat. Electric Vehicle Symp., Washington, D.C., February 1974.
178. P. C. Symons and M. J. Hammond, Energy Development Associates Report to EPRI, No. EM-249, September 1976.
179. S. Barnartt and D. A. Forejt, J. Electrochem. Soc., $\underline{111}$, 1201 (1964).
180. G. Clerici, M. de Rossi, and M. Marchetto, in Power Sources, Vol. 5, D. H. Collins, ed., Academic, New York, 1975, p. 167.

Flow-through Porous Electrodes

JOHN S. NEWMAN
*Materials and Molecular Research Division, Lawrence Berkeley Laboratory
and
Department of Chemical Engineering, University of California
Berkeley, California*

WILLIAM TIEDEMANN
*Corporate Applied Research Group
Globe-Union, Inc.
Milwaukee, Wisconsin*

1.	Introduction	355
	1.1. Applications in Electrochemical Processes	355
	1.2. Cell Configuration	358
	1.3. Fluid Flow and Pressure Drop	360
	1.4. Mass Transfer	364
	1.5. Axial Dispersion	368
	1.6. Ohmic Potential Drop	371
2.	Effect of Axial Dispersion on the Average Mass-transfer Coefficient	372
3.	The Mass-transfer Coefficient	379
4.	Ohmically Limited Reactor Capacity	389
	4.1. Copper Recovery	398
	4.2. Lead Removal	399

*Work supported by the U.S. Energy Research and Development Administration and by Globe-Union, Inc.

	4.3. Desalination	400
	4.4. Downstream Counterelectrode	402
	4.5. Matrix Resistance	404
	4.6. Fluid Flow Parallel to Separator	407
5.	Maximum Conversion	408
6.	Historical Perspective	415
7.	Recent Work	420
8.	Conclusions	426
9.	List of Symbols	426
10.	References	429

1. INTRODUCTION

The choice of method, whether electrochemical or other, in carrying out a chemical conversion or a separation of chemical constituents requires an evaluation of process alternatives, ultimately on an economic basis. We explore here certain design considerations for a class of electrochemical reactors with the hope that this process alternative will not be overlooked for a lack of understanding of how to scale up laboratory data to a technically useful equipment configuration. If an electrochemical reactor is to compete with a chemical reactor, detailed information on design and scale up must be known so that optimum reactor designs are involved in the comparison.

Flow-through porous electrodes are attractive for industrial application because they provide intimate contact between a process stream and an electrode surface and also provide an extensive interfacial area in a given volume. Furthermore, the ohmic potential drop, which may govern the specificity of the reactor, can be analyzed if not minimized.

1.1. Applications in Electrochemical Processes

Each individual process is actually quite involved if consideration is given to every detail, and its discussion occupies many volumes in the literature. We restrict ourselves here to a brief mention of some reactions where flow-through porous electrodes could be important.

Metal processing includes the purification, electrowinning, and possibly electroplating of aluminum, copper, magnesium, sodium, manganese, nickel, gold, silver, and chromium. The electrorefining of aluminum from an aluminum-manganese alloy might involve a flow-through porous anode to prevent the dissolved manganese from reaching the cathode. The leaching of a copper ore may produce a solution too dilute for economical electrowinning. From a feed stream of 670 mg/liter, an electrochemical reactor can readily yield a concentrated stream greater than 46,000 mg/liter while reducing the effluent concentration to less than 1 mg/liter.

Closely related to this process is the recovery or removal of metal ions from waste process streams. We speak of recovery when the economics are favorable. Gold can be readily recovered from plating baths that have deteriorated, and streams from processing photographic emulsions can be reduced to less than 10 p.p.m. of silver rather than the 100 p.p.m. presently accepted. Mercury cells for production of caustic and chlorine could again become feasible when it is demonstrated that mercury levels can be reduced from 10 to 0.01 p.p.m. from a solution containing 280 g/liter of NaCl. The removal of lead from waste sulfuric acid solutions is a challenge because its electrode potential is below that of hydrogen. However, the fact that a lead-acid battery can be charged gives us hope.

Removal or reduction of contaminants usually involves a large volume of dilute solution with the contaminant in the range from 1 to 1000 p.p.m. For metal removal, retention of the metal within the reactor requires special design considerations for periodic removal of the material. Oxidation of organic contaminants and cyanide ions should proceed with little or no retention of solids.

Electroorganic synthesis is a field where flow-through electrochemical reactors may be able to produce a valuable chemical while controlling the potential for optimum avoidance of side reactions. Synthesis of benzoquinone from benzene is attractive, whereas adiponitrile (hexanedioic acid, dinitrile) is presently produced commercially by an electrochemical route through dimerization of acrylonitrile (propionic acid, nitrile):

Introduction

$$2\ CH_2=CH-CN + 2H^+ + 2e^- \rightarrow NC-(CH_2)_4-CN \qquad (1)$$

The merits of a design procedure for electroorganic synthesis could be tested with the reduction of nitrobenzene to aniline:

$$C_6H_5NO_2 + 6H^+ + 6e^- \rightarrow C_6H_5NH_2 + 2H_2O \qquad (2)$$

while avoiding the evolution of hydrogen, which occurs at a more negative electrode potential, and preventing the formation of 4-aminophenol from the intermediate, hydroxylaminobenzene, which forms at a more positive potential:

$$C_6H_5NO_2 + 4H^+ + 4e^- \rightarrow C_6H_5NHOH + H_2O \qquad (3)$$

$$C_6H_5NHOH \xrightarrow{H^+} HO-C_2H_4-NH_2 \qquad (4)$$

The design and operation of cells for the preparation of organic and inorganic chemicals can differ markedly from those involved with metals because of differences in solution conductivity, solubility of reactants and products, voltage stability of solvent and solutes, competition of side reaction, and electrode reactivity and kinetics.

Flow-through electrochemical reactors have shown promise in analytical chemistry for continuous quantitative analysis of various electrolytic solutions. At low concentrations, the selectivity of the reactor approaches that of controlled-potential devices, and nearly 100% reaction can be achieved.

Energy storage and conversion systems occasionally take advantage of flow-through porous electrodes. Examples would be fuel cells with a liquid fuel such as dissolved hydrazine or methanol, flow-redox and zinc-chlorine hydrate storage systems, and primary or secondary batteries where performance would be enhanced by a fresh supply of electrolyte.

1.2. Cell Configuration

An electrochemical reactor contains at least one anode and one cathode, but there are usually many electrode pairs. In the latter case, the cells can be connected electrically in parallel or in series. However, the series arrangement requires special manifolding of the solution to reduce the stray currents that tend to bypass cells via the common solution path in the inlet and outlet manifolds.

The position of the electrodes in relation to the direction of fluid flow and collection of the current (see Fig. 1) can change the character of the reactor. In configuration a in Fig. 1 there is a common feed, and the anolyte and catholyte are maintained separate. If the solution is allowed to flow from the back to the front (configuration b), a situation is created where the ohmic potential drop in the solution is working against the favorable mass-transfer condition near the fluid inlet, and there may result a more uniform distribution of the reaction throughout the porous medium. Flowing the solution from the front face (configuration a) combines the favorable electrode potential and mass-transfer conditions to produce the maximum rate at the front and minimize the overall ohmic potential drop in the solution.

Configuration c puts the fluid flow perpendicular to the current. This is more difficult to analyze, but it has the advantage of permitting a higher flow rate while maintaining a large electrical driving force throughout the length of the reactor.

Introduction

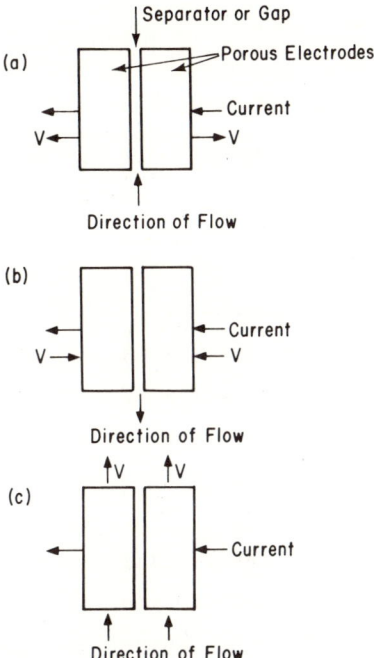

Fig. 1. Various configurations of electrode placement relative to the direction of fluid flow.

The preceding considerations apply when the solid-phase conductivity is much greater than that of the solution. For a finite electrode conductivity, the behavior can also depend on the placement of the current collector. For example, the electrical driving force (electrode potential minus solution potential) for configuration a can be made more uniform by placing the current collector on the back side instead of the front side, unless the electrode is too thick.

1.3. Fluid Flow and Pressure Drop

Figure 2 is a correlation of the pressure drop in packed beds. This is of interest because we occasionally need to estimate the cost of pumping a fluid through a porous electrode. Also, a knowledge of fluid flow helps us understand the mass-transfer process.

The fluid-flow pattern is characterized by the modified Reynolds number $Re' = d_p v/\nu(1-\varepsilon)$, where d_p is the equivalent sphere diameter of the packing and v is the superficial fluid velocity (volumetric flow rate divided by the cross-sectional area of the bed). The particle diameter is related to the specific surface area, a, of the bed by $d_p = 6(1-\varepsilon)/a$. Because of our involvement with mass transfer and interfacial processes, we prefer to use the specific surface area in forming dimensionless groups. Therefore, we define the Reynolds number as $Re = v/a\nu$.

For steady, fully developed flow, the pressure drop ΔP is proportional to the length L of the bed and can be combined into the group $\Delta P/aL\rho v^2$. Dimensional considerations show that this group can depend only on the Reynolds number for a given packing geometry, and Fig. 2 expresses this dependence.

For small Reynolds numbers, the pressure drop is directly proportional to the velocity and also proportional to the viscosity of the fluid:

$$\Delta P = 4.2 \frac{\mu v a^2 L}{\varepsilon^3} \quad (5)$$

This behavior, which prevails for $v/a\nu < 1$, implies that the local fluid velocity is everywhere doubled when the superficial velocity is doubled. For very high Reynolds numbers, the pressure

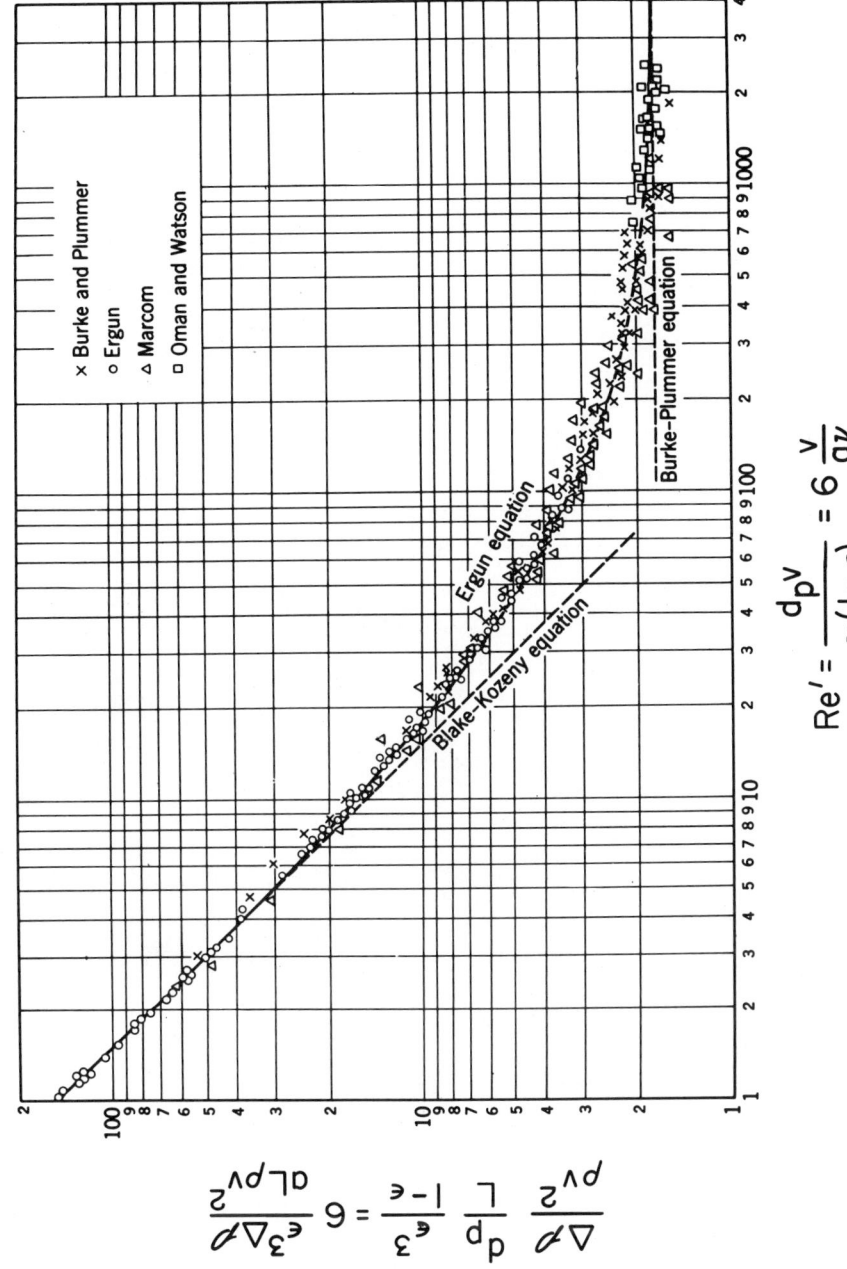

Fig. 2. Pressure-drop correlation for packed beds (9, p. 198).

drop is proportional to the square of the velocity and independent of the viscosity:

$$\Delta P = 0.29 \frac{aL\rho v^2}{\varepsilon^3} \qquad (6)$$

Here the flow is turbulent, that is, subject to random fluctuations of the local pressure and velocity. At low intermediate Reynolds numbers, the flow is still steady and laminar, but inertial effects are not negligible, and a doubling of the superficial velocity does not lead to an exact doubling of the velocity at each point in the pores. An increase in the flow rate leads first to an unsteady local velocity and finally to randomness and turbulence.

Differences in the packing geometry manifest themselves in variations in the porosity or void volume fraction ε, even for spheres, and "shape factors" must sometimes be introduced into the correlation to account for departures from spherical shape of the packing. In choosing an empirical function of porosity to correlate experimental data, bear in mind that functions of the forms A/ε^n and $A'(1-\varepsilon)^n$ may be indistinguishable in a limited range of porosity since $\varepsilon(1-\varepsilon)$ varies only between 0.21 and 0.25 for ε in the range from 0.3 to 0.7.

Instructive models have been developed in the literature to explain or correlate pressure-drop data. The fluid velocity v/ε in the interstices of the porous medium characterizes the residence time in the bed. An equivalent pore model is useful. For an equivalent pore of radius $r_0 = 2\varepsilon/a$ and length $l = \tau L$, where τ is the tortuosity factor, the average velocity in the pore is $v\tau/\varepsilon$. This pore model is in harmony with Eqs. 5 and 6 if τ takes the value 1.44 and if the tube or pipe friction factor is taken to be 0.194 at high Reynolds numbers.

Introduction

To cover the whole range of Reynolds numbers and reproduce the curve in Fig. 2, the pressure drops predicted by Eqs. 5 and 6 are added. This yields the so-called Ergun equation.

The quantities ε and a are really insufficient to characterize completely a porous medium. The effective value of the specific interfacial area is difficult to determine, but it is very important in the design of an electrochemical reactor. The BET measurement reflects all area present, including matrix, binders, and nonconductors. The electrochemically active area could be measured by charging the double-layer capacity. However, in a flow-through porous electrode it is the interfacial area which is accessible to the flowing stream that is relevant.

For a porous medium composed of particles that are themselves porous, it is helpful to think of micropores (within the particles) and macropores (between the particles). Correspondingly, one speaks of the macroporosity and the microporosity. Here we deal with an extreme case of a nonuniform pore-size distribution; pore volume occurs in two peaks at quite different pore sizes. The fluid flows through the large pores between the particles, and it is the external area of the particles which determines the pressure drop according to Eq. 5 or 6 and which is accessible directly for mass transfer with the flowing fluid. The internal area of the particles, which may be much larger, cannot be assumed to be available in general. However, when diffusion within the particle can occur and electrode kinetics are limiting, this internal area may have a favorable influence on the overall behavior of the electrochemical reactor.

1.4. Mass Transfer

We are concerned here with mass transfer of the reactant from the flowing stream to the surface of the solid matrix of the porous electrode. The porous electrode may have been selected to promote such mass transfer, and the flow through ensures a continuous supply of reactant to the electrode. In this case there is a concentration difference Δc between the solid wall and the bulk fluid flowing in a pore, and this concentration difference is the driving force for mass transfer to the surface. The apparent, average mass-transfer coefficient $\bar{k}_m$ is defined on the basis of the overall rate of mass transfer within the porous electrode $v(c_o - c_L)$ per unit of superficial electrode area and the driving force Δc_{ln} averaged appropriately through the thickness of the porous electrode. Thus

$$v(c_o - c_L) = aL\bar{k}_m \Delta c_{ln} \qquad (7)$$

where aL is the interfacial area per unit of superficial electrode area, c_o is the feed concentration, and c_L is the exit concentration. The logarithmic average is used for the driving force:

$$\Delta c_{ln} = \frac{(\Delta c)_o - (\Delta c)_L}{\ln[(\Delta c)_o/(\Delta c)_L]} \qquad (8)$$

The mass-transfer coefficient actually depends on the detailed concentration profile along the electrode surface, even for identical electrodes under identical flow conditions. Therefore, we limit our interest here to the limiting-current condition, where the concentration at the electrode surface has been effectively reduced to zero. Equations 7 and 8 now simplify to:

Introduction

$$\bar{k}_m = \frac{v}{aL} \ln\left(\frac{c_o}{c_L}\right) \tag{9}$$

As a practical matter, it is difficult to measure the mass-transfer coefficient below the limiting current because the surface concentration is not known precisely. What we do then is measure the mass-transfer coefficient at limiting current according to Eq. 9, and if necessary we use this value to predict the electrode behavior below the limiting current. Possible variations in the definition of the mass-transfer coefficient are discussed in a later section.

Consider now the quantities on which $\bar{k}_m$ can depend. First we decide to examine the dimensionless group $\varepsilon \bar{k}_m / aD_o$, called here the (average) Nusselt number, $\overline{Nu}$. The diffusion coefficient D_o is the true molecular diffusion coefficient of the reactant in the free solution, and not the effective diffusion coefficient within the porous medium. The Nusselt number depends on the electrode thickness, expressed in dimensionless form as aL. It also depends on the fluid velocity and kinematic viscosity, expressed in the dimensionless Reynolds number $Re = v/a\nu$ and Schmidt number $Sc = \nu/D_o$. In addition, the Nusselt number can depend on the details of the geometry of the porous medium. These we characterize by the porosity ε and various dimensionless shape factors that more or less successfully correlate, for example, how a stack of screens behaves differently from spherical packing material. Even spheres can be packed differently, with $\varepsilon = 0.260$ for close packing, $\varepsilon \approx 0.4$ for random packing, and $\varepsilon = 0.476$ for simple cubic packing.

To simplify this complex functional dependence, we can emphasize the asymptotic value of $\overline{Nu}$ for deep beds. This is similar to our neglect of end effects in the correlation of the pressure drop in Fig. 2. We can be sure that $\overline{Nu}$ is larger for shallow beds and drops to the asymptotic value as aL increases.

The Nusselt number depends on the flow behavior and hence on the Reynolds number, which determines the flow regime. For values of $v/a\nu$ below 1, the velocity distribution becomes independent of the value of the viscosity, and correspondingly the Nusselt number depends on the Péclet number

$$Pe = ReSc = \frac{v}{aD_o} \qquad (10)$$

and no longer has a separate dependence on the Reynolds number and Schmidt number. Figure 3 illustrates this type of dependence. For $v/a\nu$ in the range from 5×10^{-4} to 15, Wilson and Geankoplis (1) proposed the correlation

$$\frac{\varepsilon d_p \overline{k}_m}{D_o} = 1.09 \left(\frac{v d_p}{D_o}\right)^{1/3} \qquad (11)$$

Figure 3 shows a comparison with the data of Appel for a close-packed bed of spheres three layers deep (2,3). At values of $v/a\nu$ greater than 1, a separate dependence on Re and Sc is expected, and Colquhoun-Lee and Stepanek proposed the correlation (4)

$$\frac{\overline{k}_m}{aD_o} = 0.62 \left(\frac{\nu}{D_o}\right)^{1/3} \left(\frac{v}{a\nu}\right)^{0.61} \qquad (12)$$

for $14 < v/a\nu < 1400$. Figure 4 shows their plot of this equation in comparison, for $\varepsilon = 0.4$, to a number of experimental correlations from the literature.

Introduction

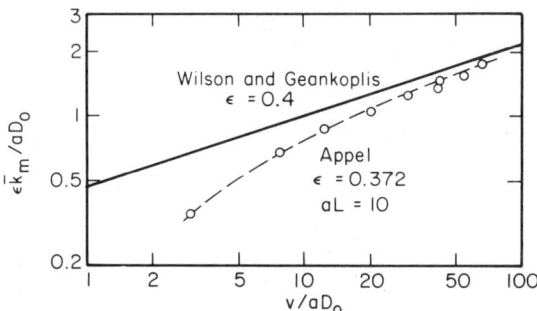

Fig. 3. Behavior of the average, overall mass-transfer coefficient in the low-Reynolds-number range where there is no dependence on the viscosity. Appel's data are for reduction of ferricyanide ions on stainless steel spheres with $a = 9.5$ cm^{-1} and $Sc = 1430$.

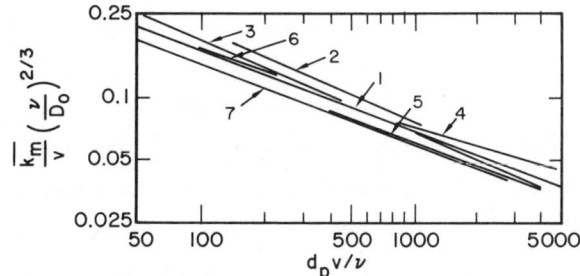

Fig. 4. Mass-transfer correlation in the high-Reynolds-number range (4). Comparison of equation 12 (line 1) with experimental results obtained by Jolls and Hanratty (line 2); Chu, Kalil, and Wetteroth (line 3); Bradshaw and Bennett (line 4); Di Cave (Line 5); Petrovic and Thodos (line 6); and Krishna et al. (line 7). See Ref. 4 for citations of the original literature.

1.5. Axial Dispersion

In addition to the local mass transfer from the flowing solution to the solid wall, there is transport in the fluid from one part of the electrode to another. First, there is molecular diffusion, which can be characterized by a diffusion coefficient

$$D_i = \frac{D_o}{\tau^2} \qquad (13)$$

where τ is again the tortuosity factor and D_o is the diffusion coefficient in the free stream outside any porous medium. Second, there is a dispersion due to a nonuniform fluid velocity at different points in a pore. Fluid near the pore wall moves more slowly than fluid toward the center of the void space. This results in a mixing of the fluid in the direction of flow (the axial direction).

Axial dispersion tends to reduce the sharpness of concentration gradients in the fluid. It can be studied, for example, by introducing at time $t = 0$ an abrupt change in the concentration of the feed to an inert packed bed while continuing the flow rate with no change. At the exit, the concentration change will have been spread over a distance that depends on the flow rate and the thickness of the bed. This phenomenon follows approximately the laws of diffusion, and one defines an axial dispersion coefficient such that the flux (per unit of superficial area of the porous medium) due to diffusion and dispersion takes the form $-\varepsilon(D_i + D_a)\nabla c_i$. However, the dispersion coefficient D_a is not a fundamental transport property. It depends on the flow rate and the packing geometry and vanishes in the absence of convective fluid motion.

Introduction

In a packed bed, dispersion can also occur in the radial[*] direction, perpendicular to the flow. This arises because the flowing stream repeatedly divides as it flows around the packing. It would be studied by introducing a solute only in a small part of the entrance cross section, while the fluid flows uniformly through the entire bed. At the exit, the solute would have spread in the direction transverse to the flow.

Dispersion should properly be represented by a tensor coefficient since dispersion is different in the axial and radial directions. Figure 5 shows a correlation of dispersion coefficients. At high flow rates, D_a becomes much greater than D_i, and we can use as an approximation:

$$D_a = \frac{vd_p}{2\varepsilon} = \frac{3v(1-\varepsilon)}{\varepsilon a} \qquad (14)$$

At low flow rates, dispersion becomes negligible compared to molecular diffusion, but the Reynolds number at which the two are comparable decreases as the Schmidt number increases.

The name of Taylor is associated with dispersion in pipe flow. For a porous medium where there is laminar flow in straight pores, we should expect D_a to take the form:

$$D_a = \frac{v^2}{12a^2 D_o} \qquad (15)$$

This form is inherently different from Eq. 14.

---

[*]A packed bed in the form of a column produces this terminology. Radial dispersion is not directly related to mass transfer between the fluid and the packing. In fact, the dispersion coefficients are best measured in an inert bed, where such mass transfer is absent.

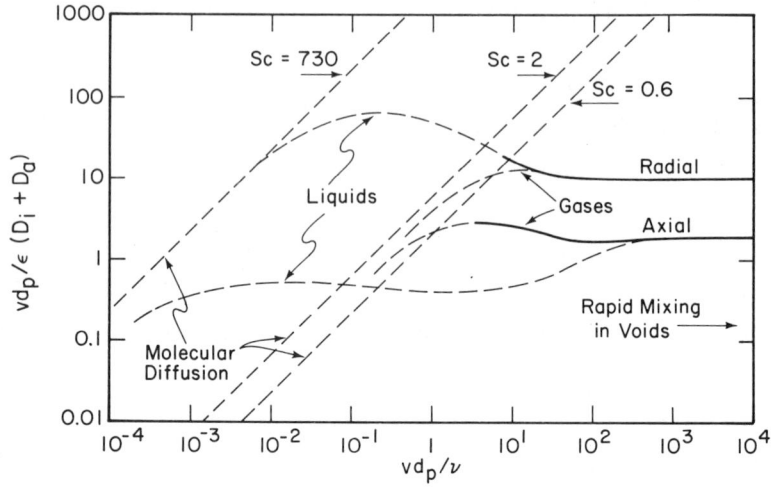

Fig. 5. Correlation of dispersion coefficients in packed beds (16). The asymptotes for molecular diffusion are based on $\varepsilon = 0.4$ and a tortuosity factor of 1.4. For spheres, $d_p = 6(1 - \varepsilon)/a$.

Axial mixing of the fluid in a packed-bed electrochemical reactor reduces the efficacy of the device because it reduces the local driving force for mass transfer between the flowing fluid and the wall. This is analyzed in more detail in a later section. Even though the dispersion coefficient increases with the flow rate, it is found that its effect is largest at lower flow rates where molecular diffusion is at least comparable to dispersion induced by the flow.

We have tacitly assumed that the fluid velocity is uniform across the face of the porous electrode. Nonuniformities in the geometric structure lead to "channeling," where the fluid preferentially flows through the less densely packed part of the bed and bypasses the rest. The consequences bear some resemblance to axial dispersion, and the effectiveness of the device for carrying out the desired reactions is seriously impaired. That is, the apparent mass-transfer coefficient is lowered. The potential for channeling is greater for particles of various sizes than for particles of nearly equal size. Extreme care should be exercised when preparing a porous electrode to minimize channeling. Electrode structures of sintered metal can be recommended.

1.6. Ohmic Potential Drop

The variation of the potential in the pore solution is given approximately by Ohm's law:

$$i_2 = -\kappa \nabla \Phi_2 \qquad (16)$$

The effective conductivity κ of the solution as it finds itself in the porous structure can be estimated from the conductivity κ_o of the free solution:

$$\kappa = \varepsilon^{1.5} \kappa_o = \frac{\varepsilon \kappa_o}{\tau^2} \qquad (17)$$

Meredith and Tobias (5) have reviewed the subject of conduction in heterogeneous media. The graph of Satterfied (6) of effective diffusion coefficients within porous catalyst pellets is also instructive because it displays the porosity dependence of experimental results. The value of τ^2 can lie in the range from 2 to 5, and a power of ε as high as 3 has been used instead of 1.5.

The effective conductivity σ of the solid matrix is highly dependent on the physical structure of the electrode. When the electrode matrix is coherent, such as a sintered metal, one can use a similar equation:

$$\sigma = (1-\varepsilon)^{1.5}\sigma_o = \frac{(1-\varepsilon)\sigma_o}{\tau^2} \qquad (18)$$

However, electrodes composed of solid particles held together by the application of an external force may exhibit a low and unpredictable value of σ because of the contact resistance between particles. The situation is further complicated when the matrix is composed of a mixture of conducting and nonconducting particles and when the structure changes as the reaction proceeds. On the other hand, many practical systems have effective matrix conductivities much greater than those of the pore solution, and accurate estimation of σ becomes unimportant.

2. EFFECT OF AXIAL DISPERSION ON THE AVERAGE MASS-TRANSFER COEFFICIENT

The preceding section has provided an overview of many of the physical processes that have important consequences in the design and behavior of flow-through porous electrodes. In the present and following sections we examine some of these phenomena in more detail.

We begin this effort with the manner in which a flow-through electrode operated at the limiting current is influenced by axial dispersion. In the absence of migration in an electric field, the superficial flux of the limiting reactant in the direction of the superficial fluid velocity is

$$N_i = -\varepsilon(D_i + D_a)\frac{dc_i}{dx} + c_i v \qquad (19)$$

and this is to be substituted into the continuity equation

$$\frac{\partial \varepsilon c_i}{\partial t} + \frac{dN_i}{dx} = aj_{in} \qquad (20)$$

[The formulation of equations for porous electrodes and the approximations therein have been discussed by Newman and Tiedemann (7).] Let us suppose that the local flux from the wall to the flowing stream is given by a local mass-transfer coefficient k_m such that

$$j_{in} = k_m(c_{iw} - c_i) = -k_m c_i \qquad (21)$$

where the wall concentration c_{iw} can be set equal to zero at the limiting current. In the steady state we have:

$$v\frac{dc_i}{dx} = \varepsilon(D_i + D_a)\frac{d^2 c_i}{dx^2} - ak_m c_i \qquad (22)$$

The selection of boundary conditions for this reactor in the presence of axial diffusion and dispersion has received extensive treatment in the literature. One feels that the feed concentration c_o establishes an upstream boundary condition and that, physically, we are not at liberty to specify anything downstream. Yet the second-order equation (Eq. 22) clearly requires two boundary conditions.

Wehner and Wilhelm (8) resolved this dilemma by analyzing a system in which the active packing between $x = 0$ and $x = L$ is supplemented by inert packing extending a great distance in both the upstream and downstream directions. This extra, inert material seems desirable since it ensures a fully developed fluid flow throughout the active portion of the bed. The differential equation (Eq. 22) again applies in the inert packing, but without the last term representing mass transfer from the packing to the fluid. In the downstream region the equation is solved to yield:

$$c_i = A_1 + A_2 \exp\left[\frac{vx}{\varepsilon(D_i + D_a)}\right] \qquad (23)$$

Since the concentration cannot increase without bound in the downstream direction, A_2 must equal zero, and the concentration is a constant in this region. This is the consequence of applying a boundary condition at the downstream end of the system.

The continuity conditions at $x = L$, where the active bed joins the downstream region, is that both the concentration and the flux should be continuous (see Eq. 19). This leads directly to the result

$$\frac{dc_i}{dx} = 0 \quad \text{at} \quad x = L \qquad (24)$$

as a boundary condition at the exit of the active part of the bed, and this remains valid even if the downstream, inert region has values of ε, D_i, and D_a different from those that apply in the active bed.

In the inert packing upstream of $x = 0$, the concentration again has the form of Eq. 23, but now there is no reason to set A_2 equal to zero. However, A_1 is clearly equal to c_o, the feed concentration. The continuity conditions at $x = 0$ again require that both the concentration and the flux be continuous. With these two

Effect of Axial Dispersion on Mass-Transfer 375

conditions it is possible to eliminate the unknown A_2 and obtain

$$c_o v = c_i v - \varepsilon (D_i + D_a) \frac{dc_i}{dx} \quad \text{at} \quad x = 0 \qquad (25)$$

as the boundary condition at the entrance to the active portion of the bed. Again it is convenient that this condition remains the same if the upstream packing has different values of ε, D_i, and D_a.

This condition (Eq. 25) has a simple physical interpretation. The total amount of material entering the active bed per unit of superficial area must be $c_o v$, the feed concentration multiplied by the flow rate per unit of superficial area. In Eq. 25 we see that the diffusive flux is properly accounted for in assessing the amount of material entering the active bed since the concentration derivative is not zero there. Under this condition, the concentration at $x = 0$ will be less than c_o because some of the reactant will have diffused ahead to the active portion without waiting to be convected.

We can say that the specification of the upstream concentration as the feed concentration has led to Eq. 25 and the requirement that the downstream concentration be well behaved has led to the milder condition (Eq. 24).*

---

*When migration effects are included, as they must be with the supporting electrolyte itself, the flux condition (Eq. 19) can be modified to read:

$$N_i = -z_i u_i \varepsilon F c_i \frac{d\Phi_2}{dx} - \varepsilon (D_i + D_a) \frac{dc_i}{dx} + c_i v$$

Reasoning similar to that in the text leads to the inlet condition:

To effect a solution to Eq. 22 subject to Eqs. 24 and 25, let us introduce the dimensionless quantities:

$$\theta = \frac{c_i}{c_o}, \quad y = \frac{ak_m x}{v}, \quad \text{and} \quad D' = \frac{\varepsilon a k_m}{v^2}(D_i + D_a) \quad (26)$$

This reduces the differential equation (Eq. 22) to

$$\frac{d\theta}{dy} = D'\frac{d^2\theta}{dy^2} - \theta \quad (27)$$

and the boundary conditions to

$$c_o v - \frac{It_i}{z_i F} = c_i v - z_i u_i \varepsilon F c_i \frac{d\Phi_2}{dx} - \varepsilon(D_i + D_a)\frac{dc_i}{dx} \quad \text{at} \quad x = 0$$

The superficial current density I enters with a minus sign on the left because it represents here the anodic current density flowing from the working electrode to an upstream counterelectrode. The transference number t_i is not modified from the free-solution value by the presence of the porous structure. If the counterelectrode is downstream, I is zero, but $d\Phi_2/dx$ is not zero at $x = 0$.

For the condition at the exit, we can still set dc_i/dx equal to zero for all the species. For the potential gradient we obtain

$$I = -\varepsilon F^2 \frac{d\Phi_2}{dx}\sum_i z_i^2 u_i c_i \quad \text{at} \quad x = L$$

since the solution is electrically neutral. Now I is the superficial anodic current density flowing from the working electrode to a downstream counterelectrode. If the only counterelectrode is upstream, I is set equal to zero here.

$$\theta - D' \frac{d\theta}{dy} = 1 \quad \text{at} \quad y = 0 \tag{28}$$

and

$$\frac{d\theta}{dy} = 0 \quad \text{at} \quad y = \alpha L \tag{29}$$

where:

$$\alpha = \frac{a k_m}{v} \tag{30}$$

The solution can now be written

$$\theta = \frac{e^{-y/B} + (D'/B^2) e^{By/D'} \exp[-\alpha L(1/B + B/D')]}{B + (D'/B^2)(1 - B) \exp[-\alpha L(1/B + B/D')]} \tag{31}$$

where

$$B = \frac{1 + (1 + 4D')^{1/2}}{2} \tag{32}$$

is a constant approximately equal to 1 if D' is small. By setting y equal to αL in Eq. 31, we can obtain the concentration ratio c_L/c_o at the exit. Substitution into Eq. 9 permits us to relate the average mass-transfer coefficient $\bar{k}_m$ to the local mass-transfer coefficient k_m:

$$\bar{k}_m = \frac{k_m}{B} + \frac{v}{\alpha L} \ln \frac{B + (D'/B^2)(1 - B) \exp[-\alpha L(1/B + B/D')]}{1 + (D'/B^2)} \tag{33}$$

Axial diffusion and dispersion have the effect of lowering $\bar{k}_m$ below the value given by k_m. To display this clearly, we have prepared Fig. 6 on the basis of Eq. 33. The magnitude of D' determines the importance of axial dispersion. To be explicit, we have assumed the Wilson-Geankoplis correlation (Eq. 11) to give k_m (rather than $\bar{k}_m$), and we have used Eq. 14 for the axial dispersion coefficient. Then D' depends on the porosity ε and the Péclet number v/aD_o. Even though D_a increases proportionally with flow rate, D' and the effect of axial dispersion decrease at high Péclet numbers.

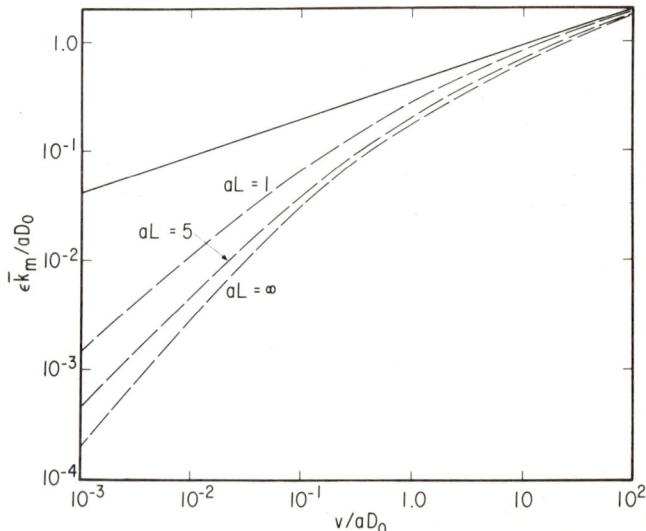

Fig. 6. Influence of axial dispersion on the apparent or overall mass-transfer coefficient. The local mass-transfer coefficient k_m is assumed to depend on the Péclet number as shown for the upper curve. Because of axial dispersion, $\bar{k}_m$ lies below k_m, and the effect becomes large at low Péclet numbers. In preparing the graph, the porosity was taken to be $\varepsilon = 0.3$, and the tortuosity factor τ was assumed to be equal to 1.

The thickness of the bed also influences the relationship. For aL approaching zero, Eq. 33 reduces to $\bar{k}_m = k_m$. However, a value of aL below 1 is unrealistic, and the value is likely to lie above 10. The effect of aL as well as Péclet number is shown in Fig. 6. For large values of aL, Eq. 33 simplifies to:

$$\bar{k}_m = \frac{k_m}{B} = \frac{2k_m}{1 + \sqrt{1 + 4D'}} \qquad (34)$$

The dimensionless variables in Eq. 26 were chosen because axial dispersion is not the dominant effect in the flow-through reactor. Simply setting D' equal to zero allows us to ignore this effect in our analysis. Equation 31 then reduces to:

$$c_i = c_o e^{-\alpha x} \qquad (35)$$

3. THE MASS-TRANSFER COEFFICIENT

It is clear that the design of porous electrodes as flow-through electrochemical reactors depends strongly on the value of the mass-transfer coefficient, and any clarification of the dependence of the Nusselt number on the Péclet number, the Schmidt number, the porosity, and the dimensionless electrode thickness aL would be useful.

First we should realize that different quantities are discussed in the literature under the term "mass-transfer coefficient." We have already used an average or overall value defined by Eq. 9 and a local value defined by Eq. 21. Bird et al. (9) give a lucid discussion of the possibilities for the analogous definition of the heat-transfer coefficient inside tubes.

The definition of a mass-transfer coefficient refers to a flux and a driving force. For the local mass-transfer coefficient, we should thus expect to see both the local flux and the local concentration difference. However, the concentration difference at the inlet is occasionally used, and the feed concentration c_o is another logical choice when operation is at the limiting current. Even the definition of the local concentration difference is a matter of choice. The bulk concentration can either be an average over the volume of the pores or an average weighted by the fluid velocity at a point in the pore cross section, the so-called cup-mixing average concentration (9).

The definition of the local mass-transfer coefficient can subtly include or exclude the effect of axial diffusion, depending on whether this effect is in Eq. 22 and the use to which the solution of Eq. 22 is put. Careful examination of the preceding section will suggest that k_m is not, in fact, a local mass-transfer coefficient but rather an average mass-transfer coefficient that has been corrected for the effect of axial dispersion by means of Eq. 33. In this view, what is measurable is $\bar{k}_m$, defined by Eq. 9 and measured at the limiting current for a particular bed geometry, bed thickness L, flow rate v, and feed concentration c_o. The correction of $\bar{k}_m$ to k_m involves a dispersion coefficient either measured in an inert bed of the same geometry or taken from a correlation.

From this reasoning, $\bar{k}_m$ may appear to be a more fundamental quantity than k_m. However, $\bar{k}_m$ can conveniently be measured only at the limiting current. Although there is no fundamental reason why a value obtained under such conditions should be applicable below the limiting current, we are inclined to correct $\bar{k}_m$ to k_m and then use this value in predicting bed behavior by means of Eq. 22, extended to include a nonzero wall concentration c_{iw}.

In the preceding discussion k_m lost its identity as a local mass-transfer coefficient because it could not be asserted that its value is constant throughout the bed. In fact, the local mass-transfer rate is influenced by upstream events and possibly by downstream events, and hence a transfer coefficient for a constant wall concentration is, in principle, different from that for a constant wall flux. This difference persists whether we are speaking of a local or an average transfer coefficient.

Some of these differences could be overcome by means of superposition integrals. For example, the wall concentration could safely be expressed as

$$c_{iw} = c_o + \frac{a}{v}\int_0^L j_{in}(x')K\left(\frac{x-x'}{\varepsilon v/a^2 D_o}\right) dx' \qquad (36)$$

where K describes the concentration distribution at the pore wall due to a localized current source j_{in} at the position x' and where K approaches 1 as x - x' approaches infinity and K approaches 0 as x - x' approaches minus infinity. If axial diffusion and dispersion are negligible, K will equal 0 whenever x is less than x'. Then the upper limit of integration can be replaced by x instead of L. In addition to the argument shown explicitly, K can depend on the Schmidt number, the Péclet number, and the geometry of the porous medium.

Experimental data may not yet be sufficiently accurate or reproducible to justify the integral-equation approach in Eq. 36. What we can expect to measure is the effluent concentration at the limiting current for beds of given thickness and operated for a certain reaction at a certain flow rate. The expected results are depicted in Fig. 7. A differential mass-transfer coefficient would be proportional to the slope of the curve at each point; theoretically, this would be approximately equal to the local mass-transfer coefficient at the corresponding point within a deeper bed operated at the same conditions. On the other hand, the average mass-transfer coefficient defined by Eq. 9 is proportional to the slope of a line drawn on Fig. 7 from the inlet condition to the exit condition. Figure 7 should illustrate that the average mass-transfer coefficient is higher for shorter bed thicknesses. Equivalently, one can say that there is an entry region where the local mass-transfer coefficient is higher than the average value. The length of this entry region may be expected to be proportional to the superficial velocity v.

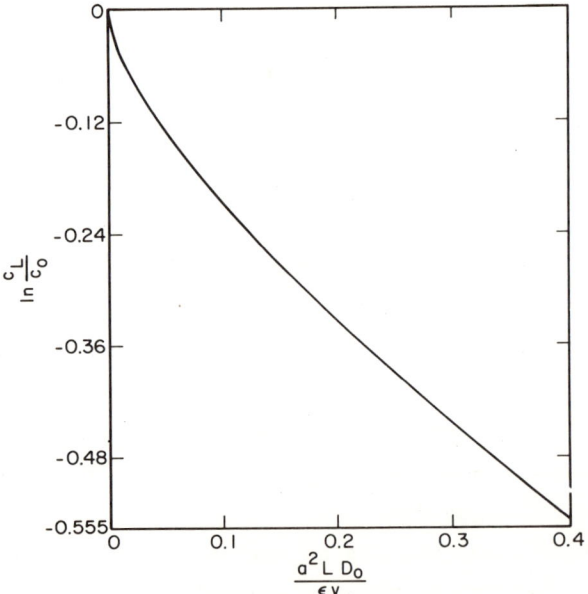

Fig. 7. Ratio of inlet and exit reactant concentrations as a function of electrode thickness. The slope of the curve, as drawn, is $-\varepsilon k_m/aD_o$. (This particular curve is actually a plot of the Graetz solution.)

On the basis of experimental results and theoretical considerations, a picture resembling Fig. 8 begins to emerge for the average mass-transfer coefficient. The solid curves represent the asymptotic result for very thick beds, derived from the final slope toward the right in Fig. 7.

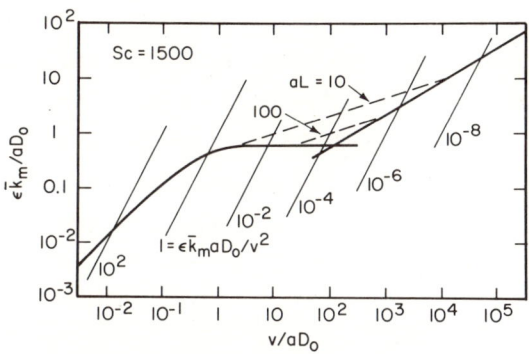

Fig. 8. Illustration of the possible behavior of the average mass-transfer coefficient for Sc = 1500. The line labeled aL = 10 is actually the Wilson-Geankoplis equation 11 for $\varepsilon = 0.4$. The line labeled aL = 100 has the same slope but is lowered by a factor of the cube root of 10. The heavy solid curves are supposed to correspond to aL = ∞. The solid line toward the right with a slope of 0.61 follows Eq. 12 and can be thought of as the high-Reynolds-number behavior where the local velocity is no longer strictly proportional to the superficial velocity. (A Reynolds number of $v/a\nu = 10$ corresponds to a Péclet number of $v/aD_o = 1.5 \times 10^4$ on the graph.) The solid curve toward the center and the left corresponds to a deep bed of spheres on a simple cubic lattice, as calculated by Sørensen and Stewart (10). This curve shows the effect of axial diffusion and dispersion toward the left and an asymptotic value of $\varepsilon \bar{k}_m/aD_o = 0.613$ at high Péclet numbers.
Light solid lines with a slope of 2 have been superposed for later calculation of the maximum velocity limited by ohmic considerations.

In the low-Reynolds-number region where the mass-transfer coefficient does not depend on the fluid viscosity, the behavior of this solid curve can be studied for both high and low ranges of the Péclet number, v/aD_o. At low Péclet numbers it can be shown that the slope of a log-log graph of the Nusselt number $\varepsilon \bar{k}_m/aD_o$ versus the Péclet number must approach unity. In Fig. 8 this behavior can be seen clearly for v/aD_o less than about 0.3, and this is the region where axial diffusion is dominant. Under these conditions the depletion of concentration in the depths of the porous electrode is essentially independent of the superficial velocity in the sense that the concentration has the form:

$$c_i \sim \exp\left(-\text{constant }\frac{ax}{\varepsilon}\right) \quad (37)$$

[For straight tubes of uniform radius, this constant is equal to 1.202τ. For spheres in a simple cubic lattice it is about 1.23, according to the results of Sørensen and Stewart (10).]

At high Péclet numbers, the slope of the curve must approach zero, as shown on Fig. 8 for values of v/aD_o between 3 and 100. For a periodic porous structure for which the velocity profile has been calculated in creeping flow, the asymptotic value of $\varepsilon \bar{k}_m/aD_o$ can be obtained by solving an eigenvalue problem similar to the Graetz problem. For straight tubes the asymptotic value is 0.914, and for a simple cubic lattice of spheres the value is about 0.613, according to the results of Sørensen and Stewart (10). Axial diffusion is of negligible importance in this region, but one could say that dispersion due to a nonuniform velocity distribution is already accounted for in the calculations.

It is interesting to note that Eq. 34 predicts the correct slopes for Fig. 8, as discussed previously, if and only if k_m

is taken to be a constant, independent of v. If k_m is proportional to the cube root of v, as suggested by Eq. 11, then the predicted slope on Fig. 8 is 1/3 at high Péclet numbers and 7/6 at low Péclet numbers.

The horizontal curve just discussed for high Péclet numbers does not persist for arbitrary values of v/aD_o. At higher Reynolds numbers the simple laminar flow breaks down and progressively becomes more turbulent. Under these conditions we are advised to use a different relationship (e.g. Eq. 12) to correlate the mass-transfer coefficient, and Fig. 8 shows that higher rates of mass transfer are predicted when the fluid-flow regime promotes mixing of the fluid across the cross section of the pores.

Our discussion of Fig. 8 has been confined to deep beds where aL is large. At low Reynolds numbers there should be an entry region where the local mass-transfer coefficient is larger than the asymptotic value for deep beds. When the entry region is appreciably greater than the bed thickness, the average mass-transfer coefficient will also be high and can be expected to be proportional to the cube root of v/L. Equation 11 of Wilson and Geankoplis (1) does not show a dependence on L, but most measurements of $\bar{k}_m$ involve thin beds for purposes of accurately determining the concentrations. Furthermore, an appreciable entry region is found for only a restricted range of flow rates. At low Péclet numbers the entry length is proportional to v, and even a shallow bed (small value of aL) behaves like a deep bed. A divergence of bed behavior with bed thickness is not observed until larger values of the Péclet number are reached. On the other hand, at high Reynolds numbers where the flow is becoming turbulent, the diffusion layer is forced to become thinner, and the entry length is reduced accordingly. Thus the entry length actually decreases markedly with increasing velocity, and the dependence of $\bar{k}_m$ on aL again disappears. This is depicted in Fig. 8 by the fact that the lines of slope equal to 1/3 eventually intersect the high-Reynolds-number line with a slope of 0.61.

In experimental results, $\bar{k}_m$ never is reported to be independent of the flow rate v. Three factors can interfere and lead to a higher slope on Fig. 8: at very low flow rates there is axial dispersion, at high flow rates there is turbulence, and at intermediate flow rates the whole bed may be in the entry region.

The graph in Fig. 8 is not accurate in quantitative detail, and additional work is required to refine it. Comparison with Appel's data in Fig. 3 suggests that axial diffusion and dispersion remain important to higher Péclet numbers than the theoretical analysis would predict. Adequate experimental results are not available to establish the behavior depicted in Fig. 8 for any particular bed geometry, such as close-packed spheres. Furthermore, it is not known exactly how the graph should be modified for beds of different geometries, although shape factors are given in chemical-engineering correlations for transfer in packed beds (11).

In discussing Fig. 8, we have tried to consider many of the complex factors that influence the mass-transfer coefficient. Models for predicting and correlating the data can be found in the literature and help us organize our thoughts on the subject. Straight tubes or periodically constricted tubes constitute useful models. The Graetz solution for the straight tube gives good insight into the entry length. The Graetz solution with periodic renewal, or transverse mixing, has been suggested to account for the cube-root dependence of the mass-transfer coefficient on fluid velocity. A particle-friction model involves: (1) the estimation of shear-stress distributions to conform with the observed pressure-drop data (see Fig. 2) and (2) application of these distributions

$$\frac{k_{m,transient}}{\bar{k}_m} = 2\left(\frac{\lambda_B}{\lambda_G}\right)^2 = 2\left(\frac{2.4048}{2.7044}\right)^2 = 1.58 \tag{39}$$

where λ_B is the first zero of the Bessel function J_o, coming from the transient solution, and λ_G is the first eigenvalue for the Graetz solution for steady mass transfer for laminar flow in a tube. The larger value of the mass-transfer coefficient obtained by the transient measurement would result in an overestimate of the performance of a proposed electrode design, so that the system would fall below acceptable performance levels.

Second, one must determine in what time domain the current should follow Eq. 38. An upper limit is perhaps 10% of the residence time, since the approach to steady state becomes pronounced at later times. At very short times, under the ideal conditions described previously, the current would be inversely proportional to the square root of time and would not display the exponential decay.

Furthermore, Posey and Morozumi (13) showed that double-layer charging of porous electrodes resulted in a linear dependence of ln I versus time. The magnitude and duration of this current depends on the specific surface area, double-layer capacity, solution conductivity, and electrode thickness. It must be possible to distinguish between this process and the current decrease due to depletion of reactant before Eq. 38 can safely be used to obtain the mass-transfer coefficient.

Finally, the ohmic potential drop within the electrode must be sufficiently low so that the concentration at the wall is virtually zero throughout the electrode at any instant in time while side reactions are simultaneously avoided. However, many practical systems are designed so that this condition is just marginally with the Lighthill transformation to yield predictions of mass-transfer rates.

The theoretical work of Sørensen and Stewart (10) stands out as an effort to calculate the mass-transfer rates rigorously. However, the problem is still formidable, and such calculations are not yet routine.

Steady-state conditions are generally used in the determina[tion] of mass-transfer coefficients. The time necessary to achieve steady state is about three times the fluid residence time εL[...]. However, for very high internal surface areas, the time requi[red] for double-layer charging may exceed the residence time.

Chu et al. (12) recently proposed a transient measurement of the mass-transfer coefficient. With the flui[d] flowing through the electrode, the current is suddenly switc[hed] on so as to maintain a limiting current. Ideally, the super[ficial] current density to the electrode would follow the form (see Eq. 14)

$$I = \frac{nFak_m c_o L}{s_R} \exp\left(-\frac{ak_m t}{\varepsilon}\right)$$

and the slope of the straight-line portion of a semilogar[ithmic] plot of I versus t is interpreted to yield ak_m/ε. The me[asure]ment should be accomplished in much less than the residen[ce time].

This transient method is subject to a number of possib[le ...]. First, it should be realized that, even when the measureme[nt is] ideal, the transient mass-transfer coefficient is inhere[ntly] different from the steady-state value. In particular, f[or an] electrode composed of straight pores with laminar flow a[nd negli]gible axial diffusion, the relationship is:

satisfied under steady-state operation. The transient currents are necessarily larger than the steady-state value, and the limiting-current condition cannot be maintained in the transient state for systems so designed.

The transient value of k_m determined for copper deposition by Chu and colleagues (12) was approximately 30 times larger than the corresponding steady-state value. This large discrepancy may be due to an overlap between double-layer and faradic charging of the porous electrode.

4. OHMICALLY LIMITED REACTOR CAPACITY

Frequently, capital costs are appreciable, and it is desirable to use the reactor to obtain the greatest possible amount of product. This is accomplished by increasing the flow rate while maintaining a desired percent of conversion. Eventually, the incremental pumping costs balance the greater utilization of the invested capital. Before this, however, the ohmic potential drop within the reactor may become so large that the side reactions occur to an unacceptable extent.

To consider a simple design problem, let us follow Bennion and Newman (14) by assuming that the reactor operates at limiting current and that the electrode potential $\Phi_1 - \Phi_2$ is confined to a certain range that ensures attainment of the limiting current but reduces side reactions to an acceptable amount. To fix ideas, examine the limiting-current curve in Fig. 9 for deposition of copper on a rotating-disk electrode from a solution containing 3175 mg/liter of copper (0.05 M in $CuSO_4$ and 1.0 M in H_2SO_4). In this diagram it appears that the limiting current can be attained at an electrode potential of −0.5 V, and evolution of hydrogen

does not become severe until the electrode potential reaches
−0.75 V. Therefore, an acceptable range of electrode potentials
might span 0.25 V.

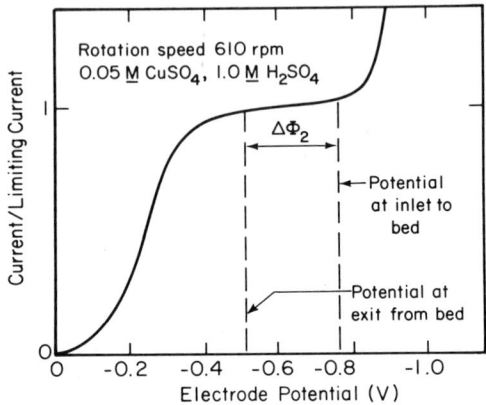

Fig. 9. Polarization of a disk electrode in an acidified cupric sulfate solution leads first to an increase of the deposition rate, next to a limiting plateau for deposition of copper, and finally to the evolution of hydrogen. The electrode potential is relative to a copper reference electrode in the bulk solution located beside the rotating disk. The curve is taken from (15). Also shown are conditions, for the inlet and exit of a flow-through porous electrode, selected to ensure a limiting current while avoiding evolution of hydrogen.

The precise amount of hydrogen evolution that can be tolerated in the porous electrode depends on how gas bubbles behave in the system. If they cause channeling of the fluid flow, it may be necessary to suppress their formation by pressurizing the system and not exceeding the solubility limit.* Korovin et al. (17-20) and Ichino et al. (21) examine the effect of gas bubbles in porous electrodes. Other side reactions, for example, reduction of thiosulfate ions in a system designed to recover silver from photographic fixing baths (22), may serve merely to reduce the current efficiency without any catastrophic effect on the overall process. The acceptable amount of side reactions is then determined by an optimization over the cost of power dissipated in the side reactions and the incremental value of recovered silver.

In some systems, such as the deposition of lead, the side reactions may occur to a significant extent before the desired reaction reaches limiting current. That is, a distinct limiting-current plateau may not exist on the current-potential curve. The reactor design must then take into account the simultaneous occurrence of side reactions. As the desired reactant becomes more dilute, the limiting-current plateau should become less distinct, as shown by the simulated current-potential curves for copper deposition in Figs. 10 and 11.

A limiting-current curve on a rotating-disk electrode or other electrode of the proper material can provide quantitative information for the design of the reactor, as illustrated previously. The interpretation of these results may be modified somewhat later. For example, the surface overpotential, which is inherent in the rotating-disk data, should be reduced toward the exit of a flow-through reactor where the reaction rate diminishes considerably with distance within the electrode.

---

*In some cases, supersaturation by a factor of 5 may be required for the formation of bubbles. On the other hand, a porous electrode should provide many sites for nucleation.

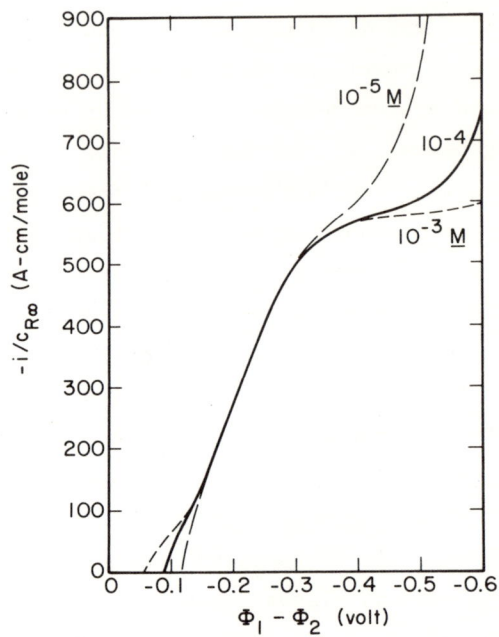

Fig. 10. Simulated current-potential curves for the deposition of copper, with simultaneous evolution of hydrogen. The mass-transfer coefficient k_m is taken to be 3×10^{-3} cm/s, corresponding to rotation of a disk electrode at about 411 rpm. The limiting-current density for a 0.1 M cupric ion concentration is then 57.9 mA/cm^2, and the current-density axis is scaled with this bulk concentration (which is given as a parameter on the curves). The potential difference $\Phi_1 - \Phi_2$ is the potential of the electrode measured relative to an adjacent reference electrode of a given kind, copper in a 0.1 M cupric ion solution.

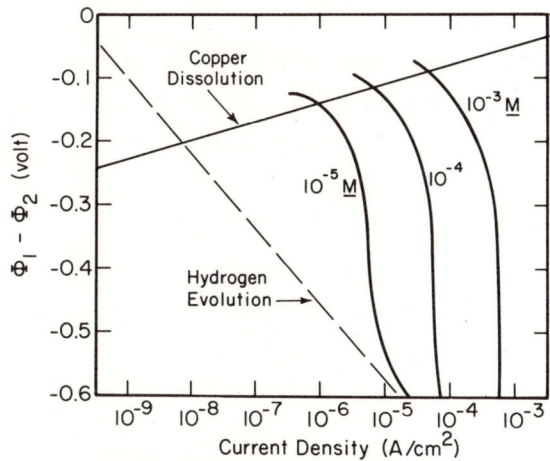

Fig. 11. The simulated current-potential curves of Fig. 10, replotted with a logarithmic scale for the current density. On this graph, the background curves for hydrogen evolution and anodic copper dissolution can be shown clearly. The limiting-current plateau becomes shorter and less distinct as the bulk cupric ion concentration is decreased. This method of plotting makes the plateaux look more distinct than for the corresponding curves on Fig. 10.

Ohm's law governs the variation of potential in the pore electrolyte:

$$i_2 = -\kappa \nabla \Phi_2 \qquad (40)$$

Diffusion potentials have been ignored here, which is a good assumption when an excess of supporting electrolyte is present. The case of a binary electrolyte requires separate treatment. In addition, the conductivity κ can be taken to be constant if the limiting reactant is a minor constituent.

With the assumption that side reactions do not occur significantly within the operating range of electrode potentials, the current can be related to the concentration through Faraday's law. We have

$$v \frac{dc_R}{dx} = \varepsilon(D_R + D_a) \frac{d^2 c_R}{dx^2} - \frac{s_R}{nF} \frac{di_2}{dx} \qquad (41)$$

Integration gives

$$v(c_R - c_L) = \varepsilon(D_R + D_a) \frac{dc_R}{dx} - \frac{s_R}{nF} i_2 \qquad (42)$$

where it has been assumed that there is no counterelectrode downstream of the working electrode. (The case of a downstream counterelectrode is considered in a later section.) Since operation is assumed to be at the limiting current, Eq. 31 expresses the concentration distribution to substitute into Eq. 42.

Integration of Eq. 40 now gives the potential distribution within the pore solution. Figure 12 indicates the nature of this variation. The potential gradient is steepest at the front of the electrode where the current enters from the counterelectrode. The potential gradient is very small within the depth of the electrode because the reaction rate diminishes rapidly. The most important aspect of the potential distribution is the potential difference in the pore solution from the front to the back of the electrode. This result can be expressed as

$$\Phi_2(L) - \Phi_2(0) = \frac{nFvc_o}{s_R \kappa} \frac{v}{ak_m} \left[B^2 \theta_L e^{\alpha L/B} - \frac{D'}{B} - (\alpha L + 1 + D')\theta_L \right] \quad (43)$$

where D' and B are defined by Eqs. 26 and 32.

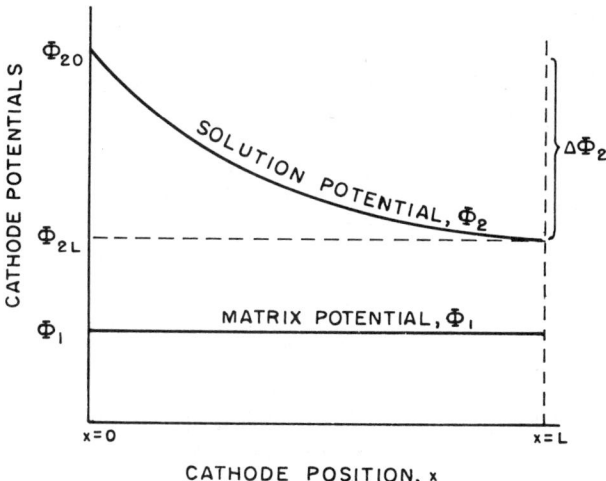

Fig. 12. Solution-phase and solid-matrix-phase potentials, as functions of cathode position.

The bracketed quantity in Eq. 43 is dimensionless and of order unity. The coefficient of this quantity, representing the magnitude of the potential difference, can be thought of as a current density, multiplied by a length through which this current flows, and divided by the conductivity κ. The current density $nFvc_o/s_R$ represents the maximum current density that would flow if there were complete conversion of all the reactant entering the reactor. The length v/ak_m is the order of magnitude of the distance through the reactor where the reaction occurs to an appreciable extent (see Eqs. 30 and 35).

The result in Eq. 43 is important because it indicates the maximum flow rate that can be permitted while maintaining a limiting-current condition at the exit of the electrode and simultaneously restricting side reactions to a tolerable level at the entrance. Thus $\Phi_2(L) - \Phi_2(0)$ can be assigned a certain maximum value $\Delta\Phi_2$, perhaps on the basis of a limiting-current curve like Fig. 9, and Eq. 43 can be solved for v.

In the absence of a significant effect of axial diffusion and dispersion, that is, as D' approaches zero, Eq. 43 can be written in the simpler form

$$\frac{\varepsilon \bar{k}_m a D_o}{v^2} = \frac{\varepsilon n F D_o c_o}{s_R \kappa \Delta \Phi_2} [1 - (\alpha L + 1)\theta_L] \qquad (44)$$

and for deep beds where θ_L becomes small, Eq. 43 becomes:

$$\frac{\varepsilon \bar{k}_m a D_o}{v^2} = \frac{\varepsilon n F D_o c_o}{s_R \kappa \Delta \Phi_2} \qquad (45)$$

Equations 44 and 45 are written in this form to facilitate a graphical solution by means of Fig. 8. The right sides of these equations can readily be computed from the properties of the feed solution, as well as the desired conversion that determines αL and θ_L. The flow rate cannot be computed directly because the mass-transfer coefficient depends on the velocity. However, when the combination of $\bar{k}_m$ and v given on the left sides of Eq. 44 and 45 has a certain value, this can be represented in Fig. 8 (or Fig. 3) as a straight line with a slope of 2. The intersection of this straight line with the experimental curve for $\varepsilon \bar{k}_m/aD_o$ yields a graphical solution for the Péclet number and hence the velocity v.

A similar graphical solution could, of course, be carried out on a plot of the Nusselt number versus the Reynolds number $v/a\nu$

or on a plot like Fig. 4.

One interesting feature of these results is that the limit of 100% conversion leads to a simple finite limit for the maximum flow rate, that given by Eq. 45. The term on the right in Eq. 45 also gives a measure of the difficulty of the process to be carried out in the flow-through reactor. There we find the ratio c_o/κ of the reactant concentration in the feed to the solution conductivity (within the porous medium). A dilute solution with a highly conducting supporting electrolyte can be processed at a higher flow rate than a concentrated solution of poor conductivity. A system with only a small range of tolerable electrode potentials must be processed at a low flow rate. A high specific interfacial area permits higher processing rates.

We sometimes become preoccupied with high degrees of conversion. A higher production rate for the reactor can be achieved at a lower conversion (although a subsequent separation of products from reactants may then be necessary). At low conversions, Eq. 44 can be approximated by:

$$v(c_o - c_L) = \sqrt{\frac{2a\bar{k}_m s_R c_o \kappa \Delta \Phi_2}{nF}} \qquad (46)$$

The maximum production rate is achieved at high velocities where $\bar{k}_m$ is higher, and it is proportional to the square root of the specific interfacial area, the allowable range of electrode potentials, and the product of the feed concentration and the solution conductivity.

We might notice that this maximum production rate is only 41% higher than the production rate at high conversions (where the 2 is absent on the right side of Eq. 46 and c_L is negligible on the left side). The difference is that at low conversions the velocity can be set at any desired value (so as to achieve a high value of $\bar{k}_m$) and Eq. 46 determines the maximum conversion at which the electrode potential can be maintained in the allowable range, whereas at high conversions c_L is negligible, and the corresponding form of Eq. 46 is used to determine the maximum value of v (through Eq. 45). This comparison is based on the production rate per unit of superficial area. The production rate per unit volume of the porous electrode is much higher when operation is at low conversions because the electrode thickness is less. At low conversions, the production rate per unit volume of the electrode is:

$$\frac{v(c_o - c_L)}{L} = a\bar{k}_m c_o \qquad (47)$$

4.1. Copper Recovery (14)

Consider now some example designs, where in each case the electrode is taken to have a porosity of 0.3 and a specific interfacial area of 25 cm^{-1}. Let the copper concentration be reduced from 667 mg/liter in the feed to 1 mg/liter in the dilute product, so that $\alpha L = 6.5$. Take the diffusion coefficient to be $D_o = 6 \times 10^{-6}$ cm^2/s and the conductivity to be $\kappa_o = 0.17$ $(\Omega\text{-cm})^{-1}$. Equation 44 then yields:

$$\frac{\varepsilon \bar{k}_m a D_o}{v^2} = 6.46 \times 10^{-4} \qquad (48)$$

Figure 8 shows that we are in the region where the entry length is important and the Nusselt number depends on aL. We tentatively assume that the bed is deep because of the high conversion, and we obtain v/aD_o = 30.8 or $v = 4.62 \times 10^{-3}$ cm/s and $\bar{k}_m = 3 \times 10^{-4}$ cm/s (a factor of 10 smaller than that used in the calculation of Figs. 10 and 11). Consequently, the electrode thickness is $L = 3.92$ cm, and $aL = 98$. Figure 8 now indicates that the mass-transfer coefficient could be somewhat higher, and a second estimate of v/aD_o could be obtained. We stop here and note that Bennion and Newman (14) obtained $v = 3.6 \times 10^{-3}$ cm/s and $L = 4.96$ cm by using a more conservative correlation (11) for the mass-transfer coefficient.

4.2. Lead Removal (7,23)

In the manufacture of lead acid batteries, it may be necessary to dispose of sulfuric acid saturated with lead sulfate. The porous electrode should reduce the concentration from 1.45 to 0.05 mg/liter for a solution having a conductivity of $\kappa_o = 0.8$ $(\Omega\text{-cm})^{-1}$. We take the allowable potential variation to be 0.1 V, although the standard electrode potential for lead deposition is actually negative to that for hydrogen evolution. The high overpotentials for hydrogen evolution on lead permit the lead acid battery to be charged and suggest that it may be possible to remove lead electrochemically.

With $D_o = 4 \times 10^{-6}$ cm^2/s, Eq. 44 leads to:

$$\frac{\varepsilon \bar{k}_m a D_o}{v^2} = 1.05 \times 10^{-7} \qquad (49)$$

Figure 8 suggests that we are in the range where Eq. 12 is applicable. Solution with that equation yields $v/aD_o = 6603$ and consequently $v = 0.66$ cm/s and $\bar{k}_m = 1.52 \times 10^{-3}$ cm/s. It follows that the electrode thickness is L = 58.4 cm and aL = 1461. The use of a more conservative correlation (11) leads to (7,23) v = 0.64 cm/s and L = 60 cm. The design shows that the low feed concentration and high solution conductivity lead to a much higher processing rate than for the copper process, despite the smaller range of allowed electrode potentials used in the calculations.

In our experiments we have not been able to achieve our design objectives, although we have (23) reduced the lead concentration from 4.32 to 0.55 mg/liter.

4.3. Desalination

For an example of a process involving a binary electrolyte, we choose a desalination method suggested by Tobias (24). Let the feed be 0.565 M sodium chloride for which the electrode reaction is:

$$Ag + Cl^- \rightarrow AgCl + e^- \qquad (50)$$

The chloride ion is to be removed at the limiting current while the sodium ion migrates to the upstream counterelectrode. The design problem calls for depletion of the NaCl concentration from 33,000 to 500 mg/liter. With neglect of axial diffusion and dispersion, the concentration is given by Eq. 35, but the diffusion coefficient of the NaCl electrolyte is to be used in the correlations for the mass-transfer coefficient instead of the diffusion coefficient of the reacting Cl^- ion. The current density distribution in the pore electrolyte is:*

*This result can be obtained from the governing relations for porous electrodes (7). Since a binary electrolyte obeys the equation of convective diffusion without the explicit appearance of a migration term, it is also possible to infer mass-transfer results for binary electrolytes from equations developed for the case of excess supporting electrolyte (25).

$$i_2 = -\frac{Fvc_o}{1-t_-}(e^{-\alpha x} - e^{-\alpha L}) \quad (51)$$

For Ohm's law in a binary 1-1 electrolyte, we replace Eq. 40 by:

$$\nabla \Phi_2 = -\frac{\vec{i}_2}{c\Lambda} - \frac{RT}{F}(t_+ - t_-)\nabla \ln c \quad (52)$$

For the binary electrolyte, we take the molar conductance $\Lambda = \kappa/c$ to be constant rather than the conductivity itself. The last term in Eq. 52 represents the gradient of the diffusion potential.

Substitution of the current-density distribution and integration yield the potential variation in the pore solution:

$$\Phi_2(L) - \Phi_2(0) = \frac{RT}{F}(t_+ - t_-)\alpha L$$

$$+ \frac{Fv^2/a\bar{k}_m \Lambda}{1-t_-}(\alpha L - 1 + \theta_L) \quad (53)$$

This result is less favorable than Eqs. 43 and 44 for the supporting-electrolyte case by the presence of αL within the parentheses and the factor $1 - t_-$ in the denominator in the last term on the right. Also leading to a larger variation of the potential within the pore solution is the fact that the ratio $c_o/\kappa = 1/\Lambda$ is inherently larger in the absence than in the presence of a supporting electrolyte. In contrast to Eq. 43 and 44, there is no simple finite limit for the maximum flow rate in the limit of 100% conversion.

With $t_+ = 0.373$, $D_o = 1.47 \times 10^{-5}$ cm^2/s, $\Lambda_o = 91.2$ cm^2/Ω-equiv, and $\Delta\Phi_2 = 0.2$ V, Eq. 53 yields

$$\frac{\varepsilon \bar{k}_m aD_o}{v^2} = 1.073 \qquad (54)$$

at 25°C. According to Fig. 8, we obtain $v/aD_o = 0.61$, $v = 2.2 \times 10^{-4}$ cm/s, $\bar{k}_m = 4.9 \times 10^{-4}$ cm/s, $L = 7.7 \times 10^{-2}$ cm, and $aL = 1.9$. The adverse conditions with a binary electrolyte lead to an extremely low flow rate and an electrode thickness so small that we hardly have a porous electrode. The cost of desalination would be correspondingly high. In addition, there would be an appreciable inventory of silver, and it would be necessary to recover any silver leaving the electrode as a consequence of the solubility of silver chloride.

4.4 Downstream Counterelectrode

For the case with supporting electrolyte, let the counterelectrode now be placed downstream of the working porous electrode, as in configuration b of Fig. 1. The integration of Eq. 41 involves a different integration constant, and Eq. 42 is replaced by

$$v(c_R - c_o) = \varepsilon(D_R + D_a)\frac{dc_R}{dx} - \frac{s_R}{nF}i_2 \qquad (55)$$

so that the current density is zero at the entrance at $x = 0$. Integration of Ohm's law, Eq. 40, leads to the potential distribution in the pore solution. With neglect of axial diffusion and dispersion, the potential difference from the entrance to the exit can be evaluated and expressed in the form:

$$\Phi_2(L) - \Phi_2(0) = -\frac{nFvc_o}{s_R\kappa} \frac{v}{a\bar{k}_m} (\alpha L - 1 + \theta_L) \tag{56}$$

Now the electric driving force is largest at the fluid exit, where it can compensate to some extent for the lower concentration driving force between the flowing fluid and the pore wall. This feature is of interest when operation is below the limiting current. Side reactions would be expected to occur first at the fluid exit because of the large electric driving force there.

However, under the design conditions where the electrode potential must be maintained within a certain range and the limiting-current condition is assumed to exist, Eq. 56 is less favorable than Eq. 44 for the upstream counterelectrode because of the presence of αL within the parentheses in Eq. 56. This means that there is no simple limit for the maximum flow rate in the limit of 100% conversion, and at high conversion Eq. 56 leads to a lower maximum flow rate than Eq. 44.

For the example of copper recovery, with flow from the back side, Eq. 56 yields:

$$\frac{\varepsilon \bar{k}_m a D_o}{v^2} = 3.59 \times 10^{-3} \tag{57}$$

From Fig. 8, we obtain $v/aD_o = 13.1$ and consequently $v = 1.96 \times 10^{-3}$ cm/s, a reduction by a factor of 2.4 from the maximum flow rate permissible with an upstream counterelectrode. The electrode thickness is now reduced to $L = 1.7$ cm, and $aL = 41.6$.

4.5. Matrix Resistance

The preceding designs were based on the variation of the potential Φ_2 in the pore solution. Thus the conductivity σ of the matrix was, in effect, taken to be much larger than the effective conductivity κ of the pore solution. If we consider the effect of a finite conductivity of the matrix, we find that the position of the current collector now influences the range of electrode potentials $\Phi_1 - \Phi_2$ encountered in the porous electrode when it is operated at the limiting current.

Two cases can be treated simply and rapidly. When the current collector is on the same side of the porous electrode as the counterelectrode, the current enters the porous electrode in the pore solution at one face, penetrates a certain distance into the electrode to the point where it is transferred by the electrochemical reaction to the matrix phase, and then flows through the matrix back to the current collector at the same face where it originally entered the electrode. Under this condition the resistivities of the pore solution and the matrix are effectively in series, and $1/\kappa + 1/\sigma$ should replace $1/\kappa$ in evaluating the variation of the electrode potential. In particular, for an upstream counterelectrode and an upstream current collector, Eq. 43 or 44 applies, but with $1/\kappa$ replaced by $1/\sigma + 1/\kappa$ and $\Phi_2(L) - \Phi_2(0)$ or $\Delta\Phi_2$ replaced by the variation $\Delta\Phi$ of the electrode potential. The electrical driving force is largest at the fluid inlet and smallest at the outlet. Furthermore, for the case of a downstream counterelectrode and a downstream current collector, Eq. 56 applies, but with the same substitutions. The electrical driving force is largest at the fluid outlet and smallest at the inlet.

The situation is more complicated when the counterelectrode is upstream and the current collector is downstream. Now the current density i_2 in the pore solution and the current density i_1 in the matrix sum to the superficial total current density $nFv(c_o - c_L)/s_R$, which is a constant through the thickness of the electrode. The electrical driving force is now smallest at an intermediate point x within the electrode, given by:

$$e^{-\alpha x} = \frac{\kappa + \sigma e^{-\alpha L}}{\kappa + \sigma} \qquad (58)$$

For

$$\frac{\kappa}{\sigma} \leq \frac{1 - (\alpha L + 1)e^{-\alpha L}}{\alpha L - 1 + e^{-\alpha L}} \qquad (59)$$

the electrical driving force is largest at the fluid inlet, and the variation of the electrode potential within the electrode is:

$$\Delta \Phi = \frac{nFv^2 c_o}{s_R a \bar{k}_m} \left(\frac{1}{\kappa} + \frac{1}{\sigma} \right) \left[1 + \frac{\kappa + \sigma e^{-\alpha L}}{\kappa + \sigma} \ln \left(\frac{\kappa + \sigma e^{-\alpha L}}{(\kappa + \sigma)e} \right) \right] \qquad (60)$$

For larger values of κ/σ, the electrical driving force is largest at the fluid outlet, and the variation of the electrode potential within the electrode is:

$$\Delta \Phi = \frac{nFv^2 c_o}{s_R a \bar{k}_m} \left(\frac{1}{\kappa} + \frac{1}{\sigma} \right) \left[e^{-\alpha L} + \frac{\kappa + \sigma e^{-\alpha L}}{\kappa + \sigma} \ln \left(\frac{\sigma + \kappa e^{\alpha L}}{(\kappa + \sigma)e} \right) \right] \qquad (61)$$

406   J. Newman and W. Tiedemann

To facilitate comprehension of these complicated relations, Fig. 13 has been prepared for $\alpha L = 6.5$, approximately the design for the copper-recovery problem. This graph shows that a finite matrix resistance lowers the variation of electrode potential until $\kappa/\sigma = 0.18$ and that the downstream current collector is superior to the upstream current collector when $\kappa/\sigma < 0.32$.

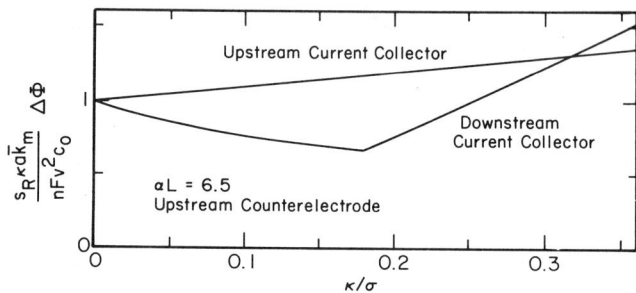

Fig. 13. Dependence of the variation of the electrode potential on the resistivity of the matrix.

The last case to treat involves a downstream counterelectrode and an upstream current collector. The preceding paragraph through Eq. 61 applies exactly to this situation if the conductivities κ and σ are interchanged. With the same substitution, Fig. 13 can also be used for this case.

The graphical solution method with Fig. 8, as developed after Eq. 45, can still be used with a finite matrix conductivity to solve for the maximum permissible flow rate. The value of $\varepsilon \bar{k}_m a D_o / v^2$ is now evaluated from the equations referred to in the present subsection. Of the four cases treated in this section, only the first, involving an upstream counterelectrode and an upstream current collector, leads to a simple finite limit for the maximum flow rate in the limit of 100% conversion.

4.6. Fluid Flow Parallel to Separator

One way to achieve high flow rates and high production rates without violating the allowed range of potentials is to stage several electrochemical reactors, so that part of the overall conversion is accomplished in each of them. A variation of this idea is to adopt configuration c of Fig. 1, so that a large conversion can be attained in a single unit. At the downstream end of this reactor, the electrical driving force is maintained as large as possible, subject to the limitation of a tolerable extent of side reactions. Near the downstream end, the reaction rate is low because the reactant has been depleted. Hence the ohmic potential drop in the solution is small, and the large electrical driving force is available for effecting a high degree of conversion. This is similar to the case of the downstream counterelectrode in that a large electrical driving force is available near the fluid outlet.

In configuration c of Fig. 1, the same electrical driving force is available over the entire length of the reactor. The higher reaction rates near the fluid inlet require a larger ohmic potential drop, so that a lower electrical driving force is actually available locally within the porous electrode.

This configuration has been used for oxidation of ferrous ions (26) and for recovery of copper (27). The difficult two-dimensional analysis of the system has been carried out by Alkire and Ng (28) for a cylindrical geometry.

It is not clear that this configuration has an inherent cost advantage over the other configurations in Fig. 1. Adams et al. (29) cite lower cost figures for configuration a for oxidation of ferrous ions (actually, they flowed the solution through the cathode before flowing it through the anode). Details of fabrication costs may have more influence than the gross flow and current geometry.

5. MAXIMUM CONVERSION

When operation is at the limiting current, so that the wall concentration is effectively zero, the reactant concentration can be reduced to an arbitrarily low level by using a sufficiently deep bed and a sufficiently low flow rate. When axial diffusion and dispersion are neglected, the relationship is $\alpha L = \ln(c_o/c_L)$. The preceding sections have indicated how this design can be achieved, with some allowance for side reactions and ohmic potential drop.

There remain two interrelated questions: (1) what cell potential would be required to achieve a given effluent concentration and (2) what is the ultimate limit below which the effluent concentration of reactant cannot be reduced? The answers require a realization that the wall concentration can be made small only with the occurrence of a substantial concentration overpotential and that this will lead to side reactions.

Let this overpotential be detected by means of a reference electrode of a given kind located in the dilute-product stream. The potential of the working electrode relative to this reference electrode is a measure of the electrode potential $\Phi_1 - \Phi_2$ at the fluid outlet. Let us assume that the electrode thickness can be increased substantially and that this same value of the electrode potential can be maintained over some distance near the fluid outlet.*

---

*In terms of the discussion of the previous section, this implies that we have an upstream counterelectrode and an upstream current collector. If the matrix conductivity σ is high, the current collector need not be placed upstream.

The wall concentration and also the effluent concentration then approach the equilibrium concentration at this electrode potential, rather than the zero value assumed in earlier sections.

In their work on copper removal, Bennion and Newman (14) used a saturated calomel electrode in the dilute-product stream. In that case, one wishes to treat the open-circuit potential of the following cell:

$$
\begin{array}{c|c|c|c|c|c|c|c}
\alpha & \beta & \delta & \epsilon & & \lambda & \phi & \alpha' \\
Pt(s) & Hg(l) & Hg_2Cl_2(s) & \text{KCl in } H_2O \text{ saturated} & \text{transition region} & CuSO_4, Na_2SO_4, H_2SO_4, H_2O & Cu(s) & Pt(s)
\end{array}
\tag{62}
$$

With neglect of liquid-junction potentials and activity-coefficient corrections, the open-circuit potential of the copper electrode relative to the calomel electrode is

$$U = U^\theta + \frac{RT}{2F} \ln \left(\frac{c_{Cu^{++}}^\lambda \left(c_{Cl^-}^\epsilon\right)^2}{\rho_o^3} \right) \tag{63}$$

where U^θ = 0.337 − 0.2676 = 0.0694 V at 25°C is the standard cell potential, ρ_o is the density of pure water (g/cm^3), and the concentrations must be expressed in moles per liter.

If U can be equated to the extreme value that can be maintained for the potential of the porous electrode relative to the reference electrode in the dilute product, then Eq. 63 can be used to estimate the minimum cupric ion concentration that can be attained in the effluent.

On the other hand, if the design calls for an effluent concentration of, say, 1 mg/liter, Eq. 63 can be used to estimate the electrode potential that must be maintained near the outlet of the reactor. To permit use of the design equations presented earlier, one would seek to maintain the wall concentration at a lower value, such as 0.1 mg/liter.

To obtain the total equilibrium concentration of copper in the effluent, we must also estimate the cuprous concentration and add this to the cupric concentration from Eq. 63. A parallel development leads to the open-circuit potential of a cell involving cuprous ions instead of cupric. Alternatively, one can develop the equilibrium relationship for the cuprous-disproportionation reaction

$$2Cu^+ \rightarrow Cu + Cu^{++} \qquad (64)$$

with the result (30):

$$c_{Cu^+} = \sqrt{0.6 \times 10^{-6} \rho_o c_{Cu^{++}}} \qquad (65)$$

Figure 14 shows these results as functions of the electrode potential U. Bennion and Newman were able to attain an electrode potential of −0.25 V at the fluid outlet. At this potential the equilibrium concentration is about 4.7×10^{-5} mg/liter and is composed predominantly of cuprous ions. The actual effluent concentration (14) at this potential was in the range from 0.06 to 0.38 mg/liter, depending on the flow rate. Since the reactor was designed to reduce the concentration only to 1 mg/liter, agreement with the equilibrium value should not be expected.

Trainham and Newman (31) have analyzed several other systems as well. Their results are summarized in Table I.

TABLE I

Experimental Minimum Effluent Concentration and Predicted Wall Concentration at that Potential for Various Systems, with Electrode Potentials Relative to a Saturated Calomel Electrode in All Cases

System	Feed Concentration (mg/liter)	Observed Effluent Concentration (mg/liter)	Electrode Potential (V)	Calculated Wall Concentration (mg/liter)
Copper	667	0.06	-0.25	4.7×10^{-5}
Silver	1000	42	-0.46	23
Lead	4.32	0.55	-0.56	0.028
Mercury	4.12	<0.005	-0.46	1.4×10^{-14}
Antimony	100	5	-0.38	1.6×10^{-13}
Ferrous oxidation	705	1	0.7	1.4

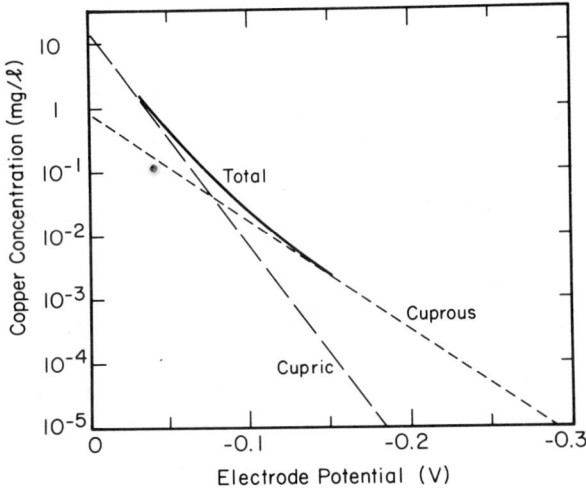

Fig. 14. Equilibrium copper ion concentrations as functions of the electrode potential relative to a saturated calomel electrode.

The silver deposition (22) is from a 1.737 M thiosulfate solution: a simulated fixing bath. The strong complex of silver with thiosulfate shifts the standard electrode potential from 0.7991 V for the reaction

$$Ag^+ + e^- \rightarrow Ag \tag{66}$$

to 0.0164 V for the reaction:

$$Ag(S_2O_3)_2^{3-} + e^- \rightarrow Ag + 2S_2O_3^{2-} \tag{67}$$

This makes it more difficult to attain a low concentration of dissolved silver, without interference from hydrogen evolution. Furthermore, the thiosulfate ion can be reduced according to the reaction:

$$S_2O_3^{2-} + 8e^- + 8H^+ \rightarrow 2HS^- + 3H_2O \tag{68}$$

This constitutes an additional side reaction that consumes current and produces a species, HS^-, that could lead to precipitation of silver sulfide.

At the potential -0.46 V used in Table I, the current efficiency is still high, and side reactions are not dominant. In the experiments (22), electrode potentials as negative as -0.55 V were used, and the silver concentration was reduced to as low as 0.8 mg/liter.

Mercury removal (32) from a brine solution 4.35 M in NaCl involves predominantly the complex $HgCl_4^{2-}$ in the calculations of the equilibrium wall concentration.

Lead deposition has a standard electrode potential of −0.126 V, which is negative to that of hydrogen. Preliminary experiments (23) showed a reduction of the concentration to within a factor of 20 of that calculated for the equilibrium wall concentration. Some further improvement may be possible.

Ferrous oxidation (29) was from a dilute sulfuric acid solution containing about 1.43 g/liter of H_2SO_4.

Kuhn and Houghton (33) reduced the antimony concentration from 100 to 5 mg/liter in 1 M sulfuric acid, by recirculation of the solution through the reactor until no further reduction in the antimony concentration was observed. No electrode potentials were reported. The equilibria in this system are complicated. The most probable species in solution is SbO^+ followed by $Sb(OH)_3$ and $HSbO_2$ in nearly equal concentrations. The standard electrode potential is 0.2075 V for the reaction

$$Sb + H_2O \rightarrow SbO^+ + 2H^+ + 3e^- \tag{69}$$

and 0.2307 V for the reaction:

$$Sb + 3H_2O \rightarrow Sb(OH)_3 + 3H^+ + 3e^- \tag{70}$$

Antimony can be reduced further to stibine (SbH_3), a gas more or less soluble in the solution (and also highly toxic). The standard electrode potential is −0.5104 V for the reaction:

$$SbH_3 \rightarrow Sb + 3H^+ + 3e^- \tag{71}$$

The equilibria thus lead to a minimum in the soluble antimony concentration at an electrode potential of −0.17 V, and this is what is reported for the calculated wall concentration in Table I. At more negative potentials, the increasing equilibrium concentration of stibine outweighs the decreasing equilibrium concentration of antimony in the +3 valence state. It would seem that designed, controlled-potential experiments could lead to a minimum effluent concentration below the 5-mg/liter level reported by Kuhn and Houghton (33).

The ohmic potential drop in the solution within the porous electrode itself also pertains to the matters under discussion. If, in the copper-recovery process, the electrode potential is −0.25 V at the fluid outlet and the flow rate is selected for $\Delta\Phi_2$ = −0.2 V, then the electrode potential is −0.45 V at the fluid inlet. This implies that the electrode potential of −0.45 V does not, in itself, lead to an intolerable level of side reactions. To achieve a larger electrical driving force at the fluid exit, one could add a second electrochemical reactor downstream of the first. If the copper concentration is depleted by, say, 99% in the first reactor, then the total current density in the second reactor will be considerably reduced, and the electrode potential could be maintained closer to −0.45 V throughout the thickness of the porous electrode. A lower effluent concentration can ultimately be achieved in this manner.

Furthermore, because of the nonzero reaction rate at the fluid outlet, an appreciable surface overpotential there produces a wall concentration higher than the calculated equilibrium value. A second reactor downstream would provide an additional opportunity for the effluent concentration to approach the equilibrium wall concentration. In this case a similar effect can be accomplished by making the electrode thicker if the counterelectrode is upstream and either the matrix conductivity is high or the current collector is also upstream.

One perceives that configuration c of Fig. 1 or recirculation of the solution through a single reactor after the manner of Kuhn and Houghton (33) accomplishes much the same purpose as one or more reactors downstream of the first. For a number of systems we should note that conversions approaching 100% can be achieved in a single reactor, and this is probably more economic than multiple reactors when it accomplishes the required conversion.

The surface overpotential at the reactor exit and the potential drop $\Delta\Phi_2$ within the solution in the porous electrode also influence the prediction of the overall cell potential. One also needs to analyze the electrode potential at the counterelectrode and the ohmic potential drop in the solution from the anode to the cathode.

6. HISTORICAL PERSPECTIVE

We must keep in mind that fluid flow in porous media and mass transfer and chemical reactions in packed beds have been studied extensively for many years in geology and in civil and chemical engineering. Much of the information concerning the characterization of flow-through systems can be transferred directly to the electrochemical system of interest from the fluid-flow, heat-transfer, and mass-transfer situations from which it was generated. For summaries of these contributions, we refer to the texts of Bird et al. (34) and Sherwood et al. (16) and the collection edited by de Wiest (35).

The potential advantages of flow-through porous electrodes over porous electrodes with no flow have been known for some time. Liebenow (36) showed how mass-transport limitations could be diminished by flowing a solution through an electrode. Heise (37) discussed in great detail the usefulness of flow-through porous

electrodes for a number of processes of industrial interest:
(1) conversion of manganate to permanganate, ferrous to ferric,
nitrate to nitrite, and arsenite to arsenate; (2) electrowinning
of copper; and (3) electroorganic oxidation and reduction. The
advantages and limitations of such electrode systems are also
discussed.

The development of theoretical models to treat flow-through
electrodes has followed a little behind that of porous electrodes
without flow (7) due to the added complexities produced by the
flowing solution. Perskaya and Zaǐdenman (38-40), Sohm (41),
Gurevich and Bagotskiǐ (42,43), and Austin et al. (44) were the
first to treat in detail the polarization behavior of flow-through
porous electrodes. The assumptions included a homogeneous porous
electrode in steady operation with uniform flow, a one-dimensional
treatment, first-order electrode kinetics with no side reactions,
negligible ohmic losses in the solid matrix, a constant solution
conductivity, negligible diffusion and dispersion in the direction
of fluid flow, and negligible mass-transfer resistance within
the pore itself. Both upstream and downstream counterelectrodes
were included in the preceding treatments. Wroblowa (45) discussed reaction orders different from one (i.e., non-first-order)
and their possible determination from measurements of flow-through
porous electrodes.

Johnson and Newman (46) examined the adsorption and desorption
of ions in a flow-through porous electrode as a means of desalting
water. Their analysis included ohmic losses in the carbon matrix
[see also Posey and Morozumi (13)] and transient operation.
Schmidt and Gidaspow (47) have presented a similar treatment.

The papers of Sioda (48-52) have shown an increasing awareness
that the overall reaction rate is determined more by mass-transfer
limitations within the pores than by the kinetics of the electrode

reaction. Bennion and Newman (14) emphasized the role of the mass-transfer coefficient in their concentration of dilute solutions containing copper. Fleischmann and Oldfield (53) have presented similar equations for fluidized-bed electrodes.

The potential distributions within flow-through porous electrodes under limiting-current conditions have been treated by Sioda (51), Bennion and Newman (14), and (for fluidized-bed electrodes) Fleischmann and Oldfield (53). The potential distributions not only contribute to the overall polarization of the electrode, but they also determine whether the electrode can operate within a narrow range of electrode potentials $\Phi_1 - \Phi_2$. Correspondingly, there will be a maximum permissible flow rate for operation of the porous electrode with controlled-potential electrolysis. These considerations of electrode design have been emphasized by Bennion and Newman (14).

Sioda (54-56) and Wroblowa and Razummey (57) have extended the steady-state analysis to include axial diffusion. Here we should also recall the early chemical-engineering contributions (8,16,58). The present review suggests that axial diffusion and dispersion do not have a dominant effect on the overall mass-transfer rate except at very low flow rates, which are likely to be outside the range of industrial interest.

A few papers have dealt with the situation below the limiting current. Alkire and Ng (28) analyzed the two-dimensional current distribution resulting from placement of the counterelectrode parallel to the direction of flow within the working electrode (see configuration c of Fig. 1). Alkire and Mirarefi (59) have analyzed the current distribution in a tubular electrode. These are both sophisticated analyses. Considerably simpler is the one-dimensional analysis of a macrohomogeneous model below the limiting current. Sioda (52) has given such an analysis for a reversible reaction whose equilibrium can be described by the Nernst equation.

Fleischmann et al. (60) have reported on fluidized beds of conducting particles in a stream of electrolyte. (This is only one of many papers on this subject, which is outside the scope of this review.) Several reports (61-62) have compared fluidized-bed electrodes with flow-through, packed-bed electrodes. The fixed-bed electrodes are simpler to operate, have a higher matrix conductivity and a higher specific interfacial area, and can achieve high superficial current densities. The distances over which the current must flow, thereby producing an ohmic potential drop in the solution, are small. Flow-through porous electrodes have also been compared (62,63) with other electrode arrangements, such as parallel-plate electrodes and geometries with stationary or fluidized turbulence promoters.

In electroanalytical chemistry, flow-through porous electrodes can effect a controlled-potential coulometric analysis because of the high conversions that can be achieved (48,49). With proper design, essentially all the reactant entering the electrode can be consumed before the solution leaves the electrode. A second electroanalytical application involves polarography at a solid electrode. With a relatively thin porous electrode or a high flow rate, not all the material reacts, but the limiting current is proportional to the concentration in the feed solution. This possibility has been explored by Blaedel and Boyer (64) as well as Sioda. A third mode of operation involves electrodeposition of metals from solutions too dilute to analyze directly. Subsequent anodic stripping (14,65) or dissolution will produce a more concentrated solution, which can be analyzed.

We have already cited a number of papers emphasizing removal of metal ions from waste streams, economic recovery of metals, or hydrometallurgy (12,14,22,23,27, 31-33, 37,60,65), and others have dealt with redox reactions (29,37,44, 48-52, 54-56). Additional papers not yet cited are concerned with copper (66-68), zinc (69), metals removal (70) or fractionation (71), the iodine-iodide couple (72), and effluent treatment by oxidation of organics (73).

Beccu (74) has reviewed the use of flow-through porous electrodes, employing redox couples, for the storage of electrical energy. On charge, the feed in configuration a of Fig. 1 contains the oxidized species of the negative couple and the reduced species of the positive couple. The effluent from the positive anode contains the oxidized species of the positive couple (as well as the oxidized species of the negative couple). The effluent from the negative cathode contains the reduced species of both couples. On discharge the flows are reversed, so that the system becomes configuration b of Fig. 1. For positive couples, Beccu suggests Fe(II)/Fe(III) and Cr(III)/Cr(IV). Possible negative couples include Cr(II)/Cr(III) and Ti(III)/Ti(IV).

Fitzjohn (75) and Davis (76) have reported on the recent renewed interest in electroorganic synthesis. The electrochemical alternative to well-established chemical processes has been well known for some time, but the technical and economic aspects of the electrochemical processes still need proper evaluation. Only when the proper treatment of design and scale-up parameters has been given can an appropriate comparison between chemical and electrochemical manufacturing processes be made. Sioda (48) has treated reactions with 9,10-diphenylanthracene, and the reduction of p-nitrobenzoate ion has been carried out by Sioda and Kemula (77). The selectivity and possible production rate for an intermediate in ec and ece mechanisms has been analyzed theoretically by Sioda (78,79).

The mass-transfer coefficient can be obtained from chemical-engineering correlations (4,11,16) or can be calculated in a few cases from fundamental principles (10). In addition to those electrochemical investigations (2,3,80) leading directly toward values of the mass-transfer coefficient, such values could also be deduced from the results of some other studies (12,14,29,56,81, 82) with somewhat different objectives, although it might be necessary first to estimate the effective specific interfacial area of the electrodes.

7. RECENT WORK

This section reviews the literature on flow-through porous electrodes since the original submission of this work to the editors.

Alkire and Gracon (82) have included a very general analysis of a single electrode reaction below the limiting current that predicts nonuniform reaction rates due to ohmic, mass-transfer, and heterogeneous kinetic limitations. Their model also has provision for treating the effects of axial diffusion. However, a simplification of the Danckwerts (58) and Wehner-Wilhelm (8) boundary condition at the reactor inlet leads to an incorrect treatment of this effect. With the authors' simplification, it would be possible for the overall reaction rate to exceed the supply of reactant in the upstream flowing stream. Results are given for upstream placement of the counterelectrode.

Agreement between the model of Alkire and Gracon (82) and their deposit-distribution data at the limiting current is poor, and no satisfactory explanation is given, as noted by Ateya and Austin (83). However, Ateya and Austin also offer no explanation for this discrepency and present a very weak model in comparison to that of Alkire and Gracon (82).

The analysis of Ateya and Austin (83) is similar to the earlier work of Sioda (52) for a single electrode reaction described by the Nernst equation, but unlike Sioda (52) they neglect the interfacial mass-transfer resistance. The main purpose of their paper is to study the effect of axial diffusion, and agreement between experimental data and the simple model is good. But as mentioned earlier, the effects of axial diffusion and dispersion are generally secondary compared to the effects of convection, reaction, and mass transfer to the wall, except at very low flow rates. Therefore, the model proposed by Ateya and Austin (83) has limited industrial application.

Trainham and Newman (84) have also developed a model that is similar to that of Alkire and Gracon (82) and includes ohmic potential drop, mass-transfer resistance, heterogeneous reaction kinetics, and the effects of axial diffusion and dispersion, without simplification of the Danckwerts (58) and Wehner-Wilhelm (8) boundary conditions as was done previously (82). The model also includes the possibility of a simultaneous side reaction, where the composition dependence of the side reaction has been ignored. In this paper Trainham and Newman (84) speculate as to why the agreement between the model of Alkire and Gracon (82) and their data is poor. This is perhaps due to the inadequate representation of the derivatives in the governing equations at the limiting current by standard central-difference approximations. It was found that these approximations were not sufficiently accurate to approach the limiting-current result, which can be calculated analytically [see Bennion and Newman (14) and Eqs. 31 and 41]. This anomalous behavior was primarily due to the high degree of nonuniformity in the current and concentration distributions near the front of the electrode. A more accurate representation of the derivatives required that more of the physics be incorporated in their dependence. New approximations were developed that reproduced the limiting-current result and also gave satisfactory agreement with the experimental data.

Alkire and Gould (85) have extended the work of Alkire and Gracon (82) to multiple-reaction sequences, which include deposition of several metals, deposition of a metal in the presence of a redox system, and an ece sequence applicable to electroorganic synthesis; results are given for a downstream counterelectrode. The overall analysis is good, but the simplification of the Danckwerts and Wehner-Wilhelm boundary conditions persists as before (82). However, as we have tried to emphasize, the effects of axial diffusion and dispersion are small, and over the parameter space chosen by Alkire and Gould (85) the oversight in the boundary condition has a negligible effect on the results.

Ateya (86) has next considered the case of two consecutive single-electron-transfer reactions for both flow-through and non-flow electrodes with use of a polarization equation derived by Vetter (87). The analysis assumes negligible ohmic losses and again neglects the interfacial mass-transfer resistance. Experimental verification of this simple analysis is accomplished using an extremely shallow reactor [aL = 1.1×10^{-4} (sic)].

Coeuret et al. (88) have introduced the idea of the effectiveness (i.e., the effective electrode length) of flow-through porous electrodes in terms of the local overpotential when linear kinetics are applicable. This analysis is analogous to treating the internal mass-transfer resistance within a catalyst pellet for a first-order chemical reaction. In this case, the local overpotential plays the same role as the concentration of a gaseous reactant.

In order to make this type of analysis of practical use (i.e., in a range of large overpotential), generalization of the method has been suggested (88). The assumption is made that the local current density is proportional to the local overpotential to the n-th power, where n is a number less than one. In another paper,

Paulin et al. (89) carry out the theoretical and experimental study of this extension.

Theoretically, the method suggested by Coeuret and co-workers (88,89) is valid only under conditions when linear kinetics are applicable. For large overpotentials below the limiting current, the local current density exhibits Tafel behavior, and at even higher values of the local overpotential the local current density becomes mass-transfer limited. Unfortunately, the empiricism suggested masks this physically meaningful behavior.

On a brighter note, the experimental potential distributions measured below the limiting current by Coeuret and co-workers (88,89) may prove useful in verifying more complete models (82, 84,85), but the lack of conversion data or data for the operating current will make this difficult.

In other work at the limiting current, Coeuret (90) (using a Bennion and Newman (14) type of analysis) has derived the expression for the potential distribution for downstream placement of the counterelectrode [Eq. 56 replaced with $L = x$ and $\theta_L = \theta_x = \exp(-\alpha x)$] and has verified this distribution experimentally with a small moveable Luggin capillary connected to a saturated calomel reference electrode. Similarly, Gaunand et al. (91) have rearranged the expression given by Bennion and Newman (14) for the potential distribution for upstream placement of the counterelectrode and have verified this distribution experimentally with the same technique used by Coeuret (90).

As mentioned previously, Sioda (56) has included the effect of axial diffusion and dispersion in his limiting-current model. Recently (92), he has used this theoretical result with limiting-current data obtained by the reduction of ferricyanide on platinum screens to obtain the axial dispersion coefficient for this system.

Coeuret (93) has apparently done a similar analysis for the effect of axial diffusion and dispersion as presented here and has used this result to correct his mass-transfer-coefficient data. However, no derivation is given, and the final result presented for the exit concentration appears to be in error. If the axial dispersion coefficient is set equal to zero in the expression given, the exit concentration goes to one instead of Eq. 9 as the author claims.

Ateya (94) has defined a simple criterion that describes the effects of diffusion on conversion in tubular pores. Entrance effects are also considered by application of Levich's discussion of the Lévêque problem (95). Unfortunately, the author reaches the wrong conclusion and states that the local current density is lower in the entrance region than in the fully developed region; the reverse is true.

Ateya et al. (96) have calculated (at the limiting current) the asymptotic concentration distribution for long straight pores with the Graetz solution. In the conclusions of this paper the authors have again misinterpreted the physics of the various mass-transfer regions. If a limiting-current condition is achieved along the entire length of a tube, not only is the entrance region mass-transfer controlled, but so is the fully developed region where the Graetz analysis is valid.

Van Zee and Newman (22) have shown the feasibility of removing silver ions from photographic fixing solution. However, before an optimum cell design can be achieved, more experimental data are needed, especially in clarifying the electrochemistry of the side reaction and an apparent current efficiency greater than unity.

Recently, Trainham and Newman (97) have used their model (84) to assess the effect of electrode placement and finite matrix conductivity on the performance of flow-through porous electrodes. Calculations show that a limiting-current distribution can only be achieved (for electrode lengths of industrial interest) for upstream placement of the counterelectrode and not for a downstream counterelectrode. However, if interference by the side reaction is substantial (because of either the chemical system or too high a superficial velocity), a limiting-current distribution cannot be achieved for any electrode configuration at any matrix conductivity.

For an upstream counterelectrode when the matrix conductivity is not large, the choice between upstream or downstream placement of the current collector depends on the chemical system and what matrix conductivity can be achieved in practice. Below the limiting current, the performance is only slightly better for an upstream counterelectrode than for downstream placement of the counterelectrode. It is evident from these results that sophisticated models are necessary to achieve an optimum design.

As mentioned earlier, the Graetz solution for straight tubes is useful conceptually in understanding mass transfer in packed beds. However, a more useful approach would be to model a packed bed as an array of periodically constricted tubes. Fedkiw and Newman (98) have taken this approach, solving the creeping-flow equations for fully developed flow in a sinusoidal tube and then using this velocity profile to solve the convective diffusion equation for the rate of mass transfer. Results are presented for the asymptotic Sherwood number of a deep-bed reactor with a large reactant Péclet number. The results depend on two dimensionless geometric variables. The bed Sherwood number exhibits different behavior in the amplitude-radius ratio (A/r_A) for small and large values of r_A (radius-period ratio). In beds of long thin tubes (small r_A), the Sherwood number increases with

A/r_A, whereas for large r_A this trend reverses itself.

8. CONCLUSIONS

Flow-through porous electrodes show promise for industrial applications in the areas of removal and recovery of metals from process streams, synthesis of organic and inorganic chemicals, and storage of electric energy. It is now possible to design and optimize such systems by consideration of the principles of fluid flow, mass transfer, ohmic potential drop, electrode kinetics, and current distribution.

In the future we can expect to find more extensive analysis of operation below the limiting current with both upstream and downstream counterelectrodes, side reactions, and transient behavior and control. Cost optimization should include conversion and current efficiency in relation to energy consumption. Process alternatives for investigation include multiple reactors as opposed to a single reactor and flow parallel to the electrodes versus flow perpendicular to the electrodes. More consideration must be given to the best choice of the reaction at the counterelectrode and the materials and method of fabrication.

9. LIST OF SYMBOLS

a	Specific surface or interfacial area (per unit volume of the porous electrode), cm^{-1}
A_1, A_2	Constants in Eq. 23, mol/cm^3
B	See Eq. 32
c	concentration of binary electrolyte, mol/cm^3

List of Symbols

c_i	Concentration of species i, mol/cm^3
c_{iw}	Concentration of species i at the wall, mol/cm^3
c_o	Feed concentration of reactant, mol/cm^3
c_L	Exit concentration of reactant, mol/cm^3
c_R	Reactant concentration, mol/cm^3
$c_{R\infty}$	Reactant concentration far from the rotating disk, mol/cm^3
Δc	Reactant concentration difference between the flowing fluid and the pore wall, mol/cm^3
d_p	Equivalent spherical particle diameter of bed packing, cm
D'	Dimensionless dispersion coefficient, $\varepsilon a k_m (D_i + D_a)/v^2$
D_a	Axial dispersion coefficient, cm^2/s
D_i	Effective diffusion coefficient within a porous electrode, cm^2/s
D_o	Diffusion coefficient of reactant in the feed solution, cm^2/s
F	Faraday's constant, 96,487 C/equiv
i_1	Superficial current density in the matrix phase, A/cm^2
i_2	Superficial current density in the pore solution phase, A/cm^2
I	Superficial current density to an electrode, A/cm^2
j_{in}	Pore-wall flux of species i, mol/(cm^2-s)
k_m	Local mass-transfer coefficient, cm/s
$\bar{k}_m$	Average mass-transfer coefficient, cm/s
K	Kernel in integral Eq. 36
l	Pore length, τL, cm
L	Electrode thickness, cm
n	Number of electrons transferred in the electrode reaction
N_i	Superficial flux of species i, mol/(cm^2-s)
$\overline{Nu}$	(Average) Nusselt number, $\varepsilon \bar{k}_m / a D_o$
P	Dynamic pressure, dyne/cm^2
Pe	Péclet number, v/aD_o

r_o	Equivalent pore radius, cm
R	Universal gas constant, 8.3143 J/(mol K)
Re	Reynolds number, $v/a\nu$
Re'	Modified Reynolds number, $d_p v/\nu(1-\varepsilon)$
s_R	Stoichiometric coefficient of reactant in the electrode reaction
Sc	Schmidt number, ν/D_o
t	Time, s
t_i	Transference number of species i
T	Absolute temperature, K
u_i	Mobility of species i, $(cm^2\text{-mol})/(J\text{-}s)$
U	Equilibrium, open-circuit potential, V
U^θ	Standard cell potential, V
v	Superficial fluid velocity, cm/s
x	Distance through the porous electrode from feed end, cm
y	Dimensionless distance, $ak_m x/v$
z_i	Valence or charge number of species i
α	Reciprocal reaction distance, ak_m/v, cm^{-1}
ε	Porosity or void volume fraction
θ	Dimensionless concentration, c_i/c_o
θ_L	Dimensionless exit concentration, c_L/c_o
κ	Effective conductivity of the solution phase, mho/cm
κ_o	Conductivity of feed solution outside of the electrode, mho/cm
λ_B, λ_G	Eigenvalues of the Bessel function and the Graetz problem
Λ	Molar conductance of a binary electrolyte, κ/c, $(mho\text{-}cm^2)/mol$
μ	Viscosity, g/(cm-s)
ν	Kinematic viscosity, μ/ρ, cm^2/s
ρ	Fluid density, g/cm^3
ρ_o	Density of the pure solvent, g/cm^3

σ	Effective conductivity of the solid matrix, mho/cm
σ_o	Conductivity of the solid materials composing the matrix, mho/cm
τ	Tortuosity factor
Φ_1	Electric potential in the matrix, V
Φ_2	Electric potential in the solution, V
$\Delta\Phi$	Variation of $\Phi_1 - \Phi_2$ within the electrode, V
$\Delta\Phi_2$	Variation of Φ_2 within the electrode, V

10. REFERENCES

1. E. J. Wilson and C. J. Geankoplis, Liquid mass transfer at very low Reynolds numbers in packed beds, Ind. Eng. Chem. Fund., 5, 9-14 (1966).

2. Peter Appel, Electrochemical systems: impedance of a rotating disk and mass transfer in packed beds, Dissertation, No. LBL-5132, University of California, Berkeley, May 1976.

3. Peter W. Appel and John Newman, Application of the limiting-current method to mass transfer in packed beds at very low Reynolds numbers, AIChE J., 22, 979-984 (1976).

4. I. Colquhoun-Lee and J. Sepanek, Mass transfer in single phase flow in packed beds, Chem. Eng. (London), No. 282, 108-111 (Feb. 1974).

5. Robert E. Meredith and Charles W. Tobias, Conduction in heterogeneous systems, in Advances in Electrochemistry and Electrochemical Engineering, Vol. 2, Charles W. Tobias, ed., Wiley Interscience, N.Y., pp. 15-47, 1962.

6. Charles N. Satterfield, Mass Transfer in Heterogeneous Catalysis, M.I.T. Press, Cambridge, Mass., 1970, p. 36.

7. John Newman and William Tiedemann, Porous-electrode theory with battery applications, AIChE J., 21, 25-41 (1975).
8. J. F. Wehner and R. H. Wilhelm, Boundary conditions of flow reactor, Chem. Eng. Sci., 6, 89-93 (1956).
9. R. Byron Bird, Warren E. Stewart, and Edwin N. Lightfoot, Transport Phenomena, Wiley, New York, 1960, pp. 297, 390-396.
10. Jan P. Sørensen and Warren E. Stewart, Computation of forced convection in slow flow through ducts and packed beds -- III. Heat and mass transfer in a simple cubic array of spheres, Chem. Eng. Sci., 29, 827-832 (1974).
11. R. Byron Bird, Warren E. Stewart, and Edwin N. Lightfoot, Transport Phenomena, Wiley, New York, 1960, pp. 411-412.
12. A. K. P. Chu, M. Fleischmann, and G. J. Hills, Packed bed electrodes. I. The Electrochemical extraction of copper ions from dilute aqueous solutions, J. Appl. Electrochem., 4, 323-330 (1974).
13. F. A. Posey and T. Morozumi, Theory of potentiostatic and galvanostatic charging of the double layer in porous electrodes, J. Electrochem. Soc., 113, 176-184 (1966).
14. Douglas N. Bennion and John Newman, Electrochemical removal of copper ions from very dilute solutions, J. Appl. Electrochem., 2, 113-122 (1972).
15. Jan Robert Selman, Measurement and interpretation of limiting currents, Dissertation, No. UCRL-20557, University of California, Berkeley (June 1971).
16. Thomas K. Sherwood, Robert L. Pigford, and Charles R. Wilke, Mass Transfer, McGraw-Hill, New York, 1975.
17. N. V. Korovin, Study of liquid-gaseous electrode, Comptes Rendus Deuxiemes Journées Internationales d'Étude des Piles a Combustible, Serai, Brussels, 1967, pp. 148-151.

18. N. V. Korovin and A. S. Chudinov, Issledovanie zhidkostno-gazovykh elektrodov. I. Kapillyarnyĭ zhidkostno-gazovyĭ elektrod v usloviyakh estestvennogo gazoudaleniya, Elektrokhimiya, 4, 426-431 (1968).
19. N. V. Korovin and A. S. Chudinov, Issledovanie zhidkostno-gazovykh elektrodov. II. Kapillyarnyĭ zhidkostno-gazovyĭ elektrod v usloviyakh prinuditel'nogo gazoudaleniya, Elektrokhimiya, 4, 636-640 (1968).
20. N. V. Korovin, A. S. Chudinov, and A. F. Feoktistov, Issledovanie zhidkostno-gazovogo elektrodova. III. Ploskaya sistema kapillyarov, Elektrokhimiya, 5, 430-434 (1969).
21. Hidetoshi Ichino, Masamichi Yamashita, and Masao Kubokawa, Cathodic reaction of dissolved oxygen at liquid flow-through porous carbon electrodes (2) effects of electrocatalysts on electrodes, Denki Kagaku, 37, 504-509 (1969).
22. John Van Zee and John Newman, Electrochemical removal of silver ions from photographic fixing solutions using a porous flow-through electrode, J. Electrochem. Soc., 124, 706-708 (1977).
23. James Trainham and John Newman, The removal of lead ions from very dilute solutions using a porous flow-through electrode, Inorganic Materials Research Division Annual Report 1973, No. LBL-2299, Lawrence Berkeley Laboratory, University of California, April 1974, pp. 51-53.
24. Charles W. Tobias, personal communication, University of California, Berkeley, 1971.
25. John S. Newman, Electrochemical Systems, Prentice-Hall, Englewood Cliffs, N. J., 1973, Section 114, problem 7 of Chapter 17 and problem 2 of Chapter 11.
26. See Ref. 1-3 of Adams et al. (29).

27. J. M. Williams and K. B. Keating, Extended-surface electrolysis removes heavy metals from waste streams, Dupont Innovation, 6 (no. 3), 6-10 (1975).
28. Richard Alkire and Patrick K. Ng, Two-dimensional current distribution within a packed-bed electrochemical flow reactor, J. Electrochem. Soc., 121, 95-103 (1974).
29. George B. Adams, Roger P. Hollandsworth, and Douglas N. Bennion, Electrochemical oxidation of ferrous iron in very dilute solutions, J. Electrochem. Soc., 122, 1043-1048 (1975).
30. John S. Newman, Electrochemical Systems, Prentice-Hall, Englewood Cliffs, N.J., 1973, p. 57.
31. James A. Trainham and John Newman, A thermodynamic estimation of the minimum concentration attainable in a flow-through porous electrode reactor, J. Appl. Electrochem., 7, 287-297 (1977).
32. James A. Trainham and John Newman, unpublished results, University of California, Berkeley, 1975.
33. A. T. Kuhn and R. W. Houghton, Antimony removal from dilute solutions using a restrained bed electrochemical reactor, J. Appl. Electrochem., 4, 69-73 (1974).
34. R. Byron Bird, Warren E. Stewart, and Edwin N. Lightfoot, Transport Phenomena, Wiley, New York, 1960, pp. 196-200, 411-412, 679.
35. R. de Wiest, ed., Flow through Porous Media, Academic, New York, 1969.
36. C. Liebenow, Über die Berechnung der Kapazität eines Bleiakkumulators bei variabeler Stromstärke, Ze. Elektrochem., 4, 58-63 (particularly p. 63) (1897).
37. George W. Heise, Porous carbon electrodes, Trans. Electrochem. Soc., 75, 147-166 (1939).
38. R. M. Perskaya and I. A. Zaĭdenman, O zhidkostnykh diffuzionnykh elektrodakh, Dok. Akad. Nauk SSSR, 115, 548-551 (195

39. I. A. Zaĭdenman and R. M. Perskaya, O zhidkostnykh diffuzionnykh elektrodakh. (O nachalnom naklone polyarizatsionnykh krivykh.), Zh. Fiz. Khim., 23, 50-57 (1959).
40. I. A. Zaĭdenman, K teorii zhidkostnykh diffuzionnykh elektrodov. II. Ob elektrodakh vtorogo tipa, Zh. Fiz. Khim., 23, 437-440 (1959).
41. J. C. Sohm, Les électrodes à circulation de liquide, Electrochim. Acta, 7, 629-652 (1962).
42. I. G. Gurevich and V. S. Bagotskiĭ, Rabota zhidkostnykh poristykh elektrodov v rezhime s vynuzhdennoĭ podacheĭ reagenta, Inzhen.-fiz. Zh., 6 (no. 5), 75-85 (1963).
43. I. G. Gurevich and V. S. Bagotzky, Porous electrodes with liquid reactants under steady-state operating conditions, Electrochim. Acta, 9, 1151-1176 (1964).
44. Leonard G. Austin, Pierre Palasi, and Richard R. Klimpel, Polarization at porous flow-through electrodes, in Fuel Cell Systems, Advances in Chemistry Series, No. 47, American Chemical Society, Washington, D.C., 1965, pp. 35-60.
45. Halina S. Wroblowa, Flow-through electrodes. I. Diagnostic mechanistic criteria, J. Electroanal. Chem. Interfacial Electrochem., 42, 321-328 (1973).
46. A. M. Johnson and John Newman, Desalting by means of porous carbon electrodes, J. Electrochem. Soc., 118, 510-517 (1971).
47. T. Schmidt and D. Gidaspow, Electrosorption with porous electrodes, J. Electrochem. Soc., 120, 110C (1973) Abstr. No. 250, Chicago meeting.
48. R. E. Sioda, Electrolysis with flowing solution, Electrochim. Acta, 13, 375-382 (1968).
49. R. E. Sioda, Electrolysis with a flowing solution on graphite packing, Electrochim. Acta, 13, 1559-1562 (1968).

50. R. E. Sioda, Electrolysis with flowing solution on porous and wire electrodes, Electrochim. Acta, 15, 783-793 (1970).
51. R. E. Sioda, Distribution of potential in a porous electrode under conditions of flow electrolysis, Electrochim. Acta, 16, 1569-1576 (1971).
52. R. E. Sioda, Current-potential dependence in the flow electrolysis on a porous electrode, J. Electroanal. Chem. Interfacial Electrochem., 34, 399-409 (1972).
53. M. Fleischmann and J. W. Oldfield, Fluidized bed electrodes. Part I. Polarization predicted by simplified models, J. Electroanal. Chem. Interfacial Electrochem., 29, 211-230 (1971).
54. R. E. Sioda, Flow electrolysis on a porous electrode composed of parallel grids, J. Electroanal. Chem. Interfacial Electrochem., 34, 411-418 (1972).
55. R. E. Sioda, Limiting current in flow electrolysis on porous electrode, Electrochim. Acta, 17, 1939-1941 (1972).
56. R. E. Sioda, Limiting current on porous graphite electrodes under flow conditions, J. Appl. Electrochem., 5, 221-228 (1975).
57. H. S. Wroblowa and G. Razummey, Axial diffusion effects on flow-through electrodes, J. Electrochem. Soc., 121, 124C (1974) (abstr. No. 355, San Francisco meeting).
58. P. V. Danckwerts, Continuous flow systems. Distribution of residence times, Chem. Eng. Sci., 2, 1-13 (1953).
59. Richard Alkire and Ali Asghar Mirarefi, The current distribution within tubular electrodes under laminar flow, J. Electrochem. Soc., 120, 1507-1515 (1973).
60. M. Fleischmann, J. W. Oldfield, and L. Tennakoon, Fluidized bed electrodes. Part IV. Electrodeposition of copper in a fluidized bed of copper-coated spheres, J. Appl. Electrochem., 1, 103-112 (1971).

61. R. D. Armstrong, O. R. Brown, R. D. Giles, and J. A. Harrison, Factors in the design of electrochemical reactors, Nature, 219, 94 (1968).
62. G. Kreysa, S. Pionteck, and E. Heitz, Comparative investigations of packed and fluidized bed electrodes with non-conducting and conducting particles, J. Appl. Electrochem., 5, 305-312 (1975).
63. A. T. Kuhn and R. W. Houghton, A comparison of the performance of electrochemical reactor designs in the treatment of dilute solutions, Electrochim. Acta, 19, 733-737 (1974).
64. W. J. Blaedel and S. L. Boyer, Electrochemical characteristics of the gold micromesh electrode, Anal. Chem., 45, 258-263 (1973).
65. A. K. P. Chu and G. J. Hills, Packed bed electrode. II. Anodic stripping and the recovery of copper from dilute solutions, J. Appl. Electrochem., 4, 331-336 (1974).
66. R. S. Wenger, D. N. Bennion, and J. Newman, Electrochemical concentrating and purifying from dilute copper solutions, J. Electrochem. Soc., 120, 109C (1973) (abstr. No. 236, Chicago meeting).
67. Robert S. Wenger, Electrochemical concentrating and purifying from dilute copper solutions, M.S. Thesis, University of California, Los Angeles, 1974.
68. F. A. Posey, Electrolytic demonstration unit for copper removal from distillation plant blowdown, Oak Ridge National Laboratory, Report No. ORNL-TM-4112, 1973.
69. Takuji Yoshimura, Hitoshi Furuta, and Masamichi Yamashita, Discharge behaviour of zinc-bed electrodes, Nippon Kagaku Kaishi, 1148-1152 (1975).

70. G. A. Carlson and E. E. Estep, Porous cathode cell for metals removal from aqueous solution, J. Electrochem. Soc., 119, 114C (1972) (abstr. No. 211, Houston meeting).
71. D. K. Roe, Electrochemical fractionation: Potentiostatic chromatography and elution voltammetry, Anal. Chem., 36, 2371-2372 (1964).
72. Halina S. Wroblowa and A. Saunders, Flow-through electrodes II. The I_3^-/I^- redox couple, J. Electroanal. Chem. Interfacial Electrochem., 42, 329-346 (1973).
73. E. C. Beck and A. P. Giannini, Basic electric cell in effluent treatment, J. Electrochem. Soc., 119, 114C (1972) (abstr. No. 212, Houston meeting).
74. Klaus-D. Beccu, Zur Deckung des zukünftigen Spitzenbedarfs elektrischer Energie mittels electrochemischer Energiespeicher, Chem. Ing. Tech., 46, 95-99 (1974).
75. J. L. Fitzjohn, Electro-organic synthesis, Chem. Eng. Prog., 71 (no. 2), 85-91 (1975).
76. John C. Davis, New spark comes to electroorganic syntheses, Chem. Eng. (New York), 82 (no. 14), 44-48 (1975).
77. R. E. Sioda and W. Kemula, Application of flow electrolysis on porous electrodes for the electroreduction of potassium p-nitrobenzoate, Electrochim. Acta, 17, 1171-1174 (1972).
78. R. E. Sioda, Certain preparative aspects of the flow electrolysis on porous electrodes, Electrochim. Acta, 19, 57-62 (1974).
79. R. E. Sioda, The ECE mechanism in flow electrolysis in porous electrodes under conditions of limiting current, Electrochim. Acta, 20, 457-461 (1975).
80. Harry Hung-Kwan Yip, Mass transfer coefficient in packed beds at low Reynolds numbers, M.S. Thesis, University of California, Berkeley, Lawrence Berkeley Laboratory Report No. LBL-1831, 1973.

81. Brian Eugene Gracon, Experimental studies of flow-through porous electrodes, M.S. Thesis, University of Illinois, Urbana, 1974.
82. Richard Alkire and Brian Gracon, Flow-through porous electrodes, J. Electrochem. Soc., 122, 1594-1601 (1975).
83. B. G. Ateya and L. G. Austin, Steady-state polarization at porous, flow-through electrodes with small pore diameter. I. Reversible kinetics, J. Electrochem. Soc., 124, 83-89 (1977).
84. James A. Trainham and John Newman, A flow-through porous electrode model: Application to metal-ion removal from dilute streams, J. Electrochem. Soc., 124, 1528-1540 (1977).
85. Richard Alkire and Ronald Gould, Analysis of multiple reaction sequences in flow-through porous electrodes, J. Electrochem. Soc., 123, 1842-1849 (1976).
86. B. Ateya, Kinetics of multiple electron transfer reactions at porous electrodes under stationary and flow conditions, J. Electroanal. Chem. Interfacial Electrochem., 76, 315-325 (1977).
87. Klaus J. Vetter, Electrochemical Kinetics, Theoretical and Experimental Aspects, Academic, New York, 1967, pp. 152, 481.
88. F. Coeuret, D. Hutin, A. Gaunand, Study of the effectiveness of fixed flow-through electrodes, J. Appl. Electrochem., 6, 417-423 (1976).
89. M. Paulin, D. Hutin, and F. Coeuret, Theoretical and experimental study of flow-through porous electrodes, J. Electrochem. Soc., 124, 180-188 (1977).
90. F. Coeuret, L'Electrode poreuse percolante (EPP)-III. Difference de potentiel metal-solution au sein de l'electrode en regime diffusionnel, Electrochim. Acta, 21, 203-213 (1976).
91. A Gaunand, D. Hutin, and F. Coeuret, Potential distribution in flow-through porous electrodes under limiting current conditions, Electrochim. Acta, 22, 93-97 (1977).

92. R. E. Sioda, Axial dispersion in flow porous electrodes, J. Appl. Electrochem., *7*, 135-137 (1977).

93. F. Coeuret, L'Electrode poreuse percolante (EPP)-I. Transpert de matiere en lit fixe, Electrochim. Acta, *21*, 185-193 (1976).

94. Badr G. Ateya, Radial diffusion and entrance effects in porous flow-through electrodes, J. Electroanal. Chem. Interfacial Electrochem., *76*, 183-190 (1977).

95. Veniamin G. Levich, Physicochemical Hydrodynamics, Prentice-Hall, Englewood Cliffs, 1962, pp. 112-115.

96. B. G. Ateya, E. A. S. Arafat, and S. A. Kafai, Hydrodynamic effects on the efficiency of porous flow-through electrodes, J. Appl. Electrochem., *7*, 107-112 (1977).

97. James A. Trainham and John Newman, The effect of electrode placement and finite matrix conductivity on the performance of flow-through porous electrodes, J. Electrochem. Soc., *125*, 58-67 (1978).

98. Peter Fedkiw and John Newman, Mass transfer at high Péclet numbers for creeping flow in a packed-bed reactor, AIChE J., *23*, 255-263 (1977).

Index

A

Absorbed, ions, 64
  molecules, 5, 64
Acetonitrile, 182
Acrylonitrile, 356
Activationless, discharge, 96
  processes, 76, 87, 88
Adiponitrile, 356
Adsorption, 416
  competitive, 149, 184
  energy, 88, 250
  layer, 49
  of organic molecules, 71
  rate, 75
  specific, 57, 64
Adsorptional Ψ-potential, 67
Alcohol, 106
  on mercury, 72
Alkaline electrolytes, 279
Alloys, 152, 187
Aluminum, 355
Amalgamation, 321
Aminophenol, 4, 357
Aniline, 357
André, Henri, 282
Anions, adsorption, 67, 94
  effects, 177
  electroreduction of, 57
  radical NO_3^{2-}, 98
  specific adsorption of, 65
Anisotropy, 210, 211, 222
Anodic, films, 318
  processes, 298
Anomalous codeposition, 191
Antimony, 413
Arsenate, 416
Arsenite, 416
Aqueous KOH, phase diagram for, 284
Autoelectronic emission, 10
Axial, diffusion, 380
  dispersion, 368, 372, 423

B

Barrierless, discharge H_3O^+, 87, 88
  process, 68, 91
Becquerel effect, 3
Benzoquinone, 356
Binding energies, 154
Bismuth, 38, 39, 41, 79, 84, 90, 102, 139

C

Cadmium, 58
  oxide, 321
Capacitance measurements, 60
Catalytic, activity, 234
  decomposition of NO_3^{2-}, 102
Cathodic processes, 305
Channeling, 370, 390
Characteristics of adsorbed layers, 71
Charge transfer reaction, 41
Charging methods, 314
Chemical, potential, 154
  reactions in solution, 73
Chemistry, analytical, 357
  electroanalytical, 418
Chloride containing solutions, 181
Chromium, 355
CO_2, 43, 96
Coadsorption, 149, 184
Conductance band in the solution, 46
Conductivity, 371, 404
Conservation of momentum, 14
Controlled potential, 357, 417
  analysis, 418
Convection, 324
Copper, 139, 355, 356, 389, 391, 398, 403, 406, 407, 409, 416, 419
Correction, Ψ', 41
Coulostatic mode, 31
Cr(II)/Cr(III), 419

Cr(III)/Cr(IV),419
Cuprous ions,410
Current distribution,323
Cyanide,356
Cyclic voltammetry,133
Cycling behavior,317

D

Daniell cell,278
Darimont cell,279
de Broglie wavelength,49
Deep beds,365,385,396
Debye screening length,51
de la Rive,279
Dendrites,313
Dendritic deposits,311
Dendrite,326
  formation,319
Densification,325
Density-of-states,198
Desalination,400
Desalting,416
Desorption,electrochemical,74,77,
    86,87,90,91,92,95
  bismuth electrode,93
  of hydrogen,6,90
  spectra,195
Differential reflectance
    spectroscopy,201,214
Diffuse,electric layer,36
  layer,49,51
  fields,57
  of emitted electron,18
Diffusion,coefficient,365,368
  equations,25
  processes,112
Diphenylanthracene,9,10,419
Dipole barrier,251
Disintegration,6
Double-layer,capacity,363
  charging,387,388
  structure of(see also electrical double-layer),78
Drumm,282
Dynamic discharge effect,23

E

ec(mechanisms),419
ece(mechanisms),419
  sequence,422
Edison and Jungner,281
Electrical double layer,9,11
  dense part of,63
  role,49
  structure of,48,94
Electrochemical reactions,
    consecutive,96
Electrode,antimony,80,94
  bismuth,80,81,91,93,95,103
  bromide,341
  impedance,30
  indium,101
  lead,80,103
  mercury,40,54,68,80,81,84,86,
    90,99,103
    adsorption of halogen ion,65
    dropping,6,29,41
  platinum,69,94,101
  porous,302
  porosity,302,325
  potential,dependence on,24,33
  reaction mechanisms,47
  rotating ring-disk,134
  zinc,277
Electrolyte,battery,283
  binary,393,400
  solid,111
  supporting,393,397,400
Electron,affinity,246
  binding energies,192
  bound,220
  capture,32
  by scavengers,50
  density-of-states,20
  drift of,50
  dry,27
  emitted,14,43
  excess,in condensed media,111
  excess,in polar solutions,43,46
  hydrated,46,55
  medium interaction,10

scavengers, H_3O^+, NO_3^-, N_2O, O_2, CO_2, 8
solvent interaction, for mercury, 46
wavelength, 49
Electronegativity, 239
Electronic polarization, 46
Electroosmosis, 324
Electroorganic, 416
  synthesis, 356, 419, 422
Electrooxidation, 97, 102
  of ethanol radicals, 108
  of CO_2^-, 108
  of NO_3^{2-}, 102
Electroreduction, 97
Electroreflectance, 221, 228
  enhanced, 225
Electrosorption valency, 160
Electrostatic model, 181
Ellipsometry, 210
Epitaxial growth, 199
Ethanol, 89, 104, 106
Excess binding energy, 243
Exchange currents, 171
  density, 306

F

Fe(II)/Fe(III), 419
Fermi, energies, 38
  surface, 20
Ferric, 416
Ferricyanide, 367, 423
Ferrous, 407, 413, 416
Five-halves law, 22, 34, 63, 76
Flash, desorption, 194
  measurements, 31, 110
  method, 32
  techniques, 108
Flow-redox, 358
Fluidized-bed electrodes, 418
Formate ion, 103, 106
Fowler's law, 12, 22
Frumkin, Ψ'-correction, 52
  isotherm, 165

relationship, 45

G

Gas bubbles, 390
Gautherot, 277
Gold, 139, 355, 356
Govy Chapman theory, 51, 55, 57
Graetz solution, 424, 425
Graphite, 144
Grove, Bunsen, 279
  cell, 279

H

H_{ads}, ionization of, 94
  recombination of, 94
Half-wave potential, 103
Halide, adsorption of, 84
  specific, 84
  solutions, 40
Halogen, specific adsorption of, 65
Heavy metal additives, 321
Helmholtz, layer, mercury electrode, 63
  plane outer, 63
Hexamethylphosphoric triamide, 48
Hexanedioic acid, dinitrile, 356
Homogeneous decomposition of NO_3^{2-}, 102
Hydrazine, 357
Hydrogen, 357
  adsorbed, 75
  adsorption, 75, 141
  atomic, 73, 74, 75, 77
    abstraction of, 110
    adsorption energy, 87
    cathodic removal of, 86
    from solutions, 73
    oxidation of, 80
    recombination of, 76
    removal of, 76, 83, 89, 94

evolution, 48, 73, 305, 315, 338
  on pure zinc in 6N and 9N
    KOH, 315
  overpotential, 68
ions, 73, 80, 87
  as a scavenger, 41, 79
  discharge, 41, 94
ionization, 88, 92
overvoltage, 278
oxidation, 78
α, 108
Hydroxylaminobenzene, 357

I

Image, forces, 11, 22
  potential, 244
Indium, 38, 41, 100
Interaction, energy of electrons, 47
  of adsorbed particles with an electron, 64
Interband transition, 220
Interface, potential barrier at, 38
Iodide anion, 99
Iodine-iodide couple, 419
Ionicity, 242
Ionization potential, 245
Iridium, 139
Iron, 318
Isotherm, 165
Iteration methods, 211

K

Kinetics, 171
  constants, 102
  electrochemical, 40, 41
  energy of electrons, 19
  of electrode reaction, 112
KOH, conductivity of, 285
Kramers-Kronig analysis, 208

L

Lead, 38, 41, 58, 102, 139, 316, 356, 391, 399, 413
  oxide, 321
Leclanché, 280
  and Chaperon, 280
Lighthill transformation, 387
Light modulation, 29
Linear, kinetics, 422
Lithium, 185

M

Macropores, 363
Magnesium, 355
Maiche cell, 281
Manganate, 416
Manganese, 355
Mass transfer, 364
  coefficient, 364, 372, 379, 419
Mean hydration length, 43
Measurement techniques, 29
Mercury, 38, 63, 79, 102, 356, 412
Metal deposition, 422
Methane, 104
Methanol, 104, 106, 110, 184, 358
Michalowski, 281
Micropores, 363
MnO_2 electrode, 331
Modulation spectroscopy, 201
Molten salts, 187
Monochloro derivitives, 104
Mono dielectric function, 205
Mössbauer spectroscopy, 233
Mossy deposit, 312
Morphology changes, 318
Multilayer system, 206

N

Nickel, 355
  oxide electrodes, 337
Nitrate, 416

Nitrite,416
Nitrobenzene,357
Nitrous oxide,55,63,80,105
Nonequilibrium,146
Nucleation,149
Nusselt number,365,379

O

OH,alcohol interactions,107
  capture of,110
  reduction,105
Ohmic potential drop,358,371,388,
  389,414
Optical studies,201
Orientation polarization,46
Oxide,formation,111
  layers,111
  surface,6,101
Oxygen reduction,305,317
  on zinc electrodes,317

P

Packed beds,360
Palladium,139
Parallel-plate electrodes,418
Partial charge,161,164
Pauling method,239
Passivation,111,319,325
Péclet number,366
Periodically constricted tubes,
  386,425
Permanganate,416
Phase change,209
pH,effect of,84
Phenol,110
Photocells,electrochemical,4
Photocurrent,adsorbate effect on,
  tetrabutylammonium cation,
  67
  anodic,7
  residual,42
  stationary,11

Photodecomposition,7
Photodesorption,7
Photodiffusion,currents,26
  process,59
Photodischarge,5
Photoeffect,inner,4
  outer,3
Photoemission,action of adsorbed
    hydrogen,70
  current,20
  energetics of,38
  into SF_6,110
  into vacuum,9
  laws,111
  of electrons,7
  theory,11,33
  threshold potential,38,39
  proper,49
  vacuum,10
  yield,108
Photoexcitation,112
  bulk,25
  surface,25
Photographic,emulsion of silver,
  356
  fixing baths,391
Photolysis of water molecule,5
Photopotential,31
Plasma edge,217
Platinum,139
  group metals,111
P-nitrobenzoate ion,419
Pogendorff,279
Polarizability,251
Polarography,418
Pore model,362
Porosity,362
Potassium hydroxide,284
Potential,effect of Ψ',53
  pH diagram(fig. 13),294
  Ψ',49,50,51,52,55,56,59,65,
    68,78,94,96
  $\phi*$,85
Pourbaixdiagram,294
Pressure drop,360
Primary cells,277,329

Propionic acid,nitrile,356
Propylene carbonate,182
Pulse,method,31
  radiolysis,107

Q

Quantum,mechanical theory,7,12
  yield,31

R

Radial dispersion,369
  chemistry,47
    reactions of,104
  frequency,dependence on,11
Radical,97
  CH_3,103
  CO_2^-,103
  electroreduction of,105
  free,72,96
  H,104,105
    reactions of,110
  OH,80,102,104,105,110
    reactions of,110
  oxidation,98
  RCHOH,105,106
  α,108
Rapid scan spectrometer,202
Rate constants,77,88
  for electron capture,44
  of homogeneous reaction,97
  of OH-alcohol interaction,107
Red Boundary,37
Redox,419
  couples,419
  system,422
  transformation of CO_2^-,103
Residence time,362,387
Resistivity,294
Reynolds number,360
Rhodium,139
Ruben,cell,283
  Samuel,283

S

Scattering length,65
Scavenger,25,26,32
  charged,50
  concentration,28
    photocurrent dependence,41
  dependence on,33
  electrode,229
  H_3O^+,96
  NO_3^-,96
  NO_2^-,96
  N_2O,69,80,96,105
Schmidt number,365
Schottky effect,10
Schrödinger equation,16
Screening effect,22
Sealed,cells,326
  operation,326
Secondary cells,277
Self-depolarization,251
Semiconductors,32
Separators,315
Shape,change,319,321,323
  factors,365
Silver,139,355,391,402,412,424
Single-crystal surfaces,172
Slow discharge,40,78
Sodium,355
Solid metals,32
Solutions,nonaqueous,111,159,184
Solvated electrons,5,8,25,26,
    32,43
Solvents,47
Source function,27,43
Stationary illumination,32
Stoichiometric coefficient,26
Storage energy,419
Sulfur hexafluoride,110
Super-position integrals,381
Superlattice,176,198
Surface,area,specific,360
  conductance,231
  molecule,239
  states,200
  treatment,147

Index

T

Tafel behavior, 423
Taylor dispersion, 369
TBA, 68
  adsorption of, 96
    on lead electrodes, 68
    on mercury electrodes, 68
  bromide, 99
  cation, adsorption of, 94
Templin isotherm, 168
Tallium oxide, 321
Theoretical specific energy, 329
Thermalization, 25, 43
  of photoelectrons, 27
Thermoemission of electrons into solution, 47, 48
Thin layer cell, 136
Thiosulfate, 391, 412
Three-phase system, 206
Threshold, approximation, 12, 17, 24, 35
  frequency, 12
  potential, 33, 34, 36, 44
Ti(III)/Ti(IV), 419
Tin, 139
Transfer coefficient, 77, 78, 82, 89, 90, 93, 95
Transference number, 376
Transient, 387
Tortuosity factor, 362
Twin electrode thin layer cells, 134

U

Ultraviolet photoemission spectroscopy, 198
Underpotential shift, 156, 157, 182
Uniaxial absorbing film, 211

V

Vacuum, 38

Volta, 277
  piles, 277

W

Walker-Wikins cell, 281
Water, interlayer, 63
  molecules, 87
Wave function, 14
Work function, 160, 243
  and the electrode potential, 24
  electron, 10, 37, 44
  photoemission, 38

X

XPS studies, 191
X-ray fluorescence, 145

Z

Zero charge potentials, 37, 45, 57, 60, 138, 181
Zinc, 419
  /air, 331
  amalgamated, 278, 280, 304
  batteries, 277
  bromide, 294
    cell, 339
    complexes, 294
    water system, phase diagram for, 294
  cells, 277, 329
  chemistry and electrochemistry of, 277
  chloride, 289
    cell, 338
    complexes, 289
    hydrate, 358
    resistivity data for, 289
    water system, phase diagram for, 289
  densification, 319

deposition, 305, 306, 314
dissolution, 300
electrode, electroplated, 327
　porous, 327
　structure and preparation, 326
electrolyte production of, 275
/KOH/air cell, 331
/KOH/nickel oxide, 335
/manganese dioxide cells, 329
/mercuric oxide cells, 331
morphology, 314, 317
mossy, 318, 327
/nickel oxide, 282
oxide, 277
passivation, 298
pastes, 328
powder, 327
redistribution, 283
refining, 275
/silver oxide, 282
　cells, 335
slurries, 328
slurry electrode, 333
utilization, 302
Zincate, diffusivity of, 288
　in 7.03 M KOH, activity coefficients for, 289
　ion, 284
ZnO, 286, 327
　in KOH, solubility of, 285